Java Concurrency and Parallelism

Master advanced Java techniques for cloud-based applications
through concurrency and parallelism

Jay Wang

‹packt›

Java Concurrency and Parallelism

Portfolio Manager: Kunal Sawant

Publishing Product Manager: Teny Thomas

Book Project Manager: Manisha Singh

Senior Editor: Aditi Chatterjee

Technical Editor: Sweety Pagaria

Copy Editor: Safis Editing

Proofreader: Aditi Chatterjee

Indexer: Tejal Soni

Production Designer: Nilesh Mohite

DevRel Marketing Coordinator: Shrinidhi Manoharan

First published: August 2024

Production reference: 1090824

Published by Packt Publishing Ltd.

Grosvenor House

11 St Paul's Square

Birmingham

B3 1RB, UK

ISBN 978-1-80512-926-4

www.packtpub.com

Contributors

About the author

Jay Wang, a trailblazer in the IT sector, boasts a career spanning over two decades, marked by leadership roles at IT powerhouses such as Accenture, IBM, and a globally renowned telecommunications firm. An expert in Java since 2001 and cloud technologies since 2018, Jay excels in transitioning projects from monolithic to microservice architectures and cloud. As founder of Digitech Edge, he guides clients through AI-driven cloud solutions. His educational background includes an MS in management of IT from the University of Virginia and an MS in information systems from George Mason University.

About the reviewers

Artur Skowroński is the head of Java/Kotlin engineering at VirtusLab. He has been in the industry for ten years. During this time, he has had the opportunity to work in various roles, such as software engineer, tech lead, architect, and even technical product manager. This diverse experience enables him to approach problems from a holistic perspective. He is also an active member of the tech community, serving as the lead of the Krakow Kotlin User Group and the author of the JVM Weekly Newsletter.

Akshay Phadke started his journey as a software engineer after graduating with a master of science in electrical and computer engineering from Georgia Institute of Technology in 2016. He has worked on building data-intensive applications and experiences across different industries such as networking and telecommunications, enterprise software, and finance. His work has spanned multiple areas of software development such as big data, data and platform engineering, CI/CD and DevOps, developer productivity and tooling, infrastructure and observability, and full stack web development. His professional interests include open source software, distributed systems, and building and scaling products in a start-up environment.

Table of Contents

4

Java Concurrency Utilities and Testing in the Cloud Era   101

Part 2: Java's Concurrency in Specialized Domains

6

Java and Big Data – a Collaborative Odyssey   173

7

Concurrency in Java for Machine Learning   207

8

Microservices in the Cloud and Java's Concurrency   241

9

Part 3: Mastering Concurrency in the Cloud – The Final Frontier

10

Synchronizing Java's Concurrency with Cloud Auto-Scaling Dynamics   313

11

Advanced Java Concurrency Practices in Cloud Computing   361

12

The Horizon Ahead   395

Appendix A

Setting up a Cloud-Native Java Environment   427

Appendix B

Preface

Welcome to *Java Concurrency and Parallelism*, which explores Java's concurrency and parallelism within the context of modern cloud computing. As technology evolves, mastering these concepts is crucial for developing robust, scalable, and efficient cloud-native applications.

This book is structured into three distinct yet interconnected parts, each aimed at providing in-depth knowledge and practical skills to enhance your proficiency in Java concurrency. *Part 1* lays the groundwork by introducing the fundamental principles of concurrency and parallelism, essential for any modern software development. This section covers the basics of Java concurrency, including threads and processes, and delves into advanced topics such as the `java.util.concurrent` package, the Fork/Join framework, and parallel streams. It also explores practical implementation strategies and concurrency patterns tailored for cloud environments, equipping you with the foundational tools needed to tackle the unique challenges of cloud-native Java applications.

Part 2 builds on the foundational knowledge from *Part 1*, focusing on how Java's concurrency capabilities address complex challenges across various specialized domains. This part examines Java's role in big data, machine learning, microservices, and serverless computing. Through practical examples, code snippets, and real-world use cases, it demonstrates how to harness Java's concurrency features to build scalable, high-performance applications in these cutting-edge fields.

Part 3 synthesizes the knowledge from the previous sections, applying it to the advanced realm of cloud computing. It covers synchronization with cloud auto-scaling, and advanced concurrency practices and explores emerging trends such as edge computing and quantum computing. This section prepares you to leverage Java's concurrency capabilities to their fullest potential, enabling you to stay at the forefront of technological advancements in cloud computing.

Who this book is for

This book is intended for Java developers, software engineers, and cloud computing professionals who aim to deepen their understanding of concurrent programming and parallel processing in Java. A foundational knowledge of Java and a basic familiarity with cloud computing concepts will be beneficial. Whether you are an aspiring developer or an experienced professional, this book will provide you with the knowledge and skills to excel in developing cloud-native Java applications.

What this book covers

Chapter 1, Concurrency, Parallelism, and the Cloud: Navigating the Cloud-Native Landscape, introduces the fundamental principles and distinctions between concurrency and parallelism in Java. It sets the stage for understanding how these concepts are critical in the development of modern cloud-native applications.

Chapter 2, Introduction to Java's Concurrency Foundations: Threads, Processes, and Beyond, explores the basics of Java concurrency, including threads and processes. This chapter provides a solid introduction to the building blocks of concurrent programming in Java.

Chapter 3, Mastering Parallelism in Java, delves into tools and frameworks that enhance Java's parallel processing capabilities. You will learn how to leverage multi-core processors and distributed systems to improve performance.

Chapter 4, Java Concurrency Utilities and Testing in the Cloud Era, focuses on practical implementation strategies and testing for cloud-based applications. You will discover how to design, implement, and validate concurrent systems that can thrive in distributed environments.

Chapter 5, Mastering Concurrency Patterns in Cloud Computing, provides a toolkit of concurrency patterns for cloud environments. These patterns serve as essential solutions for common challenges in cloud-native Java applications.

Chapter 6, Java and Big Data – a Collaborative Odyssey, examines Java's role in big data and its concurrency tools. You will learn how Java efficiently processes massive datasets through parallel and distributed computing.

Chapter 7, Concurrency in Java for Machine Learning, demonstrates how Java's concurrency features accelerate machine learning. This chapter shows how to build scalable, high-performance machine learning applications using multi-threading and parallel processing.

Chapter 8, Microservices in the Cloud and Java's Concurrency, explores Java's concurrency capabilities in microservices architectures. You will understand the strategies for managing concurrent operations in distributed environments to create resilient, scalable microservices.

Chapter 9, Serverless Computing and Java's Concurrent Capabilities, illustrates Java's adaptation to serverless architectures. You will discover how to build scalable, cost-effective solutions for varying workloads using Java's concurrency models.

Chapter 10, Synchronizing Java's Concurrency with Cloud Auto-Scaling Dynamics, covers synchronization with cloud auto-scaling. You will learn how to build applications that seamlessly adapt to the elastic nature of cloud environments.

Chapter 11, Advanced Java Concurrency Practices in Cloud Computing, delves into advanced concurrency techniques for the cloud. This chapter covers cutting-edge techniques such as GPU acceleration and implementing robust redundancy mechanisms.

Chapter 12, The Horizon Ahead, looks at emerging trends and future innovations in cloud computing. You will explore serverless Java beyond function as a service, Java's role in edge computing, and its integration with AI and machine learning in cloud ecosystems.

To get the most out of this book

You should have a few prerequisite skills or a certain level of knowledge in the following:

- Basic knowledge of Java programming
- Basic knowledge of cloud computing, especially the AWS environment, as most code examples for cloud applications are in AWS
- Familiarity with Java frameworks such as Spring Cloud
- Experience with build tools such as Maven
- Basic understanding of containerization using Docker
- Awareness of Quarkus for building cloud-native applications

We hope this book provides you with the insights and tools needed to master Java concurrency in the cloud era, transforming you from proficient developers into experts ready to tackle the challenges and opportunities of tomorrow's technological landscape.

Software/hardware covered in the book	Operating system requirements
Java, AWS account, Spring cloud, Docker, and Quarkus	Windows, macOS, or Linux

If you are using the digital version of this book, we advise you to type the code yourself or access the code from the book's GitHub repository (a link is available in the next section). Doing so will help you avoid any potential errors related to the copying and pasting of code.

Download the example code files

You can download the example code files for this book from `https://github.com/PacktPublishing/Java-Concurrency-and-Parallelism`. If there's an update to the code, it will be updated in the GitHub repository.

We also have other code bundles from our rich catalog of books and videos available at `https://github.com/PacktPublishing/`. Check them out!

> **Important note**
>
> Due to a recent tech update and page limit constraints, many code snippets in this book are shortened versions. They are used in chapters for demonstration purposes only. Some code has also been revised based on the update. For the most current, complete, and functional code, please refer to the book's accompanying GitHub repository. The repository should be considered the primary and preferred source for all code examples.

Conventions used

There are a number of text conventions used throughout this book.

`Code in text`: Indicates code words in text, database table names, folder names, filenames, file extensions, pathnames, dummy URLs, user input, and Twitter handles. Here is an example: "After submitting the tasks, we use `task1.get()` and `task2.get()` to wait for both tasks to complete. The `get()` method blocks until the task is finished and returns the result."

A block of code is set as follows:

```
import java.util.stream.IntStream;

public class ParallelKitchen {
  public static void main(String[] args) {
    IntStream.range(0, 10).parallel().forEach(i -> {
        System.out.println("Cooking dish #" + i + " in parallel...");
        // Simulate task
        try {
            Thread.sleep(600);
        } catch (InterruptedException e) {
            Thread.currentThread().interrupt();
        }
    });
  }
}
```

Any command-line input or output is written as follows:

```
aws cloudformation describe-stacks --stack-name <your-stack-name>
```

Bold: Indicates a new term, an important word, or words that you see onscreen. For instance, words in menus or dialog boxes appear in **bold**. Here is an example: "**Authentication**: Verify user credentials and issue secure tokens (e.g., JWTs) for access."

> **Tips or important notes**
> Appear like this.

Get in touch

Feedback from our readers is always welcome.

General feedback: If you have questions about any aspect of this book, email us at `customercare@packtpub.com` and mention the book title in the subject of your message.

Errata: Although we have taken every care to ensure the accuracy of our content, mistakes do happen. If you have found a mistake in this book, we would be grateful if you would report this to us. Please visit www.packtpub.com/support/errata and fill in the form.

Piracy: If you come across any illegal copies of our works in any form on the internet, we would be grateful if you would provide us with the location address or website name. Please contact us at `copyright@packt.com` with a link to the material.

If you are interested in becoming an author: If there is a topic that you have expertise in and you are interested in either writing or contributing to a book, please visit authors.packtpub.com.

Share Your Thoughts

Once you've read *Java Concurrency and Parallelism*, we'd love to hear your thoughts! Scan the QR code below to go straight to the Amazon review page for this book and share your feedback.

https://packt.link/r/1805129260

Your review is important to us and the tech community and will help us make sure we're delivering excellent quality content.

Download a free PDF copy of this book

Thanks for purchasing this book!

Do you like to read on the go but are unable to carry your print books everywhere?

Is your eBook purchase not compatible with the device of your choice?

Don't worry, now with every Packt book you get a DRM-free PDF version of that book at no cost.

Read anywhere, any place, on any device. Search, copy, and paste code from your favorite technical books directly into your application.

The perks don't stop there, you can get exclusive access to discounts, newsletters, and great free content in your inbox daily

Follow these simple steps to get the benefits:

1. Scan the QR code or visit the link below

https://packt.link/free-ebook/9781805129264

2. Submit your proof of purchase
3. That's it! We'll send your free PDF and other benefits to your email directly

Part 1:
Foundations of Java Concurrency and Parallelism in Cloud Computing

The first part of the book lays the groundwork for understanding and implementing concurrency and parallelism in Java, focusing on cloud computing environments. It introduces key concepts, tools, and patterns that form the backbone of efficient and scalable Java applications in the cloud.

This part includes the following chapters:

- *Chapter 1, Concurrency, Parallelism, and the Cloud: Navigating the Cloud-Native Landscape*
- *Chapter 2, Introduction to Java's Concurrency Foundations: Threads, Processes, and Beyond*
- *Chapter 3, Mastering Parallelism in Java*
- *Chapter 4, Java Concurrency Utilities and Testing in the Cloud Era*
- *Chapter 5, Mastering Concurrency Patterns in Cloud Computing*

1

Concurrency, Parallelism, and the Cloud: Navigating the Cloud-Native Landscape

Welcome to an exciting journey into the world of Java's **concurrency** and **parallelism** paradigms, which are crucial for developing efficient and scalable cloud-native applications. In this introductory chapter, we'll establish a solid foundation by exploring the fundamental concepts of concurrency and parallelism, as well as their significance in contemporary software design. Through practical examples and hands-on practice problems, you'll gain a deep understanding of these principles and their application in real-world scenarios.

As we progress, we'll delve into the transformative impact of cloud computing on software development and its synergistic relationship with Java. You'll learn how to leverage Java's powerful features and libraries to tackle the challenges of concurrent programming in cloud-native environments. We'll also explore case studies from industry leaders such as Netflix, LinkedIn, X (formerly Twitter), and Alibaba, showcasing how they have successfully harnessed Java's concurrency and parallelism capabilities to build robust and high-performance applications.

Throughout this chapter, you'll gain a comprehensive understanding of the software paradigms that shape the cloud era and Java's pivotal role in this landscape. By mastering the concepts and techniques presented here, you'll be well-equipped to design and implement concurrent systems that scale seamlessly in the cloud.

So, let's embark on this exciting journey together and unlock the full potential of concurrency and parallelism in Java cloud-native development. Get ready to acquire the knowledge and skills necessary to build innovative, efficient, and future-proof software solutions.

Technical requirements

Here is the minimal Java JRE/JDK setup guide for macOS, Windows, and Linux. You can follow these steps:

1. Download the desired version of Java JRE or JDK from the official Oracle website: `https://www.oracle.com/java/technologies/javase-downloads.html`.

2. Choose the appropriate version and operating system to download.

3. Install Java on your system:

    *   macOS:

        i.   Double-click the downloaded `.dmg` file.

        ii.   Follow the installation wizard and accept the license agreement.

        iii. Drag and drop the Java icon into the Applications folder.

    *   Windows:

        i.   Run the downloaded executable (`.exe`) file.

        ii.   Follow the installation wizard and accept the license agreement.

        iii. Choose the installation directory and complete the installation.

    *   Linux:

        *   Extract the downloaded `.tar.gz` archive to a directory of your choice.

        For system-wide installation, move the extracted directory to `/usr/local/java`.

4. Set the environment variables:

    *   macOS and Linux:

        i.   Open the terminal.

        ii.   Edit the `~/.bash_profile` or `~/.bashrc` file (depending on your shell).

        iii. Add the following lines (replace `<JDK_DIRECTORY>` with the actual path) in the file: `export JAVA_HOME=<JDK_DIRECTORY>` and `export PATH=$JAVA_HOME/bin:$PATH`.

        iv.   Save the file and restart the terminal.

- Windows:

    i.   Open the Start menu and search for **Environment Variables**.

    ii. Click on **Edit the system environment variables**.

    iii. Click on the **Environment Variables** button.

    iv. Under **System Variables**, click **New**.

    v.   Set the variable name as JAVA_HOME and the value as the JDK installation directory.

    vi. Find the **Path** variable, select it, and click **Edit**.

    vii. Add %JAVA_HOME%\bin to the **Path** variable.

    viii. Click **OK** to save the changes.

5. Verify the installation:

    i.   Open a new terminal or command prompt.

    ii. Run the following command: java -version.

    iii. It should display the installed Java version.

For more detailed installation instructions and troubleshooting, you can refer to the official Oracle documentation:

- macOS: https://docs.oracle.com/en/java/javase/17/install/installation-jdk-macos.html

- Windows: https://docs.oracle.com/en/java/javase/17/install/installation-jdk-microsoft-windows-platforms.html

- Linux: https://docs.oracle.com/en/java/javase/17/install/installation-jdk-linux-platforms.html

Please note that the exact steps may vary slightly depending on the specific Java version and operating system version you are using.

You need to install a Java **Integrated Development Environment** (**IDE**) on your laptop. Here are a few Java IDEs and their download URLs:

- IntelliJ IDEA

    - Download URL:https://www.jetbrains.com/idea/download/

    - Pricing: Free Community Edition with limited features, Ultimate Edition with full features requires a subscription

- Eclipse IDE:

  - Download URL: `https://www.eclipse.org/downloads/`

  - Pricing: Free and open source

- Apache NetBeans:

  - Download URL: `https://netbeans.apache.org/front/main/download/index.html`

  - Pricing: Free and open source

- **Visual Studio Code (VS Code)**:

  - Download URL: `https://code.visualstudio.com/download`

  - Pricing: Free and open source

VS Code offers a lightweight and customizable alternative to the other options on this list. It's a great choice for developers who prefer a less resource-intensive IDE and want the flexibility to install extensions that are tailored to their specific needs. However, it may not have all the features out of the box compared to the more established Java IDEs.

Further, the code in this chapter can be found on GitHub: `https://github.com/PacktPublishing/Java-Concurrency-and-Parallelism`.

> **Important note**
>
> Due to a recent tech update and page limit constraints, many code snippets in this book are shortened versions. They are used in chapters for demonstration purposes only. Some code has also been revised based on the update. For the most current, complete, and functional code, please refer to the book's accompanying GitHub repository. The repository should be considered the primary and preferred source for all code examples.

The dual pillars of concurrency versus parallelism – a kitchen analogy

Welcome to the kitchen of Java concurrency and parallelism! Here, we'll whisk you through a culinary journey, unveiling the art of multitasking and high-speed cooking in programming. Imagine juggling different tasks like a master chef – that's concurrency. Then, picture multiple chefs cooking in harmony for a grand feast – that's parallelism. Get ready to spice up your Java applications with these essential skills, from handling user interactions to crunching massive data. Bon appétit to the world of efficient and responsive Java cooking!

Defining concurrency

In Java, concurrency allows a program to manage multiple tasks such that they seem to run simultaneously, enhancing performance even on single-core systems. A **core** refers to a processing unit within a computer's CPU that is capable of executing programming instructions. While true parallel execution requires multiple cores, with each core handling a different task at the same time, Java's concurrency mechanisms can create the illusion of parallelism by efficiently scheduling and executing tasks in a way that maximizes the use of available resources. They can do this on a single- or multi-core system. This approach enables Java programs to achieve high levels of efficiency and responsiveness.

Defining parallelism

Parallelism is the simultaneous execution of multiple tasks or calculations, typically on multi-core systems. In parallelism, each core handles a separate task concurrently, leveraging the principle of dividing large problems into smaller, independently solvable subtasks. This approach harnesses the power of multiple cores to achieve faster execution and efficient resource utilization. By assigning tasks to different cores, parallelism enables true simultaneous processing, as opposed to concurrency, which creates the illusion of simultaneous execution through time-sharing techniques. Parallelism requires hardware support in the form of multiple cores or processors to achieve optimal performance gains.

The analogy of a restaurant kitchen

Imagine a restaurant kitchen as a metaphor for a Java application. From this perspective, we will understand the role of concurrency and parallelism in Java applications.

First, we'll consider concurrency. In a concurrent kitchen, there's one chef (the main thread) who can handle multiple tasks such as chopping vegetables, grilling, and plating. They do one task at a time, switching between tasks (context switching). This is similar to a single-threaded Java application managing multiple tasks asynchronously.

Next, we come to parallelism. In a parallel kitchen, there are multiple chefs (multiple threads) working simultaneously, each handling a different task. This is like a Java application utilizing multi-threading to process different tasks concurrently.

The following is a Java code example for concurrency:

```java
import java.util.concurrent.ExecutionException;
import java.util.concurrent.ExecutorService;
import java.util.concurrent.Executors;
import java.util.concurrent.Future;

public class ConcurrentKitchen {
  public static void main(String[] args) {
    ExecutorService executor = Executors.newFixedThreadPool(2);
```

```java
Future<?> task1 = executor.submit(() -> {
    System.out.println("Chopping vegetables...");
    // Simulate task
    try {
        Thread.sleep(600);
    } catch (InterruptedException e) {
        Thread.currentThread().interrupt();
    }
});

Future<?> task2 = executor.submit(() -> {
    System.out.println("Grilling meat...");
    // Simulate task
    try {
        Thread.sleep(600);
    } catch (InterruptedException e) {
        Thread.currentThread().interrupt();
    }
});

// Wait for both tasks to complete
try {
    task1.get();
    task2.get();
} catch (InterruptedException | ExecutionException e) {
    e.printStackTrace();
}
executor.shutdown();
    }
}
```

Here is an explanation of the preceding code example:

1. We create a fixed thread pool with two threads using `Executors.newFixedThreadPool(2)`. This allows the tasks to be executed concurrently by utilizing multiple threads.

2. We submit two tasks to the executor using `executor.submit()`. These tasks are analogous to chopping vegetables and grilling meat.

3. After submitting the tasks, we use `task1.get()` and `task2.get()` to wait for both tasks to complete. The `get()` method blocks until the task is finished and returns the result (in this case, there is no result since the tasks have a void return type).

4. Finally, we shut down the executor using `executor.shutdown()` to release the resources.

Next, we will look at a Java code example for parallelism:

```java
import java.util.stream.IntStream;

public class ParallelKitchen {
    public static void main(String[] args) {
        IntStream.range(0, 10).parallel().forEach(i -> {
            System.out.println("Cooking dish #" + i + " in parallel...");
            // Simulate task
            try {
                Thread.sleep(600);
            } catch (InterruptedException e) {
                Thread.currentThread().interrupt();
            }
        });
    }
}
```

Explanation of the preceding code example is as follows.

This Java code demonstrates parallel processing using `IntStream` and the parallel method, which is ideal for simulating tasks in a `Parallel Kitchen`. The main method utilizes an integer stream to create a range of 0 to 9, representing a range of different dishes.

By invoking `.parallel()` on `IntStream`, the code ensures that the processing of these dishes happens in parallel, leveraging multiple threads. Each iteration simulates cooking a dish, identified by the index, `i`, and is executed concurrently with other iterations.

The `Thread.sleep(600)` inside the `forEach` lambda expression mimics the time taken to cook each dish. The sleep duration is set for simulation purposes and is not indicative of actual cooking times.

In the case of `InterruptedException`, the thread's interrupt flag is set again with `Thread.currentThread().interrupt()`, adhering to best practices in handling interruptions in Java.

Having seen the two examples, let us understand the key differences between concurrency and parallelism:

- **Focus**: Concurrency is about managing multiple tasks, while parallelism is about executing tasks simultaneously for performance gains
- **Execution**: Concurrency can work on single-core processors, but parallelism benefits from multi-core systems

Both concurrency and parallelism play crucial roles in building efficient and responsive Java applications. The right approach for you depends on the specific needs of your program and the available hardware resources.

When to use concurrency versus parallelism – a concise guide

Armed with the strengths of concurrency and parallelism, let's dive into picking the perfect tool. We'll weigh up complexity, environment, and task nature to ensure that your Java applications sing. Buckle up, master chefs, as we unlock optimal performance and efficiency!

Concurrency

Concurrency is essential for effectively managing multiple operations simultaneously, particularly in three key areas:

- **Simultaneous task management**: This is ideal for efficiently handling user requests and **Input/Output (I/O)** operations, especially with the use of non-blocking I/O. This technique allows programs to execute other tasks without waiting for data transfer to complete, significantly enhancing responsiveness and throughput.

- **Resource sharing**: Through synchronization tools such as locks, concurrency ensures safe access to shared resources among multiple threads, preserving data integrity and preventing conflicts.

- **Scalability**: Scalability is crucial in developing systems capable of expansion such as microservices in cloud environments. Concurrency facilitates the execution of numerous tasks across different servers or processes, improving the system's overall performance and capacity to handle growth.

Let's look at some examples to illustrate each of the three key areas where concurrency is essential.

The first example is related to simultaneous task management. Here is a web server handling multiple client requests concurrently using non-blocking I/O:

```
import java.io.IOException;
import java.net.InetSocketAddress;
import java.nio.ByteBuffer;
import java.nio.channels.ServerSocketChannel;
import java.nio.channels.SocketChannel;

public class NonBlockingWebServer {
  public static void main(String[] args) throws IOException {
    ServerSocketChannel serverSocket = ServerSocketChannel.open();
    serverSocket.bind(new InetSocketAddress(
        "localhost", 8080));
    serverSocket.configureBlocking(false);

    while (true) {
        SocketChannel clientSocket = serverSocket.accept();
        if (clientSocket != null) {
            clientSocket.configureBlocking(false);
```

```
            ByteBuffer buffer = ByteBuffer.allocate(1024);
            clientSocket.read(buffer);
            String request = new String(
                buffer.array()).trim();
            System.out.println(
                "Received request: " + request);

        // Process the request and send a response
            String response = "HTTP/1.1 200 OK\r\nContent-Length:
                12\r\n\r\nHello, World!";
            ByteBuffer responseBuffer = ByteBuffer.wrap(
                response.getBytes());
            clientSocket.write(responseBuffer);
            clientSocket.close();
        }
    }
}
}
```

In this simplified example, the following happened:

1. We created a `ServerSocketChannel` and bound it to a specific address and port.

2. We configured the server socket to be non-blocking using `configureBlocking(false)`.

3. Inside an infinite loop, we accepted incoming client connections using `serverSocket.accept()`. If a client is connected, we will proceed to handle the request.

4. We configured the client socket to be non-blocking as well.

5. We allocated a buffer to read the client request using `ByteBuffer.allocate()`.

6. We read the request from the client socket into the buffer using `clientSocket.read(buffer)`.

7. We processed the request and sent a response back to the client.

8. Finally, we closed the client socket.

This simplified example demonstrates the key concept of handling multiple client requests concurrently using non-blocking I/O. The server can accept and process requests from multiple clients without blocking, allowing for efficient utilization of system resources and improved responsiveness.

Note that this example has been simplified for illustration purposes and may not include all the necessary error handling and edge case considerations of a production-ready web server.

The second example is resource sharing. Here is an example of multiple threads accessing a shared counter using synchronization:

```java
public class SynchronizedCounter {
    private int count = 0;
    public synchronized void increment() {
        count++;
    }
    public synchronized int getCount() {
        return count;
    }
}


public class CounterThread extends Thread {
    private SynchronizedCounter counter;
    public CounterThread(SynchronizedCounter counter) {
        this.counter = counter;
    }
    @Override
    public void run() {
        for (int i = 0; i < 1000; i++) {
        counter.increment();
        }
    }

    public static void main(String[] args) throws InterruptedException
    {
        SynchronizedCounter counter = new SynchronizedCounter();
        CounterThread thread1 = new CounterThread(counter);
        CounterThread thread2 = new CounterThread(counter);
        thread1.start();
        thread2.start();
        thread1.join();
        thread2.join();
        System.out.println(
            "Final count: " + counter.getCount());
    }
}
```

In this example, multiple (CounterThread) threads accessed a shared SynchronizedCounter object. The increment() and getCount() methods of the counter were synchronized to ensure that only one thread could access them at a time, preventing race conditions and maintaining data integrity.

Now, let us see an example of scalability. Here is a code example of microservice architecture using concurrency to handle a large number of requests:

```
import java.util.concurrent.ExecutorService;
import java.util.concurrent.Executors;

public class MicroserviceExample {
    private static final int NUM_THREADS = 10;

    public static void main(String[] args) {
        ExecutorService executorService = Executors.
        newFixedThreadPool(NUM_THREADS);

        for (int i = 0; i < 100; i++) {
            executorService.submit(() -> {
                // Simulate processing a request
                try {
                    Thread.sleep(1000);
                } catch (InterruptedException e) {
                    e.printStackTrace();
                }
        System.out.println("Request processed by " + Thread.
        currentThread().getName());
            });
        }
        executorService.shutdown();
    }
}
```

In this example, a microservice uses an `ExecutorService` with a fixed thread pool to handle a large number of requests concurrently. Each request is submitted as a task to the executor, which distributes them among the available threads. This allows the microservice to process multiple requests simultaneously, improving scalability and overall performance.

These examples demonstrate how concurrency is applied in different scenarios to achieve simultaneous task management, safe resource sharing, and scalability. They showcase the practical applications of concurrency in building efficient and high-performing systems.

Parallelism

Parallelism is a powerful concept used to enhance computing efficiency across various scenarios:

- **Compute-intensive tasks**: It excels at deconstructing elaborate calculations into smaller, autonomous sub-tasks that can be executed in parallel. This method significantly streamlines complex computational operations.

- **Performance optimization**: By engaging multiple processor cores at once, parallelism substantially shortens the time needed to complete tasks. This simultaneous utilization of cores ensures a quicker, more efficient execution process.

- **Large data processing**: Parallelism is key in swiftly handling, analyzing, and modifying vast datasets. Its capability to process multiple data segments concurrently makes it invaluable for big data applications and analytics.

Now let's look at some short demo code examples to illustrate the concepts of parallelism in each of the mentioned scenarios.

First, let's explore how parallelism can be applied to compute-intensive tasks, such as calculating Fibonacci numbers, using the `Fork/Join` framework in Java:

```java
import java.util.Arrays;
import java.util.concurrent.ForkJoinPool;
import java.util.concurrent.RecursiveAction;

public class ParallelFibonacci extends RecursiveAction {
    private static final long THRESHOLD = 10;
    private final long n;

    public ParallelFibonacci(long n) {
        this.n = n;
    }

    @Override
    protected void compute() {
        if (n <= THRESHOLD) {
        // Compute Fibonacci number sequentially
        int fib = fibonacci(n);
        System.out.println(
            "Fibonacci(" + n + ") = " + fib);
        } else {
        // Split the task into subtasks
        ParallelFibonacci leftTask = new ParallelFibonacci(
            n - 1);
        ParallelFibonacci rightTask = new ParallelFibonacci(n - 2);

        // Fork the subtasks for parallel execution
        leftTask.fork();
        rightTask.fork();

        // Join the results
        leftTask.join();
```

```
            rightTask.join();
            }
        }
    public static int fibonacci(long n) {
            if (n <= 1)
            return (int) n;
            return fibonacci(n - 1) + fibonacci(n - 2);
        }
    public static void main(String[] args) {
            long n = 40;
            ForkJoinPool pool = new ForkJoinPool();
            ParallelFibonacci task = new ParallelFibonacci(n);
            pool.invoke(task);
        }
    }
}
```

In this example, we used parallelism to compute the Fibonacci number of a given value n. The computation is split into subtasks using the Fork/Join framework. The ParallelFibonacci class extends RecursiveAction and overrides the compute() method. If the value of n is below a certain threshold, the Fibonacci number is computed sequentially. Otherwise, the task is split into two subtasks, which are forked for parallel execution. Finally, the results are joined to obtain the final Fibonacci number.

Next is performance optimization. Parallelism can significantly optimize performance, especially when dealing with time-consuming operations such as sorting large arrays. Let's compare the performance of sequential and parallel sorting in Java:

```
import java.util.Arrays;
import java.util.Random;

public class ParallelArraySort {
    public static void main(String[] args) {
            int[] array = generateRandomArray(100000000);
            long start = System.currentTimeMillis();
            Arrays.sort(array);
            long end = System.currentTimeMillis();
            System.out.println("Sequential sorting took " + (
                end - start) + " ms");
            start = System.currentTimeMillis();
            Arrays.parallelSort(array);
            end = System.currentTimeMillis();
            System.out.println("Parallel sorting took " + (
                end - start) + " ms");
        }
```

```
private static int[] generateRandomArray(int size) {
    int[] array = new int[size];
    Random random = new Random();
    for (int i = 0; i < size; i++) {
        array[i] = random.nextInt();
    }
    return array;
}
}
```

In this example, we demonstrated the performance optimization achieved by using parallelism for sorting a large array. We generated a random array of size 100,000,000 and measured the time taken to sort the array using both sequential sorting (`Arrays.sort()`) and parallel sorting (`Arrays.parallelSort()`). Parallel sorting utilizes multiple processor cores to sort the array concurrently, resulting in faster execution compared to sequential sorting.

Now, let's turn to large data processing. Processing large datasets can be greatly accelerated by leveraging parallelism. In this example, we'll demonstrate how parallel streams in Java can efficiently calculate the sum of a large number of elements:

```
import java.util.ArrayList;
import java.util.List;

public class ParallelDataProcessing {
    public static void main(String[] args) {
        List<Integer> data = generateData(100000000);

        // Sequential processing
        long start = System.currentTimeMillis();
        int sum = data.stream().mapToInt(
            Integer::intValue).sum();
        long end = System.currentTimeMillis();
        System.out.println("Sequential sum: " + sum + ",
            time: " + (end - start) + " ms");

        // Parallel processing
        start = System.currentTimeMillis();
        sum = data.parallelStream().mapToInt(
            Integer::intValue).sum();
        end = System.currentTimeMillis();
        System.out.println("Parallel sum: " + sum + ",
            time: " + (end - start) + " ms");
    }
```

```
private static List<Integer> generateData(int size) {
    List<Integer> data = new ArrayList<>(size);
    for (int i = 0; i < size; i++) {
        data.add(i);
    }
    return data;
}
}
```

In this code, we generated a large list of 100,000,000 integers using the `generateData` method. We then calculated the sum of all elements using both sequential and parallel streams.

The sequential processing is performed using `data.stream()`, which creates a sequential stream from the data list. The `mapToInt(Integer::intValue)` operation converts each `Integer` object to its primitive `int` value, and the `sum()` method calculates the sum of all elements in the stream.

For parallel processing, we use `data.parallelStream()` to create a parallel stream. The parallel stream automatically splits the data into multiple chunks and processes them concurrently using available processor cores. The same `mapToInt(Integer::intValue)` and `sum()` operations are applied to calculate the sum of all elements in parallel.

We measure the execution time of both sequential and parallel processing using `System.currentTimeMillis()` before and after each operation. By comparing the execution times, we can observe the performance improvement achieved by using parallelism.

Choosing the right approach

So, you've mastered the power of both concurrency and parallelism. Now comes the key question: how do you choose the right tool for the job? It's a dance between performance gains and complexity, where environment and task characteristics play their part. Let's dive into this crucial decision-making process:

- **Complexity versus benefit**: Weigh the performance gain of parallelism against its increased complexity and potential debugging challenges
- **Environment**: Consider your cloud infrastructure's capability for parallel processing (number of available cores)
- **Task nature and dependencies**: Independent, CPU-intensive tasks favor parallelism, while tasks with shared resources or I/O operations may benefit from concurrency

We've just equipped you with the culinary secrets of concurrency and parallelism, the dynamic duo that powers efficient Java applications. Remember, concurrency juggles multiple tasks like a master chef, while parallelism unleashes the power of multi-core machines for lightning-fast performance.

Why is this culinary wisdom so crucial? In the cloud-native world, Java shines as a versatile chef, adapting to diverse tasks. Concurrency and parallelism become your essential tools, ensuring responsiveness to user requests, handling complex calculations, and processing massive data – all on the ever-evolving canvas of the cloud.

Now let's take this culinary expertise to the next level. In the next section, we'll explore how these concurrency and parallelism skills seamlessly blend with cloud technologies to build truly scalable and high-performance Java applications. So sharpen your knives and get ready to conquer the cloud-native kitchen!

Java and the cloud – a perfect alliance for cloud-native development

Java's journey with cloud computing is a testament to its adaptability and innovation. The fusion of their capabilities has created a powerful alliance for cloud-native development. Imagine yourself as an architect, wielding Java's toolkit at the forefront of cloud technology. Here, Java's versatility and robustness partner with the cloud's agility and scalability to offer a canvas for innovation and growth. We're not just discussing theoretical concepts – we're stepping into a realm where Java's pragmatic application in the cloud has revolutionized development, deployment, and application management. Let's uncover how Java in the cloud era is not just an option, but a strategic choice for developers seeking to unlock the full potential of cloud-native development.

Exploring cloud service models and their impact on software development

Java development has entered a new era with cloud computing. Imagine having instant access to a vast pool of virtual resources, from servers to storage to networking. Cloud services unlock this magic, empowering Java developers to build and scale applications faster and more efficiently.

Three distinct service models dominate the cloud, each impacting development needs and Java application architecture. Let us explore each one of them.

Infrastructure as a service

Infrastructure as a Service (IaaS) offers foundational cloud computing resources such as virtual machines and storage. For Java developers, this means complete control over the operating environment, allowing for customized Java application setups and optimizations. However, it requires a deeper understanding of infrastructure management.

Code example – Java on IaaS (Amazon EC2)

This code snippet showcases how to create and launch an Amazon **Elastic Compute Cloud (EC2)** instance using the AWS SDK for Java:

```
// Create EC2 client
AmazonEC2Client ec2Client = new AmazonEC2Client();

// Configure instance details
RunInstancesRequest runRequest = new RunInstancesRequest();
runRequest.setImageId("ami-98760987");
runRequest.setInstanceType("t2.micro");
runRequest.setMinCount(1);
runRequest.setMaxCount(3);
// Launch instance
RunInstancesResult runResult = ec2Client.runInstances(runRequest);
// Get instance ID
String instanceId = runResult.getReservations().get(0).getInstances().
get(0).getInstanceId();
// ... Configure Tomcat installation and web application deployment
...
```

Let's break it down step by step:

1. **Create the EC2 client**:

    *   `// Create EC2 client`

    *   `AmazonEC2Client ec2Client = new AmazonEC2Client();`

2. **Configure the details**: Configure the details of the instance we want to launch. This includes the following:

    *   `setImageId`: The **Amazon Machine Image (AMI)** ID specifies the pre-configured operating system and software stack for the instance

    *   `setInstanceType`: We define the instance type, such as `t2.micro`, for a small, cost-effective option

    *   `setMinCount`: We specify the minimum number of instances to launch (1 in this case)

    *   `setMaxCount`: We specify the maximum number of instances to launch (3 in this case, allowing the system to scale up if needed)

3. **Launch the instance**: We call the `runInstances` method on the `ec2Client` object, passing the configured `runRequest` object. This sends the request to AWS to launch the desired EC2 instances.

4. **Get instance ID**: The `runInstances` method returns a `RunInstancesResult` object containing information about the launched instances. We will extract the instance ID of the first instance (assuming a successful launch) for further use in the deployment process.

5. **Configure Tomcat and deploy application**: This comment indicates that the next steps will involve setting up Tomcat on the launched EC2 instance and deploying your web application. The specific code for this would depend on your chosen Tomcat installation method and application deployment strategy.

This example demonstrates launching instances with a minimum and maximum count. You can adjust these values based on your desired level of redundancy and scalability.

Platform as a service

Platform as a Service (**PaaS**) provides a higher-level environment with ready-to-use platforms including operating systems and development tools. This is beneficial for Java developers as it simplifies deployment and management, though it might limit lower-level control.

Code example – Java on AWS Lambda

This code snippet defines a simple Java Lambda function that processes an S3 object upload event:

```
public class S3ObjectProcessor implements RequestHandler<S3Event,
String> {
    @Override
    public String handleRequest(S3Event event,
        Context context) {
            for (S3Record record : event.getRecords()) {
            String bucketName = record.getS3().getBucket().getName();
            String objectKey = record.getS3().getObject().getKey();
            // ...process uploaded object ...
        }
    return "Processing complete";
    }
}
```

This code is a listener for Amazon S3 object uploads. It's like a robot that watches for new files in a specific bucket (such as a folder) and automatically does something with them when they arrive:

- It checks for new files in the bucket: for each new file, it gets the filename and the bucket it's in

- You need to fill in the missing part: write your own code here to say what you want the robot to do with the file (e.g., download it, analyze it, or send an email)

Once the robot finishes with all the files, it sends a message back to Amazon saying **Processing complete**.

I hope this simplified explanation makes things clearer!

Software as a service

Software as a service (SaaS) delivers complete application functionality as a service. For Java developers, this often means focusing on building the application's business logic without worrying about the deployment environment. However, customization and control over the platform are limited.

Code example – Java on SaaS (AWS Lambda)

This code snippet defines a Lambda function for processing event data:

```java
public class LambdaHandler {
    public String handleRequest(Map<String, Object> event,
        Context context) {
        // Get data from event
        String message = (String) event.get("message");
        // Process data
        String result = "Processed message: " + message;
        // Return result
        return result;
    }
}
```

This code defines a class called `LambdaHandler` that listens for events in a serverless environment such as AWS Lambda. Here's a breakdown:

1. Listening for events:

    - The class is named `LambdaHandler`, signifying its role as a handler for Lambda events.
    - The `handleRequest` method is the entry point for processing incoming events.
    - The event parameter holds the data received from the Lambda invocation.

2. Processing data:

    - Inside the `handleRequest` method, the code retrieves the message key from the event data using `event.get("message")`.
    - This assumes that the event format includes a key named `message` containing the actual data to be processed.
    - The code then processes the message and combines it with a prefix to generate a new string stored in the result variable.

3. Returning result:

    - Finally, the `handleRequest` method returns the processed message stored in the result variable. This is the response that is sent back to the caller of the Lambda function.

In simpler terms, this code acts like a small service that takes in data (messages) through an event, processes it (adds a prefix), and returns the updated version. It's a simple example of how Lambda functions can handle basic data processing tasks in a serverless environment.

Understanding the strengths and weaknesses of each cloud service model is crucial for Java developers to be able to make the best decisions for their projects. By choosing the right model, they can unlock the immense potential of cloud computing and revolutionize the way in which they build and deploy Java applications.

Java's transformation in the cloud – a story of innovation

Imagine a world transformed by the cloud, where applications soar among the constellations of data centers. This is the landscape Java navigates today, not as a relic of the past, but as a language reborn in the fires of innovation.

At the heart of this evolution lies the **Java Virtual Machine** (**JVM**), the engine that powers Java applications. Once again, it has transformed, shedding layers of inefficiency to become lean and mean, ready to conquer the resource-constrained world of the cloud.

But power alone is not enough. Security concerns loom large in the vastness of the cloud. Java, ever vigilant, has donned the armor of robust security features, ensuring that its applications remain unbreachable fortresses in the digital realm.

Yet size and security are mere tools without purpose. Java has embraced the new paradigm of microservices, breaking down monolithic structures into nimble, adaptable units. Frameworks such as Spring Boot and MicroProfile stand as a testament to this evolution, empowering developers to build applications that dance with the dynamism of the cloud.

As the cloud offers its vast array of services, Java stands ready to embrace them all. Its vast ecosystem and robust APIs act as bridges, connecting applications to the boundless resources at their fingertips.

This is not just a story of technical advancement, it's a testament to the power of adaptability and of embracing change and forging a new path in the ever-evolving landscape of the cloud.

Java – the cloud-native hero

Java sits comfortably on the throne of cloud-native development. Here's why:

- **Platform agnostic**: *Write once, run anywhere* is a feature of Java applications. These cloud-agnostic Java applications effortlessly dance across platforms, simplifying deployment across diverse cloud infrastructures.

- **Scalability and performance**: Java pairs perfectly with the cloud's inherent scalability, handling fluctuating workloads with ease. Built-in garbage collection and memory management further optimize resource utilization and drive high performance.

- **Security first**: Java's robust security features, such as sandboxing and strong type checking, shield applications from common vulnerabilities, making them ideal for the security-conscious cloud environment.

- **Rich ecosystem**: A vast and mature ecosystem of libraries, frameworks, and tools caters specifically to cloud-native development, empowering developers to build faster and with less effort.

- **Microservices champion**: Java's modularity and object-oriented design perfectly align with the growing trend of microservices architecture, allowing developers to build and scale independent services easily.

- **CI/CD ready**: Java integrates seamlessly with popular **continuous integration** (**CI**) and **continuous deployment** (**CD**) tools and methodologies, enabling automated builds, tests, and deployments for rapid cloud-native application delivery.

- **Concurrency king**: Java's built-in concurrency features, such as threads and thread pools, empower developers to create highly concurrent applications that leverage the parallel processing capabilities of cloud computing.

- **Community and support**: Java boasts a vibrant community and a wealth of online resources and documentation, providing invaluable support for developers working with cloud-native Java applications.

In conclusion, Java's inherent characteristics and compatibility with modern cloud architectures make it the natural hero for cloud-native development. With its rich ecosystem and robust security features, Java empowers developers to build and deploy high-performing, scalable, and secure cloud-native applications.

Java's cloud-focused upgrades – concurrency and beyond

The cloud demands efficient and scalable applications, and Java continues to evolve to meet this need. Here's a spotlight on key updates for cloud-native development, focusing on concurrency and parallelism.

Project Loom – virtual threads for efficient concurrency

Imagine handling a multitude of concurrent tasks without worrying about resource overhead. Project Loom introduces lightweight virtual threads, enabling efficient management of high concurrency. This is ideal for cloud environments where responsiveness and resource efficiency are paramount.

Enhanced garbage collection for high throughput

Say goodbye to long garbage collection pauses impacting performance. Recent Java versions introduce low-pause, scalable garbage collectors such as ZGC and Shenandoah GC. These handle large heaps with minimal latency, ensuring smooth operation and high throughput even in demanding cloud environments.

Record types – simplifying data modeling

Cloud applications frequently deal with data transfer objects and messaging between services. Record types, introduced in Java 16, simplify immutable data modeling, offering a concise and efficient way to represent data structures. This improves code readability, reduces boilerplate code, and ensures data consistency in cloud-based microservices.

Sealed classes – controlled inheritance hierarchies

Have you ever wanted to enforce specific inheritance rules in your cloud application? Sealed classes, finalized in Java 17, allow you to restrict which classes or interfaces can extend or implement others. This promotes clarity, maintainability, and predictable behavior within cloud-based domain models.

Other notable updates for cloud development

In addition to these key updates related to concurrency and parallelism, there are many other improvements. Here are a few:

- **Pattern matching for instanceof**: Offers a cleaner and more concise solution for checking and casting object types, improving code readability and reducing boilerplate

- **Foreign-memory access API**: Allows Java programs to safely and efficiently access memory outside the Java heap, unlocking performance potential and facilitating seamless integration with native libraries

- **HTTP client API**: Simplifies HTTP and WebSocket communication for cloud applications, enabling developers to build robust and high-performance clients for effective communication within the cloud ecosystem

- **Microbenchmark suite**: Helps in accurately measuring the performance of code snippets, allowing for precise performance tuning and ensuring your cloud applications run at their peak potential

These advancements demonstrate Java's commitment to empowering developers to build robust, scalable, and high-performing cloud applications. By leveraging these features, developers can unlock the full potential of Java in the cloud and create innovative solutions for the evolving digital landscape.

Real-world examples of successful cloud-native Java applications

Java isn't just a programming language; it's a powerhouse fueling some of the world's most innovative companies. Let's take a peek behind the scenes of four industry leaders and see how Java drives their success.

Netflix – microservices maestro

Imagine millions of people streaming movies and shows simultaneously, without a hiccup. That's the magic of Netflix's microservices architecture, which was meticulously crafted with Java. Spring Boot and Spring Cloud act as the architects, building individual services that work together seamlessly. When things get bumpy, Hystrix, a Netflix-born Java library, acts as the knight in shining armor, isolating issues and keeping the show running. Zuul, another Java gem, stands guard at the edge, routing traffic and ensuring everything flows smoothly.

LinkedIn – data's real-time river

LinkedIn's vibrant network thrives on real-time data. And who keeps this information flowing like a mighty river? Apache Kafka, a Java-powered stream processing platform. Kafka's lightning-fast speed and fault tolerance ensure connections are always live, allowing for instant updates and personalized experiences. Plus, Kafka seamlessly integrates with other Java-based systems at LinkedIn, creating a powerful data processing symphony.

X – from Ruby on Rails to the JVM's soaring heights

Remember the days of slow-loading tweets? X (formerly Twitter) does! To conquer the challenge of scale, they made a bold move: migrating from Ruby on Rails to the JVM. This switch, powered by Java and Scala, unlocked a new era of performance and scalability. Finagle, a Twitter-built RPC system for the JVM, further boosted concurrency. This allowed millions of tweets to take flight simultaneously.

Alibaba – the e-commerce titan forged in Java

When it comes to online shopping, Alibaba reigns supreme. And what's their secret weapon? Java! From handling massive spikes in traffic to managing complex data landscapes, Java's ability to handle high concurrency is Alibaba's golden ticket to success. They've even optimized Java's garbage collection to efficiently manage their immense heap size, ensuring that their platform runs smoothly even when billions of items are flying off the virtual shelves.

These are just a few examples of how Java empowers industry leaders. From streaming giants to social media havens and e-commerce titans, Java's versatility and power are undeniable. So, next time you watch a movie, share a post, or click *buy*, remember – there's a good chance Java is quietly pulling the strings, making your experience seamless and magical.

We've explored Java's hidden superpower – its seamless integration with the cloud! From concurrency and parallelism to microservices architecture, Java empowers developers to build robust, scalable, and high-performing cloud-native applications. We've seen how Netflix, LinkedIn, X, and Alibaba leverage Java's diverse capabilities to achieve their cloud goals.

But the cloud journey isn't without its challenges. Security, cost optimization, and efficient resource management all come knocking at the door of your cloud-native development. In the next section, we'll dive deep into these modern challenges, equipping you with the knowledge and tools to navigate them like a seasoned cloud explorer. So buckle up, fellow Java adventurers, as we venture into the exciting realm of modern challenges in cloud-native development!

Modern challenges in cloud-native concurrency and Java's weapons of choice

The cloud's concurrency challenges loom, but Java's not backing down. We'll tackle these challenges in transactions, data consistency, and microservices states, all while wielding tools such as Akka, Vert.x, and reactive programming. Choose your weapons wisely, for the cloud-native concurrency challenge is yours to conquer!

Wrangling distributed transactions in Java – beyond classic commits

In the wild jungle of distributed systems, managing transactions across services and databases can be a daunting task. Traditional methods stumble over network delays, partial failures, and diverse systems. But fear not, Java warriors! We've got your back with a robust arsenal of solutions:

- **Two-phase commit (2PC)**: This classic protocol ensures all parties in a transaction commit or roll back together. While not ideal for high-speed environments due to its blocking nature, 2PC remains a reliable option for more controlled transactions.

- **Saga pattern**: Think of this as a choreographed dance, where each local transaction is linked to others through a sequence of events. Java frameworks such as Axon and Eventuate help you orchestrate this graceful ballet, ensuring data consistency even when things get messy.

- **Compensating transactions**: Imagine a safety net for your saga. If a step goes wrong, compensating transactions swoop in, reversing the effects of previous operations and keeping your data safe. Java services can implement this strategy with service compensations, which are ready to clean up any spills.

Maintaining data consistency in cloud-native Java applications

Data consistency in the cloud can be a tricky tango, especially with NoSQL's eventual rhythm. But Java's got the notes to keep it harmonious:

- **Kafka's eventual beat**: Updates become rhythmic pulses, sent out and listened to by services. It's not immediate, but everyone eventually grooves to the same tune.

- **Caching whispers**: Tools such as Hazelcast and Ignite act as quick assistants, keeping data consistent across nodes, even when the main database takes a break.

- **Entity versioning**: When two updates waltz in at once, versioning helps us track who came first and resolve conflicts gracefully. No data mosh pits here!

With these moves and a bit of Java magic, your cloud applications will keep your data safe and sound, moving in perfect rhythm.

Handling state in microservices architectures

Microservices are a beautiful dance of independent services, but what about their state? Managing it across this distributed landscape can feel like wrangling a herd of wild cats. But fear not, Java offers a map and a torch to guide you through:

- **Stateless serenity**: When possible, design microservices as stateless citizens of the cloud. This keeps them lightweight, scalable, and effortlessly resilient.

- **Distributed session sherpas**: For those services that crave a bit of state, distributed session management tools such as Redis and ZooKeeper come to the rescue. They keep track of state across nodes, ensuring everyone's on the same page.

- **CQRS and event sourcing – the stateful waltz**: For truly complex state dances, patterns such as **Command Query Responsibility Segregation** (**CQRS**) and Event Sourcing offer a graceful solution. Java frameworks such as Axon provide the perfect shoes for this intricate choreography.

With these strategies in your arsenal, you can navigate the stateful microservices maze with confidence, building resilient and scalable systems that thrive in the ever-changing cloud landscape.

Cloud database concurrency – Java's dance moves for shared resources

Imagine a crowded dance floor – that's your cloud database with multiple clients vying for attention. It's a delicate tango of multi-tenancy and resource sharing. Keeping everyone in step requires some fancy footwork.

The **Atomicity, Consistency, Isolation, and Durability** (**ACID**) test adds another layer of complexity. Messy concurrency can easily trip up data integrity, especially in distributed environments. Java's got your back with a few fancy footwork moves:

- **Sharing (with manners)**: Multi-tenancy and resource sharing are no problem with Java's locking mechanisms, which include synchronized blocks and ReentrantLock. They act as bouncers, ensuring everyone gets their turn without stepping on toes (or data).

- **Optimistic versus pessimistic locking**: Think of these as different dance styles. Optimistic locking assumes that everyone plays nice, while pessimistic locking keeps a watchful eye, preventing conflicts before they happen. Java frameworks such as JPA and Hibernate offer both styles, letting you choose the perfect rhythm.

- **Caching the craze**: Frequently accessed data gets its own VIP lounge: the distributed cache. Java solutions such as Hazelcast and Apache Ignite keep this lounge stocked, reducing database load and ensuring smooth data access for everyone.

With these moves in your repertoire, your Java applications can waltz gracefully through cloud database concurrency, ensuring data consistency and smooth performance even when the dance floor gets crowded.

Parallelism in big data processing frameworks

Imagine waves of data, rushing in like a flood. You need a way to analyze it all – fast. That's where parallel processing comes in, and Java's got the tools to tackle it:

- **MapReduce**: Java is extensively used in MapReduce programming models, as seen in Hadoop. Developers write Map and Reduce functions in Java to process large datasets in parallel across a Hadoop cluster.

- **Apache Spark**: Although it is written in Scala, Spark provides Java APIs. It enables parallel data processing by distributing data across **Resilient Distributed Datasets** (**RDDs**) and executing operations in parallel.

- **Stream processing**: Java Stream API, along with tools such as Apache Flink and Apache Storm, supports parallel stream processing for real-time data analytics.

So, when data gets overwhelming, remember Java. It's got the tools to keep you informed and in control, even when the beast roars!

Here, we will kick off the thrilling journey into concurrency and parallelism in the cloud-native Java world. Get ready to transform these challenges into opportunities. In the pages ahead, you'll acquire the tools to master concurrency and parallelism. This will empower you to build robust, future-proof Java applications that thrive in the cloud.

Cutting-edge tools for conquering cloud-native concurrency challenges

The intricate dance of concurrency in cloud-native applications can be daunting, but fear not! Cutting-edge tools and techniques are here to help. Let's explore some tools to address the challenges we discussed earlier.

Cloud-native concurrency toolkits

The following tools fit well into this category:

- **Akka**: This powerful toolkit leverages the actor model for building highly scalable and fault-tolerant applications. It provides features such as message passing, supervision, and location transparency, simplifying concurrent programming and addressing challenges such as distributed locks and leader election.

- **Vert.x**: This lightweight toolkit focuses on reactive programming and non-blocking I/O, making it ideal for building highly responsive and performant applications. Vert.x's event-driven architecture can handle high concurrency effectively and simplifies asynchronous programming.

- **Lagom**: This framework is built on top of Akka and offers a high-level API for building microservices. Lagom provides features such as service discovery, load balancing, and fault tolerance, making it suitable for building complex, distributed systems.

Distributed coordination mechanisms

Tools in this category include the following:

- **ZooKeeper**: This open source tool provides distributed coordination primitives such as locking, leader election, and configuration management. ZooKeeper's simplicity and reliability make it a popular choice for coordinating distributed applications.

- **etcd**: This distributed key-value store provides a high-performance and scalable means to store and manage configuration data across nodes. etcd's features, including watches and leases, make it suitable for maintaining consistency and coordinating state changes in distributed systems.

- **Consul**: This service mesh solution offers a comprehensive set of features for service discovery, load balancing, and distributed coordination. Consul's web UI and rich API make it easy to manage and monitor distributed systems.

Modern asynchronous programming patterns

These modern asynchronous patterns enable efficient non-blocking data processing and scalable, resilient applications:

- **Reactive Streams**: This specification provides a standard way to write asynchronous, non-blocking programs. Reactive Streams improves responsiveness and scalability by ensuring that data is processed efficiently and that backpressure is managed effectively.

- **Asynchronous messaging**: This technique utilizes message queues to decouple components and handle tasks asynchronously. Asynchronous messaging can improve scalability and resilience by enabling parallel processing and handling failures gracefully.

Choosing the right tool for the job

Each toolkit, mechanism, and pattern has its own strengths and weaknesses, making them suitable for different scenarios. Here are some considerations:

- **Complexity**: Akka offers rich features but can be complex to learn and use. Vert.x and Lagom provide a simpler starting point.

- **Scalability**: All three toolkits are highly scalable but Vert.x excels for high-performance applications due to its non-blocking nature.

- **Coordination needs**: ZooKeeper is well-suited for basic coordination tasks, while etcd's key-value store offers additional flexibility. Consul provides a complete service mesh solution.

- **Programming style**: Reactive Streams requires a shift in thinking toward asynchronous programming, while asynchronous messaging can be integrated with traditional synchronous approaches.

By understanding the available solutions and their trade-offs, developers can choose the right tools and techniques to address specific concurrency challenges in their cloud-native applications. This, in turn, leads to building more scalable, responsive, and resilient systems that thrive in the dynamic cloud environment.

Conquering concurrency – best practices for robust cloud-native applications

Building cloud apps that juggle multiple tasks at once? It's like managing a bustling zoo of data and operations! But fear not, because we've got the best practices to tame the concurrency beasts and build robust, scalable cloud apps. Here are the best practices to embed:

- **Early identification**: Proactively identify and address concurrency challenges through early analysis, modeling, and code review:

  - **Analyze application requirements**: Identify critical sections, shared resources, and potential points of contention early in the design phase

  - **Use concurrency modeling tools**: Utilize modeling tools such as statecharts or Petri nets to visualize and analyze potential concurrency issues

  - **Review existing code for concurrency bugs**: Conduct code reviews and static analysis to identify potential race conditions, deadlocks, and other concurrency problems

- **Embrace immutable data**: Embrace unchangeable data to simplify concurrent logic and eliminate race conditions:

    - **Minimize mutable state**: Design data structures and objects to be immutable by default. This simplifies reasoning about their behavior and eliminates potential race conditions related to shared state modifications.

    - **Utilize functional programming principles**: Leverage functional programming techniques such as immutability, pure functions, and laziness to create inherently thread-safe and predictable concurrent code.

- **Ensure thread safety**: Secure concurrent access to shared resources through synchronized blocks, thread-safe libraries, and focused thread confinement

    - **Use synchronized blocks or other locking mechanisms**: Protect critical sections of code that access shared resources to prevent concurrent modifications and data inconsistencies

    - **Leverage thread-safe libraries and frameworks**: Choose libraries and frameworks that are specifically designed for concurrent programming and utilize their thread-safe functionalities

    - **Employ thread confinement patterns**: Assign threads to specific tasks or objects to limit their access to shared resources and simplify reasoning about thread interactions

- **Design for failure**: Build resilience against concurrency failures through fault tolerance mechanisms, proactive monitoring, and rigorous stress testing

    - **Implement fault tolerance mechanisms**: Design your application to handle and recover from concurrency-related failures gracefully. This includes retry mechanisms, circuit breakers, and fail-over strategies.

    - **Monitor and observe concurrency behavior**: Employ monitoring tools and observability practices to identify and diagnose concurrency issues in production environments.

    - **Conduct stress testing**: Perform rigorous stress testing to evaluate how your application behaves under high load and identify potential concurrency bottlenecks.

- **Leverage cloud-native tools**: Harness the power of cloud-native tools such as asynchronous patterns, distributed coordination, and dedicated frameworks to conquer concurrent challenges and build robust, scalable cloud applications

    - **Utilize asynchronous programming patterns**: Embrace asynchronous programming models such as reactive streams and asynchronous messaging to improve scalability and responsiveness in concurrent applications

    - **Adopt distributed coordination mechanisms**: Utilize distributed coordination tools such as ZooKeeper, etcd, or Consul to manage distributed state and ensure consistent operation across multiple nodes

- **Choose appropriate concurrency frameworks**: Leverage cloud-native concurrency frameworks such as Akka, Vert.x, or Lagom to simplify concurrent programming and address specific concurrency challenges effectively

With a solid understanding of best practices, let's turn our attention to code.

Code examples illustrating best practices

Let's look at some code examples.

Asynchronous programming with reactive streams

We can leverage reactive streams such as RxJava to implement an asynchronous processing pipeline. This allows for the concurrent execution of independent tasks, improving responsiveness and throughput. Here is a code example using reactive streams:

```
// Define a service interface for processing requests
public interface UserService {
    Mono<User> getUserById(String userId);
}


// Implement the service using reactive streams
public class UserServiceImpl implements UserService {

    @Override
    public Mono<User> getUserById(String userId) {
        return Mono.fromCallable(() -> {
        // Simulate fetching user data from a database
            Thread.sleep(600);
            return new User(userId, "Jack Smith");
        });
    }
}
// Example usage
Mono<User> userMono = userService.getUserById("99888");
userMono.subscribe(user -> {
    // Process user data
    System.out.println("User: " + user.getName());
});
```

This code defines a service for handling user requests in a reactive way. Think of it as a waiter at a restaurant who takes your order (user ID) and brings back your food (user information).

The following are the key points to be noted from the preceding code:

- **Interface**: `UserService` defines the service contract, promising to get a user by ID.

- **Implementation**: `UserServiceImpl` provides the actual logic for fetching the user.

- **Reactive**: It uses `Mono` from reactive streams, meaning the user data is delivered asynchronously, like a waiter who tells you your food is coming later.

- **Fetching data**: The code simulates fetching user data (sleeping for `600` milliseconds) and then returns a `User` object with the ID and name.

- **Usage**: You call `getUserById` with a user ID, and it returns `Mono` containing the user data. You can then `subscribe` to `Mono` to receive the user information later when it's ready.

In short, this code shows how to define and implement a reactive service in Java using an interface and `Mono` to handle asynchronous data retrieval.

Cloud-native concurrency frameworks

Akka is a popular cloud-native concurrency framework that provides powerful tools for building highly scalable and resilient applications. It offers features such as actor-based message passing, fault tolerance, and resource management. Here is an example of handling user requests asynchronously:

```
public class UserActor extends AbstractActor {
  public static Props props() {
        return Props.create(UserActor.class);
    }
    @Override
    public Receive createReceive() {
        return receiveBuilder()
            .match(GetUserRequest.class, this::handleGetUserRequest)
            .build();
    }
    private void handleGetUserRequest(GetUserRequest request) throws
    InterruptedException {
        // Simulate fetching user data
        Thread.sleep(600);
        User user = new User(request.getUserId(), "Jack Smith");
        getSender().tell(new GetUserResponse(user), getSelf());
    }
}

// Example usage
public class ActorManager {
    private ActorSystem system;
    private ActorRef userActor;
```

```java
    private ActorRef printActor;

    public ActorManager() {
        system = ActorSystem.create("my-system");
        userActor = system.actorOf(UserActor.props(), "user-actor");
        printActor = system.actorOf(PrintActor.props(), "print-
        actor");
    }

    public void start() {
        // Send request to UserActor and expect PrintActor to handle
        the response
        userActor.tell(new GetUserRequest("9986"), printActor);
    }

    public void shutdown() {
        system.terminate();
    }

    public static void runActorSystem() {
        ActorManager manager = new ActorManager();
        manager.start();

        // Ensure system doesn't shutdown immediately
        try {
            // Wait some time before shutdown to ensure the response
            is processed
            Thread.sleep(5000);
        } catch (InterruptedException e) {
            Thread.currentThread().interrupt();
        }

        manager.shutdown();
    }

    public static void main(String[] args) {
        // Start the actor system
        runActorSystem();
    }
}
```

In the example provided, `UserActor` class in Akka defines an actor that handles user requests asynchronously. The class provides a static `props()` method for actor instantiation, encapsulating its creation logic. In the `createReceive()` method, the actor defines its behavior by using the `receiveBuilder()` to match messages of type `GetUserRequest`. When such a message is received, it delegates the handling to the private `handleGetUserRequest()` method.

The `handleGetUserRequest()` method simulates a delay of 600 milliseconds to represent fetching user data. After the delay, it creates a `User` object with the provided user ID and a hardcoded name `"Jack Smith"`. The actor then sends a `GetUserResponse` message containing the `User` object back to the sender using `getSender()` and `getSelf()`. This design ensures that each request is processed independently, allowing the system to handle multiple concurrent requests efficiently.

In short, this code uses an actor model to handle user requests asynchronously. The actor receives tasks, works on them, and sends back results, making your application more responsive and efficient.

Distributed coordination with ZooKeeper

Imagine that our application now scales to multiple nodes. To maintain a consistent state across the nodes and prevent conflicts, we can utilize a distributed coordination tool such as ZooKeeper. The following is an example of using ZooKeeper:

```
// Connect to ZooKeeper server
CuratorFramework zkClient = CuratorFrameworkFactory.
newClient(zkConnectionString);
zkClient.start();


// Create a persistent node to store the latest processed request ID
String zkNodePath = "/processed-requests";
zkClient.create().creatingParentsIfNeeded().forPath(zkNodePath);


// Implement request processing logic
public void processRequest(String requestId) {
  // Check if the request has already been processed
  if (zkClient.checkExists().forPath(zkNodePath + "/" + requestId) !=
  null) {
     System.out.println("Request already processed: " + requestId);
     return;
  }
  // Process the request
  // ...
  // Mark the request as processed in ZooKeeper
  zkClient.create().forPath(zkNodePath + "/" + requestId);
}
```

This code snippet sets up a simple system to track processed requests using ZooKeeper, a distributed coordination service. Here's a breakdown:

- **Connect and create a node**: It connects to ZooKeeper and creates a persistent node called / `processed-requests` to store processed request IDs.

- **Check for existing requests**: Before processing a new request (identified by `requestId`), the code checks whether a node with that ID already exists under the called/processed-requests node.

- **Process and mark**: If the request hasn't been processed before, the actual processing logic is executed. Then, a new node with `requestId` is created under the called or processed-requests to mark it as processed.

Think of it like a checklist in ZooKeeper. Each request has its own checkbox. Checking it means it's been dealt with. This ensures that requests are not processed multiple times, even if the connection drops or the server restarts.

By integrating these best practices into your development process, you can build cloud-native applications that are not only highly functional but also robust and resilient to the complexities of concurrency. Remember, embracing concurrency-first design is not just about solving immediate problems. It's also about building a foundation for future scalability and sustainable growth in the dynamic cloud environment.

Ensuring consistency – the bedrock of robust concurrency strategies

In the dynamic realm of cloud-native applications, concurrency is an ever-present companion. Achieving consistent concurrency strategies throughout your application is crucial for ensuring its reliability, scalability, and performance. This consistent approach offers several key benefits.

Predictability and stability

Consistent concurrency strategies unify your code, simplifying development and boosting stability through predictable behavior:

- **Uniformity**: Utilizing consistent concurrency strategies across the application promotes predictability and stability. Developers can rely on established patterns and behaviors, leading to easier code comprehension, maintenance, and debugging.

- **Reduced complexity**: By avoiding a patchwork of ad hoc solutions, developers can focus on core functionalities instead of constantly reinventing the wheel for concurrency management.

Leveraging standard libraries and frameworks

Leverage established libraries and frameworks for built-in expertise, optimized performance, and reduced development overhead in concurrent projects:

- **Reliability and expertise**: Utilizing established libraries and frameworks designed for concurrent programming leverages the expertise and best practices embedded within them. These tools often offer built-in thread safety, error handling, and performance optimizations.

- **Reduced overhead**: Standard libraries often offer optimized implementations for common concurrency tasks, reducing development time and overhead compared to building custom solutions from scratch.

Pitfalls of ad hoc solutions

Consider the potential issues with using ad hoc concurrency solutions:

- **Hidden bugs and pitfalls**: Ad hoc concurrency solutions can introduce subtle bugs and performance issues that are difficult to detect and debug. These problems may surface only under specific conditions or high loads, leading to unexpected failures.

- **Maintainability challenges**: Implementing and maintaining ad hoc solutions can become cumbersome and error-prone over time. This complexity can hinder future development and collaboration efforts.

Shared standards and reviews for robust code

Shared guidelines and reviews prevent concurrency chaos, ensuring consistent, reliable code through teamwork:

- **Establish guidelines and standards**: Define clear guidelines and standards for concurrency management within your development team. This should include preferred libraries, frameworks, and coding practices to be followed.

- **Utilize code reviews and peer programming**: Encourage code reviews and peer programming practices to identify potential concurrency issues early and ensure adherence to established guidelines. Consider using checklists or specific review techniques tailored for concurrency concerns.

Emphasize testing and quality assurance

Concurrency in cloud-native Java applications introduces unique testing challenges. To ensure robust and resilient applications, address these challenges head-on with targeted testing strategies:

- **Concurrency-focused unit testing**: Use unit tests to isolate and examine the behavior of individual components under concurrent scenarios. This includes testing for thread safety and handling of shared resources.

- **Integration testing for distributed interactions**: Conduct integration tests to ensure that different components interact correctly under concurrent conditions, especially in microservices architectures common in cloud environments.

- **Performance and stress testing**: Stress test your application under high load to uncover issues such as deadlocks or livelocks that only emerge under specific conditions or heavy concurrent access.

- **Automated testing for efficiency**: Implement automated tests using frameworks such as JUnit, focusing on scenarios that mimic concurrent operations. Use mock testing frameworks to simulate complex concurrency scenarios and dependencies.

- **Concurrency testing tools**: Leverage tools such as JMeter, Gatling, Locust, or Tsung to test how your application handles high concurrent loads. This helps you identify performance bottlenecks and scalability issues in cloud-native environments.

- **Ongoing commitment**: Maintaining consistent concurrency strategies is an ongoing commitment. Regularly review and revise your approach as your application evolves and new libraries, frameworks, and best practices emerge. By fostering a culture of consistency and continuous improvement, you can build reliable, scalable, and performant cloud-native applications that thrive in the ever-changing digital landscape.

Maintaining consistent concurrency strategies is an ongoing commitment. Regularly review and revise your approach as your application evolves and new libraries, frameworks, and best practices emerge. By fostering a culture of consistency and continuous improvement, you can build reliable, scalable, and performant cloud-native applications that thrive in the ever-changing digital landscape.

Summary

Chapter 1 introduced the fundamental concepts of Java cloud-native development, focusing on concurrency and parallelism. It distinguished between managing tasks on single-core (concurrency) versus multi-core processors (parallelism), with practical Java examples. The chapter highlighted Java's role in cloud computing, emphasizing its scalability, ecosystem, and community. Practical applications, including the Java AWS SDK and Lambda functions, illustrated Java's adaptability across cloud models.

Significant Java updates such as Project Loom and advanced garbage collection methods were discussed for optimizing performance. Java's effectiveness in complex environments was showcased through case studies of Netflix and X (formerly Twitter), among others. These focused on microservices, real-time data processing, and scalability.

The narrative then shifted to practical strategies for distributed transactions, data consistency, and microservices state management. The chapter advocated for consistent concurrency strategies in cloud-native applications. It concluded with resources for further exploration and tools for mastering Java concurrency and parallelism, equipping developers to build scalable cloud-native applications. The foundation that has been set here will lead to deeper explorations of Java's concurrency mechanisms in subsequent chapters.

Next, we will transition to a new chapter that delves into the foundational principles of concurrency within the Java ecosystem.

Exercise – exploring Java executors

Objective: In this exercise, you will explore different types of executors provided by the Java Concurrency API. You will refer to the Java documentation, use a different executor implementation, and observe its behavior in a sample program.

Instructions:

- Visit the Java documentation for the `Executors` class: `https://docs.oracle.com/javase/8/docs/api/java/util/concurrent/Executors.html`.

- Read through the documentation and familiarize yourself with the different factory methods provided by the `Executors` class for creating `Executor` instances.

- Choose a different executor implementation other than the fixed thread pool used in the previous examples. Some options include the following:

  - `Executors.newCachedThreadPool()`

  - `Executors.newSingleThreadExecutor()`

  - `Executors.newScheduledThreadPool(int corePoolSize)`

- Create a new Java class called `ExecutorExploration` and replace the Executor creation line with the chosen executor implementation. For example, if you chose `Executors.newCachedThreadPool()`, your code would look like this:Top of Form

  ```
  ExecutorService executor = Executors.newCachedThreadPool();
  ```

- Modify the task creation and submission logic to create and submit a larger number of tasks (e.g., 100 tasks) to the executor. Here's an example of how you can modify the code to create and submit 100 tasks to the executor:

  ```
  public class ExecutorExploration {
      public static void main(String[] args) {
          ExecutorService executor = Executors.
          newCachedThreadPool();
          // Create and submit 100 tasks to the Executor
          for (int i = 0; i < 100; i++) {
              int taskId = i;
              executor.submit(() -> {
                  System.out.println("Task " + taskId + " executed
                  by " + Thread.currentThread().getName());
                  // Simulating task execution time
  ```

```
                              try {
                              Thread.sleep(1000);
                              } catch (InterruptedException e) {
                              e.printStackTrace();
                          }
                      });
                  }
                  // Shutdown the Executor
                  executor.shutdown();
                  }
          }
```

- Run the program and observe the behavior of the chosen executor. Take note of how it handles the submitted tasks and any differences compared to the fixed thread pool.

- Experiment with different executor implementations and observe how they behave differently in terms of task execution, thread creation, and resource utilization.

- Consider the following questions:

  - How does the chosen executor handle the submitted tasks compared to the fixed thread pool?

  - Are there any differences in the order or concurrency of task execution?

  - How does the executor manage threads and resource allocation?

- Feel free to refer back to the Java documentation to understand the characteristics and use cases of each executor implementation.

By completing this exercise, you will gain hands-on experience with different types of executors in Java and understand their behavior and use cases. This knowledge will help you make informed decisions when choosing an appropriate executor for your specific concurrency requirements in Java applications.

Remember to review the Java documentation, experiment with different executor implementations, and observe their behavior in action. Happy exploring!

Questions

1. What is the primary advantage of using microservices in cloud-based Java applications?

    A. Increased security through monolithic architecture

    B. Easier to scale and maintain individual services

    C. Eliminating the need for databases

    D. Unified, single-point configuration for all services

2. In Java concurrency, which mechanism is used to handle multiple threads trying to access a shared resource simultaneously?

    A. Inheritance

    B. Synchronization

    C. Serialization

    D. Polymorphism

3. Which of the following is NOT a feature of Java's `java.util.concurrent` package?

    A. Fork/join framework

    B. `ConcurrentHashMap`

    C. `ExecutorService`

    D. Stream API

4. In serverless computing, which feature is a key benefit when using Java?

    A. Static typing

    B. Manual scaling

    C. Automatic scaling and management of resources

    D. Low-level hardware access

5. What is a common challenge when managing distributed data in Java cloud applications?

    A. Graphics rendering

    B. Data consistency and synchronization

    C. Single-thread execution

    D. User interface design

2

Introduction to Java's Concurrency Foundations: Threads, Processes, and Beyond

Welcome to *Chapter 2*, where we embark on a culinary-inspired exploration of Java's concurrency model, likening it to a bustling kitchen. In this dynamic environment, **threads** are akin to nimble sous chefs, each skillfully managing their specific tasks with speed and precision. They work in unison, seamlessly sharing the kitchen space and resources. Imagine each thread whisking through their assigned recipes, contributing to the overall culinary process in a synchronized dance.

On the other hand, processes are comparable to larger, independent kitchens, each equipped with their unique menus and resources. These processes operate autonomously, handling complex tasks in their self-contained domains without the interference of neighboring kitchens.

In this chapter, we delve into the nuances of these two essential components of Java's concurrency. We'll explore the life cycle of a thread and understand how it wakes up, performs its duties, and eventually rests. Similarly, we'll examine the independent freedom and resource management of processes. Our journey will also take us through the `java.util.concurrent` package, a well-stocked pantry of tools designed for orchestrating threads and processes with efficiency and harmony. By the end of this chapter, you'll gain a solid understanding of how to manage these concurrent elements, enabling you to build robust and efficient Java applications.

Technical requirements

You need to install a Java **integrated development environment** (**IDE**) on your laptop. Here are a few Java IDEs and their download URLs:

- **IntelliJ IDEA**:
  - **Download URL**: `https://www.jetbrains.com/idea/download/`
  - **Pricing**: Free Community Edition with limited features, Ultimate Edition with full features requires a subscription

- **Eclipse IDE**:
  - **Download URL**: `https://www.eclipse.org/downloads/`
  - **Pricing**: Free and open source

- **Apache NetBeans**:
  - **Download URL**:`https://netbeans.apache.org/front/main/download/index.html`
  - **Pricing**: Free and open source

- **Visual Studio Code (VS Code)**:
  - **Download URL**: `https://code.visualstudio.com/download`
  - **Pricing**: Free and open source

VS Code offers a lightweight and customizable alternative to the other options on this list. It's a great choice for developers who prefer a less resource-intensive IDE and want the flexibility to install extensions tailored to their specific needs. However, it may not have all the features out of the box compared to the more established Java IDEs.

Further, the code in this chapter can be found on GitHub:

`https://github.com/PacktPublishing/Java-Concurrency-and-Parallelism`

Java's kitchen of concurrency – unveiling threads and processes

Mastering Java's concurrency tools, threads and processes, is akin to acquiring the skills of a culinary master. This section equips you with the knowledge to design efficient and responsive Java applications, ensuring your programs run smoothly even when juggling multiple tasks like a Michelin-starred kitchen.

What are threads and processes?

In the realm of Java concurrency, **threads** are like sous chefs in a kitchen. Each sous chef (thread) is assigned a particular task, working diligently to contribute to the overall meal preparation. Just as sous chefs share a common kitchen space and resources, threads operate in parallel within the same Java process, sharing memory and resources.

Now, picture a large restaurant with separate kitchens for different specialties, such as a pizza oven room, a pastry department, and a main course kitchen. Each of these is a **process**. Unlike threads that share a single kitchen, processes have their own dedicated resources and operate independently. They're like separate restaurants, ensuring that complex dishes such as intricate pastries get the dedicated attention they deserve, without interfering with the main course preparation.

In essence, threads are like nimble sous chefs sharing the kitchen, while processes are like independent restaurant kitchens with dedicated chefs and resources.

Similarities and differences

Imagine our bustling restaurant kitchen once again, this time buzzing with both threads and processes. While they both contribute to a smooth culinary operation, they do so in distinct ways, like skilled chefs with different specialties. Let's dive into their similarities and differences.

Both threads and processes share the following similarities:

- **Multitasking masters**: Both threads and processes allow Java applications to handle multiple tasks concurrently. Imagine serving multiple tables simultaneously, with no single dish left waiting.

- **Resource sharing**: Both threads within a process and processes themselves can share resources, such as files or databases, depending on their configuration. This allows for efficient data access and collaboration.

- **Independent execution**: Both threads and processes have their own independent execution paths, meaning they can run their own instructions without interrupting each other. Think of separate chefs working on different dishes, each following their own recipe.

Threads and processes are different in the following areas:

- **Scope**: Threads exist within a single process, sharing their memory space and resources like ingredients and cooking tools. Processes, on the other hand, are completely independent, each with its own isolated kitchen and resources.

- **Isolation**: Threads share the same memory space, making them susceptible to interference and data corruption. Processes, with their separate kitchens, offer greater isolation and security, preventing accidental contamination and protecting sensitive data.

- **Creation and management**: Creating and managing threads is simpler and more lightweight within a process. Processes, as independent entities, require more system resources and are more complex to control.

- **Performance**: Threads offer finer-grained control and can be switched quickly, potentially faster execution for smaller tasks. Processes, with their separate resources, can be more efficient for larger, independent workloads.

Both threads and processes are valuable tools in the Java chef's toolbox, each fulfilling specific needs. By understanding their similarities and differences, we can choose the right approach to create culinary masterpieces or, rather, masterful Java applications!

The life cycle of threads in Java

In exploring the life cycle of a thread, akin to the work shift of a sous chef in our kitchen metaphor, we focus on the pivotal stages that define a thread's existence within a Java application:

- **New state**: When a thread is created using the new keyword or by extending the Thread class, it enters the **new state**. It is akin to a sous chef arriving at the kitchen, ready but not yet engaged in cooking.

- **Runnable state**: The thread transitions to the **runnable state** when the `start()` method is called. Here, it's akin to the sous chef prepped and waiting for their turn to cook. The thread scheduler decides when to allocate **central processing unit** (**CPU**) time to the thread, based on thread priorities and system policies.

- **Running state**: Once the thread scheduler allocates CPU time to a thread, it enters the **running state** and begins executing the `run()` method. This is similar to the sous chef actively working on their assigned tasks in the kitchen. At any given moment, only one thread can be in the Running state on a single processor core.

- **Blocked/waiting state**: Threads enter the **blocked/waiting state** when unable to proceed without certain conditions just as a sous chef pauses their work when waiting for ingredients. This includes situations where threads wait for resources to be freed by other threads, such as when calling the `wait()`, `join()`, or `sleep()` methods.

- **Timed waiting state**: Threads can also enter a **timed waiting state** by invoking methods with a specified timeout, such as `sleep(long milliseconds)` or `wait(long milliseconds)`. This is comparable to a sous chef taking a scheduled break during their shift, knowing they will resume work after a certain time has elapsed.

- **Terminated state**: A thread reaches the **terminated state** when it completes the execution of its `run()` method or is interrupted using the `interrupt()` method. This is comparable to a sous chef finishing their tasks and ending their shift. Once terminated, a thread cannot be restarted.

This life cycle is crucial to understanding how threads are managed within a Java program. It dictates how threads are born (created), run (`start()` and `run()`), pause (`wait()`, `join()`, `sleep()`), wait with a timeout (`sleep(long)`, `wait(long)`), and ultimately end their execution (completing `run()` or being interrupted). Understanding these key methods and their impact on thread states is essential for effective concurrent programming.

Now, let's take this knowledge to the real world and explore how threads are used in everyday Java applications!

Activity – differentiating threads and processes in a practical scenario

In the vibrant kitchen of Java concurrency, the following Java code demonstrates how threads (chefs) perform tasks (preparing dishes) within a process (the kitchen). This analogy will help illustrate the concepts of thread and process in a real-world scenario:

```java
import java.util.concurrent.ExecutorService;
import java.util.concurrent.Executors;
public class KitchenSimulator {
    private static final ExecutorService kitchen = Executors.
    newFixedThreadPool(3);
    public static void main(String[] args) {
        String dishToPrepare = "Spaghetti Bolognese";
        String menuToUpdate = "Today's Specials";
        kitchen.submit(() -> {
            prepareDish(dishToPrepare);
        });
        kitchen.submit(() -> {
            searchRecipes("Italian");
        });
        kitchen.submit(() -> {
            updateMenu(menuToUpdate, "Risotto alla Milanese");
        });
        kitchen.shutdown();
    }
    private static void prepareDish(String dish) {
        System.out.println("Preparing " + dish);
        try {
            Thread.sleep(1000);
        } catch (InterruptedException e) {
            Thread.currentThread().interrupt();
        }
    }
```

```
private static void searchRecipes(String cuisine) {
    System.out.println("Searching for " + cuisine + " recipes");
    try {
            Thread.sleep(1000);
    } catch (InterruptedException e) {
            Thread.currentThread().interrupt();
    }
}
private static void updateMenu(String menu, String dishToAdd) {
    System.out.println("Updating " + menu + " with " + dishToAdd);
    try {
            Thread.sleep(1000);
    } catch (InterruptedException e) {
            Thread.currentThread().interrupt();
    }
}
}
```

Here's a breakdown of the roles of threads in the preceding Java code:

- **Chefs**: Each chef in the code represents a thread. The `ExecutorService` kitchen creates a pool of three threads to simulate three chefs working concurrently.

- **Tasks**: The `submit()` method assigns tasks (preparing a dish, searching for recipes, updating the menu) to individual threads within the pool.

- **Concurrent execution**: Threads enable these tasks to run simultaneously, potentially improving performance and responsiveness.

- **Simulation of work**: After each task is executed (printing a message), the thread sleeps for 1000 milliseconds (1 second) using `Thread.sleep(1000)`. This simulates the time taken by the chef to perform the task. During this sleep period, other threads can continue their execution, demonstrating the concurrent nature of the program.

- **Exception handling**: Since `Thread.sleep()` can throw `InterruptedException`, each task is wrapped in a try-catch block. If an interruption occurs during sleep, the exception is caught, and the thread's interrupted status is restored using `Thread.currentThread().interrupt()`. This ensures proper handling of interruptions.

The following points present a discussion on the roles of processes in the preceding Java code:

- **Java runtime**: The entire Java program, including the kitchen simulation, runs within a single operating system process

- **Resource allocation**: The process has its own memory space, allocated by the operating system, to manage variables, objects, and code execution
- **Environment**: It provides the environment for threads to exist and operate within

The key takeaways from the code example we just saw are as follows:

- **Threads within a process**: Threads are lightweight execution units that share the same process's memory and resources
- **Concurrency**: Threads enable multiple tasks to be executed concurrently within a single process, taking advantage of multiple CPU cores if available
- **Process management**: The operating system manages processes, allocating resources and scheduling their execution

Now, let's shift gears and explore the tools that unlock their full potential: the `java.util.concurrent` package. This treasure trove of classes and interfaces provides the building blocks for crafting robust and efficient concurrent programs, ready to tackle any multitasking challenge your Java app throws at them!

The concurrency toolkit – java.util.concurrent

Think of your Java application as a bustling restaurant. Orders stream in, ingredients need prepping, and dishes must be cooked and delivered seamlessly. Now, imagine managing this chaos without efficient systems – it's a recipe for disaster! Fortunately, the `java.util.concurrent` package acts as your restaurant's high-tech equipment, streamlining operations and preventing chaos. With sophisticated tools such as thread pools for managing chefs (threads), locks and queues for coordinating tasks, and powerful concurrency utilities, you can orchestrate your Java application like a Michelin-starred chef. So, dive into this toolkit and unlock the secrets of building smooth, responsive, and efficient Java programs that truly wow your users.

Let's take a glimpse at the key elements within this package.

Threads and executors

Both `ExecutorService` and `ThreadPoolExecutor` play crucial roles in orchestrating concurrent tasks:

- `ExecutorService`: A versatile interface for managing thread pools:

  - **Abstract interface**: It defines a high-level **application programming interface** (**API**) for managing thread pools and executing tasks asynchronously

  - **Focus on task management**: It encapsulates thread pool creation and management, providing methods for submitting tasks, controlling execution, and handling results

- **Flexibility**: It offers various implementations for different thread pool behaviors such as the following:

  - `FixedThreadPool` for a fixed number of threads

  - `CachedThreadPool` for a pool that grows as needed

  - `SingleThreadExecutor` for sequential execution

  - `ScheduledThreadPool` for delayed or periodic tasks

- `ThreadPoolExecutor`: A concrete implementation of `ExecutorService`:

  - **Concrete implementation**: It's the core implementation of `ExecutorService`, providing fine-grained control over thread pool behavior.

  - **Granular control**: It allows you to customize thread pool parameters such as the following:

    - Core pool size (initial threads)

    - Maximum pool size (maximum threads)

    - Keep-alive time (idle thread timeout)

    - Queue capacity (waiting tasks)

  - **Direct usage**: It involves instantiating it directly in your code. This approach gives you complete control over the thread pool's behavior, as you can specify parameters such as core pool size, maximum pool size, keep-alive time, queue capacity, and thread factory. This method is suitable when you need fine-grained control over the thread pool characteristics. Here's an example of direct usage:

```
import java.util.concurrent.ArrayBlockingQueue;
import java.util.concurrent.ThreadPoolExecutor;
import java.util.concurrent.TimeUnit;
import java.util.stream.IntStream;

public class DirectThreadPoolExample {
    public static void main(String[] args) {
        int corePoolSize = 2;
        int maxPoolSize = 4;
        long keepAliveTime = 5000;
        TimeUnit unit = TimeUnit.MILLISECONDS;
        int taskCount = 15; // Make this 4, 10, 12, 14, and
        finally 15 and observe the output.
        ArrayBlockingQueue<Runnable> workQueue = new
        ArrayBlockingQueue<>(10);
```

```
ThreadPoolExecutor executor = new
ThreadPoolExecutor(corePoolSize, maxPoolSize,
keepAliveTime, unit, workQueue);
IntStream.range(0, taskCount).forEach(
    i -> executor.execute(
        () -> System.out.println(
            String.format("Task %d executed. Pool size
            = %d. Queue size = %d.", i,
            executor.getPoolSize(),
            executor. getQueue().size())
        )
    )
);

executor.shutdown();
executor.close();
    }
}
```

In this example, `ThreadPoolExecutor` is directly instantiated with specific parameters. It creates a thread pool with a core pool size of 2, a maximum pool size of 4, a keep-alive time of 5000 milliseconds, and a work queue capacity of 10 tasks.

The code uses `IntStream.range()` and `forEach` to submit tasks to the thread pool. Each task prints a formatted string containing the task number, current pool size, and queue size.

You need to select the right tool for your tasks based on the requirements. You may keep the following in mind:

- **General task management**: Use `ExecutorService` for most cases, benefiting from its simplicity and flexibility in choosing appropriate implementations

- **Specific requirements**: Use `ThreadPoolExecutor` when you need precise control over thread pool configuration and behavior

Understanding their strengths and use cases, you can expertly manage thread pools and unlock the full potential of concurrency in your Java projects.

The next group of elements in this package is synchronization and coordination. Let us explore this category in the next section.

Synchronization and coordination

Synchronization and coordination are crucial in multi-threaded environments to manage shared resources and ensure thread-safe operations. Java provides several classes and interfaces for this purpose, each serving specific use cases in concurrent programming:

- **Lock**: A flexible interface for controlling access to shared resources:

  - **Exclusive access**: Assert fine-grained control over shared resources, ensuring only one thread can access a critical section of code at a time

  - **Use cases**: Protecting shared data structures, coordinating access to files or network connections, and preventing race conditions

- `Semaphore`: A class for managing access to a limited pool of resources, preventing resource exhaustion:

  - **Resource management**: This regulates access to a pool of resources, allowing multiple threads to share a finite number of resources concurrently

  - **Use cases**: Limiting concurrent connections to a server, managing thread pools, and implementing producer-consumer patterns

- `CountDownLatch`: This is also a class in the `java.util.concurrent` package, which allows threads to wait for a set of operations to complete before proceeding:

  - **Task coordination**: Synchronize threads by requiring a set of tasks to complete before proceeding. Threads wait at the latch until a counter reaches zero.

  - **Use cases**: Waiting for multiple services to start before launching an application, ensuring initialization tasks finish before starting a main process, and managing test execution order.

- `CyclicBarrier`: This is another class in the `java.util.concurrent` package, used for synchronizing threads that perform interdependent tasks. Unlike `CountDownLatch`, `CyclicBarrier` can be reused after the waiting threads are released:

  - **Barrier for synchronization**: Gather a group of threads at a common barrier point, allowing them to proceed only when all threads have reached that point

  - **Use cases**: Dividing work among threads and then regrouping, performing parallel computations followed by a merge operation, and implementing rendezvous points

Each of these tools serves a distinct purpose in coordinating threads and ensuring harmonious execution.

The last group in the package is concurrent collections and atomic variables.

Concurrent collections and atomic variables

Concurrent collections are designed specifically for thread-safe storage and retrieval of data in multi-threaded environments. Key members include `ConcurrentHashMap`, `ConcurrentLinkedQueue`, and `CopyOnWriteArrayList`. These collections offer thread-safe operations without the need for external synchronization.

Atomic variables provide thread-safe operations for simple variables (integers, longs, references), eliminating the need for explicit synchronization in many cases. Key members include `AtomicInteger`, `AtomicLong`, and `AtomicReference`.

For a more detailed discussion on the advanced uses and optimized access patterns of these concurrent collections, refer to the *Leveraging thread-safe collections to mitigate concurrency issues* section later in this chapter.

Next, we will look at a code example to see how java.util.concurrent is implemented.

Hands-on exercise – implementing a concurrent application using java.util.concurrent tools

For this hands-on exercise, we'll create a simulated real-world application that demonstrates the use of various `java.util.concurrent` elements.

Scenario: Our application will be a basic order processing system where orders are placed and processed in parallel, and various concurrent elements are utilized to manage synchronization, coordination, and data integrity. Here is the Java code example:

```
import java.util.concurrent.*;
import java.util.concurrent.atomic.AtomicInteger;
import java.util.concurrent.locks.Lock;
import java.util.concurrent.locks.ReentrantLock;


public class OrderProcessingSystem {

    private final ExecutorService executorService = Executors.
    newFixedThreadPool(10);
    private final ConcurrentLinkedQueue<Order> orderQueue = new
    ConcurrentLinkedQueue<>();
    private final CopyOnWriteArrayList<Order> processedOrders = new
    CopyOnWriteArrayList<>();
    private final ConcurrentHashMap<Integer, String> orderStatus = new
    ConcurrentHashMap<>();
    private final Lock paymentLock = new ReentrantLock();
    private final Semaphore validationSemaphore = new Semaphore(5);
    private final AtomicInteger processedCount = new AtomicInteger(0);
```

```java
public void startProcessing() {
    while (!orderQueue.isEmpty()) {
        Order order = orderQueue.poll();
        executorService.submit(() -> processOrder(order));
    }
    executorService.close();
}

private void processOrder(Order order) {
    try {
        validateOrder(order);
        paymentLock.lock();
        try {
            processPayment(order);
        } finally {
            paymentLock.unlock();
        }
        shipOrder(order);
        processedOrders.add(order);
        processedCount.incrementAndGet();
        orderStatus.put(order.getId(), "Completed");
    } catch (InterruptedException e) {
        Thread.currentThread().interrupt();
    }
}

private void validateOrder(Order order) throws
InterruptedException {
    validationSemaphore.acquire();
    try {
        Thread.sleep(100);
    } finally {
        validationSemaphore.release();
    }
}

private void processPayment(Order order) {
    System.out.println("Payment Processed for Order " + order.
    getId());
}

private void shipOrder(Order order) {
    System.out.println("Shipped Order " + order.getId());
}
```

```java
public void placeOrder(Order order) {

    orderQueue.add(order);
    orderStatus.put(order.getId(), "Received");
    System.out.println("Order " + order.getId() + " placed.");
}

public static void main(String[] args) {

    OrderProcessingSystem system = new OrderProcessingSystem();

    for (int i = 0; i < 20; i++) {
        system.placeOrder(new Order(i));
    }
    system.startProcessing();
    System.out.println("All Orders Processed!");
}

static class Order {
    private final int id;
    public Order(int id) {
        this.id = id;
    }
    public int getId() {
        return id;
    }
}
}
```

The preceding code example uses many concurrency elements such as the following:

- `ExecutorService` is used to handle multiple tasks (order processing) in a thread pool, enabling parallel execution

- `ConcurrentLinkedQueue` is a thread-safe queue used to hold and manage orders efficiently in a concurrent environment

- `CopyOnWriteArrayList` provides a thread-safe list implementation, suitable for storing processed orders where iteration is more frequent than modification

- `ConcurrentHashMap` offers a high-performance, thread-safe map to track the status of each order

- `ReentrantLock` is used to ensure exclusive access to the payment processing section of the code, thus avoiding concurrency issues

- `Semaphore` controls the number of concurrent validations, preventing resource exhaustion

- `AtomicInteger` is a thread-safe integer, used for counting processed orders safely in a concurrent context

Each of these classes and interfaces plays a vital role in ensuring thread safety and efficient concurrency management in the `OrderProcessingSystem`.

The key points we have learned are as follows:

- **Efficient threading**: This uses a thread pool to handle multiple orders concurrently, potentially improving performance

- **Synchronization**: This employs locks and semaphores to coordinate access to shared resources and critical sections, ensuring data consistency and preventing race conditions

- **Thread-safe data**: This manages orders and statuses with thread-safe collections to support concurrent access

- **Status tracking**: This maintains order statuses for monitoring and reporting

This example demonstrates how these concurrent utilities can be combined to build a multi-threaded, synchronized, and coordinated application for order processing. Each utility serves a specific purpose, from managing concurrent tasks to ensuring data integrity and synchronization among threads.

Next, we will explore how synchronization and locking mechanisms are used in Java applications.

Synchronization and locking mechanisms

Imagine a bakery where multiple customers place orders simultaneously. Without proper synchronization, two orders could be mixed up, ingredients double counted, or payments processed incorrectly. This is where locking steps in, acting like a *hold, please* sign, allowing one thread to use the oven or cash register at a time.

Synchronization and locking mechanisms are the guardians of data integrity and application stability in concurrent environments. They prevent race conditions, ensure atomic operations (complete or not, never partial), and guarantee predictable execution order, ultimately creating a reliable and efficient multi-threaded process.

Let's delve into the world of synchronization and locking mechanisms, explore why they're crucial, and how to wield them effectively to build robust and performant concurrent applications.

The power of synchronization – protecting critical sections for thread-safe operations

In Java, the keyword **synchronized** acts as a gatekeeper for sensitive code blocks. When a thread enters a synchronized block, it acquires a lock on the associated object, preventing other threads from entering the same block until the lock is released. This ensures exclusive access to shared resources and prevents data corruption. There are three different locks:

- **Object-level locks**: When a thread enters a synchronized block, it acquires a lock on the instance of the object associated with the block. This lock is released when the thread exits the block.

- **Class-level locks**: For static methods and blocks, the lock is acquired on the class object itself, ensuring synchronization across all instances of the class.

- **Monitor object**: The **Java virtual machine** (**JVM**) employs a monitor object for each object and class to manage synchronization. This monitor object tracks the thread holding the lock and coordinates access to the locked resource.

In cloud environments, locking mechanisms find their primary applications in several critical areas: coordinating distributed services, accessing shared data, and managing state – specifically maintaining and updating internal state information securely across multiple threads. Beyond traditional synchronization, there exist various alternative and sophisticated locking techniques. Let's delve into these together.

Beyond the gatekeeper – exploring advanced locking techniques

In our exploration of Java's concurrency tools, we've seen basic synchronization methods. Now, let's delve into advanced locking techniques that offer greater control and flexibility for complex scenarios. These techniques are particularly useful in high-concurrency environments or when dealing with intricate resource management challenges:

- **Reentrant locks**: Unlike intrinsic locks, `ReentrantLock` provides the ability to attempt a lock with a timeout, preventing threads from getting indefinitely blocked.

- **Practical scenario**: Imagine managing access to a shared printer in an office. `ReentrantLock` can be used to ensure that if a document is taking too long to print, other jobs can be processed in the meantime, avoiding a bottleneck.

- **Read/write locks**: `ReadWriteLock` allows multiple threads to read a resource concurrently but requires exclusive access for writing.

- **Example use case**: In a stock trading application, where many users are reading stock prices but few are updating them, `ReadWriteLock` optimizes performance by allowing concurrent reads while maintaining data integrity during updates.

- **Stamped locks**: Introduced in Java 8, `StampedLock` offers a mode where a lock can be acquired with an option to convert it to a read or write lock.

- **Application example**: Consider a GPS application where the location is frequently read but occasionally updated. StampedLock allows for more concurrency with the flexibility to upgrade a read lock to a write lock when an update is necessary.

- **Condition objects**: A condition object is a Java class. Normally, it is used with ReentrantLock, which allows threads to communicate about the lock status. A condition object is essentially a more advanced and flexible version of the traditional wait-notify object mechanism.

Let's look at a Java code example demonstrating the use of ReentrantLock with a condition object:

```java
import java.util.concurrent.locks.Condition;
import java.util.concurrent.locks.ReentrantLock;
import java.util.concurrent.TimeUnit;
public class PrinterManager {
    private final ReentrantLock printerLock = new ReentrantLock();
    private final Condition readyCondition = printerLock.
    newCondition();
    private boolean isPrinterReady = false;
    public void makePrinterReady() {
        printerLock.lock();
        try {
            isPrinterReady = true;
            readyCondition.signal(); // Signal one waiting thread that
            the printer is ready
        } finally {
            printerLock.unlock();
        }
    }
    public void printDocument(String document) {
        printerLock.lock();
        try {
            // Wait until the printer is ready
            while (!isPrinterReady) {
                System.out.println(Thread.currentThread().getName() +
                " waiting for the printer to be ready.");
                if (!readyCondition.await(
                    2000, TimeUnit.MILLISECONDS)) {
                    System.out.println(
                        Thread.currentThread().getName()
                            + " could not print. Timeout while waiting
                            for the printer to be ready.");
                    return;
                }
            }
            // Printer is ready. Proceed to print the document
```

```
            System.out.println(Thread.currentThread().getName() + " is
            printing: " + document);
            Thread.sleep(1000); // Simulates printing time

            // Reset the printer readiness for demonstration purposes
            isPrinterReady = false;
        } catch (InterruptedException e) {
            Thread.currentThread().interrupt();
        } finally {
            printerLock.unlock();
        }
    }
}
    public static void main(String[] args) {
        PrinterManager printerManager = new PrinterManager();
        // Simulating multiple threads (office workers) trying to use
           the printer
        Thread worker1 = new Thread(() -> printerManager.
        printDocument("Document1"), "Worker1");
        Thread worker2 = new Thread(() -> printerManager.
        printDocument("Document2"), "Worker2");
        Thread worker3 = new Thread(() -> printerManager.
        printDocument("Document3"), "Worker3");
        worker1.start();
        worker2.start();
        worker3.start();
        // Simulate making the printer ready after a delay
        new Thread(() -> {
            try {
                Thread.sleep(2000); // Simulate some delay
                printerManager.makePrinterReady();
            } catch (InterruptedException e) {
                Thread.currentThread().interrupt();
            }
        }).start();
    }
}
```

In this code, the `PrinterManager` class includes a condition object, `readyCondition`, created from `printerLock`:

- The `printDocument` method makes threads wait if the printer is not ready (`isPrinterReady` is false). Threads call `await()` on `readyCondition`, which suspends them until they are signaled or the timeout occurs.

- The new `makePrinterReady` method simulates an event where the printer becomes ready. When this method is called, it changes the `isPrinterReady` flag to true and calls `signal()` on `readyCondition` to wake up one waiting thread.

- The `main` method simulates the scenario where the printer becomes ready after a delay, and multiple worker threads are trying to use the printer.

- The code assumes a simplistic representation of a printer using a Boolean variable (`isPrinterReady`). In reality, you would need to integrate with the actual printer's API, library, or driver to communicate with the printer and determine its readiness state.

The provided code is a simplified example to demonstrate the concept of thread synchronization and waiting for a condition (in this case, the printer being ready) using locks and conditions in Java. While it illustrates the basic principles, it may not be directly applicable to a real-world scenario without further modifications and enhancements.

By understanding and applying these advanced locking techniques, you can enhance the performance and reliability of your Java applications. Each technique serves a specific purpose and choosing the right one depends on the specific requirements and characteristics of your application.

In the realm of Java's advanced locking techniques, we delve deeper into the mechanics and use cases of tools such as `ReentrantLock`, `ReadWriteLock`, and `StampedLock`. For instance, `ReentrantLock` offers a higher level of control compared to intrinsic locks, with features such as fairness policies and the ability to interrupt lock waiting threads. Consider a scenario where multiple threads are competing to access a shared database. Here, `ReentrantLock` with a fairness policy ensures that threads gain database access in the order they requested it, preventing resource hogging and enhancing system fairness.

Similarly, `ReadWriteLock` splits the lock into two parts: a read lock and a write lock. This separation allows multiple threads to read data simultaneously, but only one thread can write at a time, thereby increasing read efficiency in scenarios where write operations are less frequent, such as in caching systems.

`StampedLock`, on the other hand, provides lock modes that support both read and write locks and also offers a method for lock conversion. Imagine a navigation application where map data is read frequently but updated less often. `StampedLock` can initially grant a read lock to display the map and then convert it to a write lock when an update is needed, minimizing the time during which other threads are prevented from reading the map.

In the next section, we'll explore some common pitfalls to avoid.

Understanding and preventing deadlocks in multi-threaded applications

As we explore the bustling kitchen of Java concurrency, where threads work like sous chefs in a harmonious rhythm, we come across a notorious kitchen hitch – the **deadlock**. Much like sous chefs vying for the same kitchen appliance, threads in Java can find themselves in a deadlock when they wait on each other to relinquish shared resources. Preventing such deadlocks is vital to ensure that our multi-threaded applications, akin to our kitchen operations, continue to run smoothly without any disruptive standstills.

To prevent deadlocks, we can employ several strategies:

- **Avoid circular wait**: We can design our application to prevent the circular chain of dependencies. One way is to impose a strict order in which locks are acquired.

- **Minimize hold and wait**: Try to ensure that a thread requests all the required resources at once, rather than acquiring one and waiting for others.

- **Resource allocation graphs**: Use these graphs to detect the possibility of deadlocks in the system.

- **Timeouts**: Implementing timeouts can be a simple yet effective way. If a thread cannot acquire all its resources within a given timeframe, it releases the acquired resources and retries later.

- **Thread dump analysis**: Regularly analyze thread dumps for signs of potential deadlocks.

After delving into the theoretical aspects of locking mechanisms in cloud environments, we shift gears to practical application. In this next section, we dive into hands-on activities focused on deadlocks, a pivotal challenge in concurrent programming. This hands-on approach aims not just to understand but to develop efficient Java applications in the face of these complex issues.

Hands-on activity – deadlock detection and resolution

We simulate a real-world scenario involving two processes trying to access two database tables. We'll represent the tables as shared resources and the processes as threads. Each thread will try to lock both tables to perform some operations. We'll then demonstrate a deadlock and refactor the code to resolve it.

First, let's create a Java program that simulates a deadlock when two threads try to access two tables (resources):

```
public class DynamoDBDeadlockDemo {
    private static final Object Item1Lock = new Object();
    private static final Object Item2Lock = new Object();
    public static void main(String[] args) {
        Thread lambdaFunction1 = new Thread(() -> {
            synchronized (Item1Lock) {
                System.out.println(
```

```
                        "Lambda Function 1 locked Item 1");
                try { Thread.sleep(100);
                } catch (InterruptedException e) {}
                System.out.println("Lambda Function 1 waiting to lock
                Item 2");
                synchronized (Item2Lock) {
                    System.out.println("Lambda Function 1 locked Item
                    1 & 2");
                }
            }
        });
        Thread lambdaFunction2 = new Thread(() -> {
            synchronized (Item2Lock) {
                System.out.println("Lambda Function 2 locked Item 2");
                try { Thread.sleep(100);
                } catch (InterruptedException e) {}
                System.out.println("Lambda Function 2 waiting to lock
                Item 1");
                synchronized (Item1Lock) {
                    System.out.println("Lambda Function 2 locked Item
                    1 & 2");
                }
            }
        });
        lambdaFunction1.start();
        lambdaFunction2.start();
    }
}
```

In this code, each thread (representing a Lambda function) tries to lock two resources (Item1Lock and Item2Lock) in a nested manner. However, each thread locks one resource and then attempts to lock the other resource that may already be locked by the other thread. This scenario creates a deadlock situation because of the following reasons:

- lambdaFunction1 locks Item1 and waits to lock Item2, which might already be locked by Lambda Function 2

- lambdaFunction2 locks Item2 and waits to lock Item1, which might already be locked by Lambda Function 1

- Both Lambda functions end up waiting indefinitely for the other to release the lock, causing a deadlock

- **Simulated processing delay**: Thread.sleep(100) in each thread is crucial as it simulates a delay, allowing time for the other thread to acquire a lock on the other resource, thus increasing the likelihood of a deadlock

This example illustrates a basic deadlock scenario in a concurrent environment, similar to what might occur in distributed systems involving multiple resources. To resolve the deadlock, we ensure that both threads acquire locks in a consistent order; it prevents a situation where each thread holds one lock and waits for the other. Let us look at this refactoring code:

```java
        // Thread representing Lambda Function 1
 public class DynamoDBDeadlockDemo {
  private static final Object Item1Lock = new Object();
  private static final Object Item2Lock = new Object();

    public static void main(String[] args) {

        Thread lambdaFunction1 = new Thread(() -> {
            synchronized (Item1Lock) {
                System.out.println(
                    "Lambda Function 1 locked Item 1");
                try { Thread.sleep(100);
                } catch (InterruptedException e) {}
                System.out.println("Lambda Function 1 waiting to lock
                Item 2");
                synchronized (Item2Lock) {
                    System.out.println("Lambda Function 1 locked Item
                    1 & 2");

                }

            }
        });

        Thread lambdaFunction2 = new Thread(() -> {
            synchronized (Item1Lock) {
                System.out.println(
                    "Lambda Function 2 locked Item 1");
                try { Thread.sleep(100);
                    } catch (InterruptedException e) {}
                System.out.println("Lambda Function 2 waiting to lock
                Item 2");
                // Then, attempt to lock Item2
                synchronized (Item2Lock) {
                    System.out.println("Lambda Function 2 locked Item
                    1 & 2");

                }

            }
        });
```

```
            lambdaFunction1.start();
            lambdaFunction2.start();
        }
    }
```

Both `lambdaFunction1` and `lambdaFunction2` now acquire the locks in the same order, first `Item1Lock` and then `Item2Lock`. By ensuring that both threads acquire locks in a consistent order, we prevent a situation where each thread holds one lock and waits for the other. This eliminates the deadlock condition.

Let's look at another real-world scenario where two processes are waiting for file access, we can simulate file operations using locks. Each process will try to lock a file (represented as `ReentrantLock`) for exclusive access.

Let's demonstrate this scenario:

```java
import java.util.concurrent.locks.ReentrantLock;
import java.util.concurrent.TimeUnit;

public class FileDeadlockDetectionDemo {
    private static final ReentrantLock fileLock1 = new
    ReentrantLock();
    private static final ReentrantLock fileLock2 = new
    ReentrantLock();

    public static void main(String[] args) {
        Thread process1 = new Thread(() -> {
            try {
                acquireFileLocksWithTimeout(
                    fileLock1, fileLock2);
            } catch (InterruptedException e) {
                if (fileLock1.isHeldByCurrentThread()) fileLock1.
                unlock();
                if (fileLock2.isHeldByCurrentThread()) fileLock2.
                unlock();
            }
        });
        Thread process2 = new Thread(() -> {
            try {
                acquireFileLocksWithTimeout(
                    fileLock2, fileLock1);
            } catch (InterruptedException e) {
                if (fileLock1.isHeldByCurrentThread()) fileLock1.
                unlock();
                if (fileLock2.isHeldByCurrentThread()) fileLock2.
                unlock();
```

```
            }
        });
        process1.start();
        process2.start();
        try {
            Thread.sleep(2000);
            if (process1.isAlive() && process2.isAlive()) {
                System.out.println("Deadlock suspected, interrupting
                process 2");
                process2.interrupt();
            }
        } catch (InterruptedException e) {
            e.printStackTrace();
        }
    }
private static void acquireFileLocksWithTimeout(
    ReentrantLock firstFileLock, ReentrantLock secondFileLock) throws
    InterruptedException {
        if (!firstFileLock.tryLock(1000, TimeUnit.MILLISECONDS)) {
            throw new InterruptedException("Failed to acquire first
            file lock");
        }
        try {
            if (!secondFileLock.tryLock(
                1000, TimeUnit.MILLISECONDS)) {
                throw new InterruptedException(
                    "Failed to acquire second file lock");
            }
            System.out.println(Thread.currentThread().getName() + "
            acquired both file locks");
            try { Thread.sleep(500);
                } catch (InterruptedException e) {}
        } finally {
            if (secondFileLock.isHeldByCurrentThread())
            secondFileLock.unlock();
            if (firstFileLock.isHeldByCurrentThread()) firstFileLock.
            unlock();
        }
    }
}
```

This code demonstrates a technique for detecting and preventing deadlocks when working with concurrent processes that require access to shared resources – in this case, two files represented by `ReentrantLock`. Let's break down how the deadlock occurs and how it is prevented:

- **Deadlock scenario:**

  - **Locks for shared resources:** `fileLock1` and `fileLock2` are `ReentrantLock` objects that simulate locks on two shared files.

  - **Two concurrent processes:** The program creates two threads (`process1` and `process2`), each trying to access both files. However, they attempt to acquire the locks in opposite orders. `process1` tries to lock `fileLock1` first, then `fileLock2`; `process2` does the opposite.

  - **Potential deadlock:** If `process1` locks `fileLock1` and `process2` locks `fileLock2` at the same time, they will each wait indefinitely for the other lock to be released, creating a deadlock situation.

- **Deadlock prevention and recovery:**

  - **Timeout on lock acquisition:** The `acquireFileLocksWithTimeout` method attempts to acquire each lock with a timeout (`tryLock(1000, TimeUnit.MILLISECONDS)`). This timeout prevents a process from waiting indefinitely for a lock, reducing the chance of a deadlock.

  - **Interrupting processes:** The main thread waits for a certain duration (`Thread.sleep(2000)`) and checks whether both processes are still active. If they are, it suspects a deadlock and interrupts one of the processes (`process2.interrupt()`), helping to recover from the deadlock situation.

  - **Releasing locks on interruption:** In the catch block for `InterruptedException`, the program checks whether the current thread holds either lock and, if so, releases it. This ensures that resources are not left in a locked state, which could perpetuate the deadlock.

  - **Ensuring lock release:** The final block in the `acquireFileLocksWithTimeout` method guarantees that both locks are released, even if an exception occurs or the thread is interrupted. This is crucial for preventing deadlocks and ensuring resource availability for other processes.

- **Key takeaways:**

  - **Deadlock detection:** The program actively checks for deadlock conditions and takes measures to resolve them

  - **Resource management:** Careful management of lock acquisition and release is essential in concurrent programming to avoid deadlocks

- **Timeouts as a preventive measure**: Using timeouts when attempting to acquire locks can prevent processes from being indefinitely blocked

This approach demonstrates effective strategies for handling potential deadlocks in concurrent processes, especially when dealing with shared resources such as files or database connections in a multi-threaded environment.

In our culinary world of Java concurrency, deadlocks are like kitchen gridlocks where sous chefs find themselves stuck, unable to access the tools they need because another chef is using them. Mastering the art of preventing these kitchen standstills is a crucial skill for any adept Java developer. By understanding and applying strategies to avoid these deadlocks, we ensure that our multi-threaded applications, much like a well-organized kitchen, operate smoothly, deftly handling the intricate dance of concurrent tasks.

Next, we will discuss task management and data sharing in concurrency in Java; it involves understanding how to effectively handle asynchronous tasks and ensuring data integrity across concurrent operations. Let's delve into this topic.

Employing Future and Callable for result-bearing task execution

In Java, Future and Callable are used together to execute tasks asynchronously and obtain results at a later point in time:

- **Callable interface**: A functional interface that represents a task capable of producing a result:

  - `call()`: This method encapsulates the task's logic and returns the result

- **Future interface**: This represents the eventual completion (or failure) of a Callable task and its associated result:

  - `get()`: This method retrieves the result, blocking if necessary until completion

  - `isDone()`: This method checks whether the task is finished

  - **Submitting tasks**: `ExecutorService` accepts Callables, returning Futures for tracking completion and results

Here is an example of Callable and Future interfaces:

```
ExecutorService executor = Executors.newFixedThreadPool(2);
Callable<Integer> task = () -> {
    // perform some computation
    return 42;
};
Future<Integer> future = executor.submit(task);
```

```
// do something else while the task is executing
Integer result = future.get(); // Retrieves the result, waiting if
necessary
// Check if the task is completed
    if (!future.isDone()) {
        System.out.println("Calculation is still in progress...");
    }
executor.shutdown();
```

The Callable interface defines the task that produces a result. The Future interface acts as a handle for managing and retrieving that result, enabling asynchronous coordination and result-bearing task execution.

The key points in this code are as follows:

- **Asynchronous execution**: Callable and Future enable the task to execute independently of the main thread, potentially improving performance

- **Result retrieval**: The Future object allows the main thread to retrieve the task's result when it becomes available, ensuring synchronization

- **Flexible coordination**: Futures can be used for dependency management and creating complex asynchronous workflows

Safe data sharing between concurrent tasks

Immutable data and thread-local storage are fundamental concepts for concurrency and can greatly simplify thread-safe programming. Let's explore them in detail.

Immutable data

Immutable data is a fundamental concept where an object's state cannot be changed once it is created. Any attempt to modify such objects results in the creation of new ones, leaving the original untouched. This is in stark contrast to **mutable data**, where the state of an object can be directly altered after its creation.

Its benefits are as follows:

- **It eliminates the need for synchronization**: When immutable data is shared across threads, there is no need for synchronization mechanisms such as locks or semaphores

- **Enhances thread safety**: Immutability by its very nature guarantees thread-safe operations

- **Simplifies reasoning**: With immutability, there's no concern about unexpected changes from other threads, making the code more predictable and easier to debug

Some examples of immutable data types are as follows:

- **Strings**: In Java, string objects are immutable

- **Boxed primitives**: These include integers and Boolean

- **Date objects**: An example of such an object would be `LocalDate` in Java 8

- **Final classes with immutable fields**: Custom classes designed to be immutable

- **Tuples**: Often used in functional programming languages. Tuples are data structures that store a fixed set of elements where each element can be of a different type. Tuples are immutable, meaning that once created, the values inside them cannot be changed. While Java does not have a built-in tuple class like some other languages (Python, for instance), you can simulate tuples using custom classes or available classes from libraries.

Here is a simple example of how you might create and use a tuple-like structure in Java:

```
public class Tuple<X, Y> {
    public final X first;
    public final Y second;
    public Tuple(X first, Y second) {
        this.first = first;
        this.second = second;
    }
    public static void main(String[] args) {
        // Creating a tuple of String and Integer
        Tuple<String, Integer> personAge = new Tuple<>(
            "Joe", 30);
    }
}
```

Let us now explore thread local storage.

Thread local storage

Thread local storage or **TLS** is a method of storing data that is local to a thread. In this model, each thread has its own separate storage, which is not accessible to other threads.

Its benefits are as follows:

- **Simplifies data sharing**: TLS provides a straightforward approach to storing data specific to each thread, and each thread can access its data independently without the need for coordination

- **Reduces contention**: By keeping data separate for each thread, TLS minimizes potential conflicts and bottlenecks

- **Improves maintainability**: Code that utilizes TLS is often clearer and easier to understand

Some examples of using TLS are discussed in the following points:

- **User session management**: In web applications, storing user-specific data such as sessions
- **Counters or temporary variables**: Keeping track of thread-specific computations
- **Caching**: Storing frequently used, thread-specific data for performance optimization

While both immutable data and TLS contribute significantly to thread safety and simplify concurrency management, they serve different purposes and scenarios:

- **Scope**: Immutable data ensures consistency and safety of the data itself across multiple threads. In contrast, TLS is about providing a separate data storage space for each thread.
- **Use cases**: Use immutable data for shared structures and values that are read-only. TLS is ideal for managing data that is specific to each thread and not meant for cross-thread sharing.

The choice between immutable data and TLS should be based on the specific requirements of your application and the nature of the data access patterns involved. Leveraging both immutable data and TLS can further enhance the safety and simplicity of your concurrent systems, harnessing the strengths of each approach.

Leveraging thread-safe collections to mitigate concurrency issues

Having already explored the basics of concurrent collections and atomic variables, let's focus on advanced strategies for utilizing these thread-safe collections to further mitigate concurrency issues in Java.

The following are the advanced uses of concurrent collections:

- **Optimized access patterns**: Optimized access patterns refer to using specific concurrent collection classes that are designed for particular usage scenarios in multi-threaded environments. These classes provide efficient ways to handle common patterns of concurrent access, such as frequent reads and writes, queue processing, or read-mostly data structures:

  - `ConcurrentHashMap`: Ideal for scenarios with a high volume of concurrent read and write operations. Utilize its advanced functions such as `computeIfAbsent` for atomic operations combining checking and adding elements.

  - `ConcurrentLinkedQueue`: Best for queue-based data processing models, especially in producer-consumer patterns. Its non-blocking nature is essential for high-throughput scenarios.

  - `CopyOnWriteArrayList`: Use when the list is largely read-only but needs occasional modifications. Its iterator provides a stable snapshot view, making it reliable for iterations even when concurrent modifications occur.

- **Combining with streams**: Leverage Java streams for concurrent processing of collections, particularly with `ConcurrentHashMap`. This combination can lead to highly efficient parallel algorithms.

- **Strategic synchronization**: Even with thread-safe collections, some scenarios require additional synchronization. For instance, when iterating over `ConcurrentHashMap` and performing multiple related operations that need to be atomic as a whole.

The following are the advantages of atomic variables in addition to their basic use cases:

- **Complex atomic operations**: They utilize advanced atomic operations such as `updateAndGet` or `accumulateAndGet` in `AtomicInteger` or `AtomicLong`, which allow complex calculations in a single atomic step.

- **Memory consistency features**: They understand the memory consistency guarantees provided by atomic variables and concurrent collections. For instance, operations on `AtomicInteger` are guaranteed to have immediate visibility to other threads, which is crucial for ensuring up-to-date data visibility.

Choosing between concurrent collections and atomic variables

Understanding when to choose concurrent collections and when to use atomic variables is crucial for developing efficient, robust, and thread-safe Java applications. This knowledge allows you to tailor your choice of data structures and synchronization mechanisms to the specific needs and characteristics of your application. Making the right choice between these two options can significantly impact the performance, scalability, and reliability of your concurrent applications. This section delves into the considerations for selecting between concurrent collections, which are ideal for complex data structures, and atomic variables, which are best suited for simpler, single-value scenarios:

- **Data complexity**: Choose concurrent collections for managing complex data structures with multiple elements and relationships. Use atomic variables when dealing with single values requiring atomic operations without the overhead of a full collection structure.

- **Performance considerations**: Balance the choice based on performance implications. `ConcurrentHashMap` has excellent scalability for concurrent access, whereas atomic variables are lightweight and efficient for simpler use cases.

By deepening your understanding of when and how to use these advanced features of thread-safe collections and atomic variables, you can optimize your Java applications for concurrency, ensuring both data integrity and exceptional performance.

Concurrent best practices for robust applications

While *Chapter 5, Mastering Concurrency Patterns in Cloud Computing*, of our book delves into Java concurrency patterns specifically tailored for cloud environments, it is crucial to lay the groundwork with some best practices and general strategies for concurrent programming.

Best practices in concurrent programming include the following:

1. **Master concurrency primitives**: Master the basics of concurrency primitives in Java, such as synchronized, volatile, lock, and condition. Understanding their semantics and usage is crucial for writing correct concurrent code.

2. **Minimize shared state**: Limit the amount of shared state between threads. The more data shared, the higher the complexity and potential for concurrency issues. Aim for immutability where feasible.

3. **Handle thread interruption**: When catching `InterruptedException`, restore the interrupt status by calling `Thread.currentThread().interrupt()`.

- **Avoid deadlocks**: Be vigilant about potential deadlocks. This can be achieved by always acquiring locks in a consistent order and considering timeouts when trying to acquire locks.

- **Use high-level concurrency utilities**: Utilize Java's high-level concurrency utilities, such as `ExecutorService`, `CountDownLatch`, and `CyclicBarrier` to manage threads and synchronization.

- **Use thread pools**: Manage threads efficiently by using thread pools. They help in reusing and managing threads, reducing the overhead of thread creation and destruction.

- **Prefer non-blocking algorithms**: Where performance and scalability are critical, consider non-blocking algorithms. These algorithms, such as those based on `AtomicInteger`, can be more scalable than lock-based approaches.

- **Cautious with lazy initialization**: Lazy initialization in a concurrent setting can be tricky. Double-checked locking with a volatile variable is a common pattern but requires careful implementation to be correct.

- **Test concurrency thoroughly**: Concurrent code should be rigorously tested under conditions that simulate real-world scenarios. This includes testing for thread safety, potential deadlocks, and race conditions.

- **Document concurrency assumptions**: Clearly document the assumptions and design decisions related to concurrency in your code. This helps maintainers understand the concurrency strategies employed.

- **Optimize thread allocation**: Balance the number of threads with the workload and the system's capabilities. Overloading a system with too many threads can lead to performance degradation due to excessive context switching.

- **Monitor and tune performance**: Regularly monitor the performance of your concurrent applications and tune parameters such as thread pool sizes or task partitioning strategies for optimal results.
- **Avoid blocking threads unnecessarily**: Design tasks and algorithms to avoid keeping threads in a blocked state unnecessarily. Utilize concurrent algorithms and data structures that allow threads to progress independently.

These best practices form the bedrock of robust, efficient, and maintainable concurrent applications, irrespective of their specific domain, such as cloud computing.

Summary

As we conclude *Chapter 2*, let's reflect on the essential concepts and best practices we've uncovered in our exploration of Java's concurrency. This summary, akin to a chef's final review of a successful banquet, will encapsulate the crucial learnings and strategies for effective concurrent programming in Java.

We learned about threads and processes. Threads, like nimble sous chefs, are the fundamental units of execution, working in shared environments (kitchens). Processes are like independent kitchens, each with its resources, operating in isolation. We journeyed through a thread's life cycle, from creation to termination, highlighting the critical stages and how they are managed within the Java environment.

Like coordinating a team of chefs, we've explored various synchronization techniques and locking mechanisms essential for managing access to shared resources and preventing conflicts. Next, we tackled the challenge of deadlocks, understanding how to detect and resolve these standstills in concurrent programming, much like resolving bottlenecks in a busy kitchen.

Then, we delved into advanced tools such as `StampedLock` and condition objects. We equipped you with sophisticated methods for specific concurrency scenarios.

A pivotal part of this chapter was the discussion on concurrent best practices for robust applications. We discussed best practices in concurrent programming. These practices are akin to the golden rules in a professional kitchen, ensuring efficiency, safety, and quality. We emphasized the importance of understanding concurrency patterns, proper resource management, and the judicious use of synchronization techniques to build robust and resilient Java applications.

Moreover, through hands-on activities and real-world examples, we've seen how to apply these concepts and practices, enhancing our understanding of when and how to utilize different synchronization strategies and locking mechanisms effectively.

This chapter gave you the tools and best practices to conquer concurrency's complexities. You're now primed to design robust, scalable applications that thrive in the multi-threaded world. However, our culinary journey isn't over! In *Chapter 3, Mastering Parallelism in Java*, we ascend to the grand hall of **parallel processing**, where we'll learn to harness multiple cores for even more potent Java magic. Prepare to leverage your concurrency expertise as we unlock the true power of parallel programming.

Questions

1. What is the primary difference between threads and processes in Java's concurrency model?

    A. Threads and processes are essentially the same.

    B. Threads are independent, while processes share a memory space.

    C. Threads share a memory space, while processes are independent and have their own memory.

    D. Processes are used only in web applications, while threads are used in desktop applications.

2. What is the role of the `java.util.concurrent` package in Java?

    A. It provides tools for building graphical user interfaces.

    B. It offers a set of classes and interfaces for managing threads and processes efficiently.

    C. It is used exclusively for database connectivity.

    D. It enhances the security features of Java applications.

3. Which scenario best illustrates the use of `ReadWriteLock` in Java?

    A. Managing user sessions in a web application.

    B. Allowing multiple threads to read a resource concurrently but requiring exclusive access for writing.

    C. Encrypting sensitive data before sending it over a network.

    D. Serializing objects for saving the state of an application.

4. How does `CountDownLatch` in Java's concurrency model function?

    A. It dynamically adjusts the priority of thread execution.

    B. It allows a set of threads to wait for a series of events to occur.

    C. It provides a mechanism for threads to exchange data.

    D. It is used for automatic memory management in multi-threaded applications.

5. What is the main advantage of using `AtomicInteger` over traditional synchronization techniques in Java?

    A. It offers enhanced security features for web applications.

    B. It allows for lock-free thread-safe operations on a single integer value.

    C. It is used for managing database transactions.

    D. It provides a framework for building graphical user interfaces.

3

Mastering Parallelism in Java

Embark on an exhilarating journey into the heart of Java's parallel programming landscape, a realm where the combined force of multiple threads is harnessed to transform complex, time-consuming tasks into efficient, streamlined operations.

Picture this: an ensemble of chefs in a bustling kitchen or a symphony of musicians, each playing a vital role in creating a harmonious masterpiece. In this chapter, we delve deep into the Fork/Join framework, your maestro in the art of threading, skillfully orchestrating a myriad of threads to collaborate seamlessly.

As we navigate through the intricacies of parallel programming, you'll discover its remarkable advantages in boosting speed and efficiency akin to how a well-coordinated team can achieve more than the sum of its parts. However, with great power comes great responsibility. You'll encounter unique challenges such as thread contention and race conditions, and we'll arm you with the strategies and insights needed to master these obstacles.

This chapter is not just an exploration; it's a toolkit. You'll learn how to employ the Fork/Join framework effectively, breaking down daunting tasks into manageable sub-tasks, much like a head chef delegating components of a complex recipe. We'll dive into the nuances of `RecursiveTask` and `RecursiveAction`, understanding how these elements work in unison to optimize parallel processing. Additionally, you'll gain insights into performance optimization techniques and best practices, ensuring that your Java applications are not just functional but are also performing at their peak like a well-oiled machine.

By the end of this chapter, you'll be equipped with more than just knowledge; you'll possess the practical skills to implement parallel programming effectively in your Java applications. You'll emerge ready to enhance functionality, optimize performance, and tackle the challenges of concurrent computing head-on.

So, let's begin this exciting adventure into the dynamic world of Java's parallel capabilities. Together, we'll unlock the doors to efficient, concurrent computing, setting the stage for you to craft high-performance applications that stand out in the world of modern computing.

Technical requirements

You will need **Visual Studio Code** (**VS Code**), which you can download here: `https://code.visualstudio.com/download`.

VS Code offers a lightweight and customizable alternative to the other available options. It's a great choice for developers who prefer a less resource-intensive **Integrated Development Environment** (**IDE**) and want the flexibility to install extensions tailored to their specific needs. However, it may not have all the features out of the box compared to the more established Java IDEs.

Furthermore, the code in this chapter can be found on GitHub:

`https://github.com/PacktPublishing/Java-Concurrency-and-Parallelism`

Unleashing the parallel powerhouse – the Fork/Join framework

The **Fork/Join framework** unlocks the power of parallel processing, turning your Java tasks into a symphony of collaborating threads. Dive into its secrets, such as work-stealing algorithms, recursive conquers, and optimization strategies, to boost performance and leave sequential cooking in the dust!

Demystifying Fork/Join – a culinary adventure in parallel programming

Imagine stepping into a grand kitchen of parallel computing in Java. This is where the Fork/Join framework comes into play, transforming the art of programming much like a bustling kitchen brimming with skilled chefs. It's not just about adding more chefs; it's about orchestrating them with finesse and strategy.

At the heart of this bustling kitchen lies the Fork/Join framework, a masterful tool in Java's arsenal that automates the division of complex tasks into smaller, more manageable bites. Picture a head chef breaking down a complicated recipe into simpler tasks and delegating them to sous chefs. Each chef focuses on a part of the meal, ensuring that no one is waiting idly, and no task is overwhelming. This efficiency is akin to the work-stealing algorithm, the framework's secret ingredient, where chefs who finish early lend a hand to those still busy, ensuring a harmonious and efficient cooking process.

In this culinary orchestra, `ForkJoinPool` plays the role of an adept conductor. It's a specialized thread pool tailored for the Fork/Join tasks, extending both the `Executor` and `ExecutorService` interfaces introduced in *Chapter 2, Introduction to Java's Concurrency Foundations: Threads, Processes, and Beyond*. The `Executor` interface provides a way to decouple task submission from the mechanics of how each task will be run, including details of thread use, scheduling, and so on. The `ExecutorService` interface supplements this with methods for life cycle management and tracking the progress of one or more asynchronous tasks.

`ForkJoinPool`, built on these foundations, is designed for work that can be broken down into smaller pieces recursively. It employs a technique called work-stealing, where idle threads can *steal* work from other busy threads, thereby minimizing idle time and maximizing CPU utilization.

Like a well-orchestrated kitchen, `ForkJoinPool` manages the execution of tasks, dividing them into sub-recipes, and ensuring no chef—or thread—is ever idle. When a task is complete, much like a sous chef presenting their dish, `ForkJoinPool` expertly combines these individual efforts to complete the final masterpiece. This process of breaking down tasks and combining the results is fundamental to the Fork/Join model, making `ForkJoinPool` an essential tool in the concurrency toolkit.

The Fork/Join framework revolves around the `ForkJoinTask` abstract class, which represents a task that can be split into smaller subtasks and executed in parallel using `ForkJoinPool`. It provides methods for splitting the task (fork), waiting for subtask completion (join), and computing the result.

Two concrete implementations of `ForkJoinTask` are **RecursiveTask** and **RecursiveAction**. `RecursiveTask` is used for tasks that return a result, while `RecursiveAction` is used for tasks that don't return a value.

Both allow you to break down tasks into smaller chunks for parallel execution. You need to implement the compute method to define the base case and the logic to split the task into subtasks. The framework handles the distribution of subtasks among the threads in `ForkJoinPool` and the aggregation of results.

The key difference between `RecursiveTask` and `RecursiveAction` lies in their purpose and return type. `RecursiveTask` computes and returns a result, while `RecursiveAction` performs an action without returning a value.

To illustrate how `RecursiveTask` and `RecursiveAction` are used within the Fork/Join framework, consider the following code example. `SumTask` demonstrates summing a data array, while `ActionTask` shows processing data without returning a result:

```
import java.util.concurrent.RecursiveTask;
import java.util.concurrent.RecursiveAction;
import java.util.ArrayList;
import java.util.concurrent.ForkJoinPool;

public class DataProcessor{
    public static void main(String[] args) {
        // Example dataset
        int DATASET_SIZE = 500;
        ArrayList<Integer> data = new ArrayList<Integer> (
DATASET_SIZE);
        ForkJoinPool pool = new ForkJoinPool();
        // RecursiveAction for generating large dataset
        ActionTask actionTask = new ActionTask(data, 0, DATASET_SIZE);
```

```
            pool.invoke(actionTask);
            // RecursiveTask for summing large dataset
            SumTask sumTask = new SumTask(data,0,DATASET_SIZE);
            int result = pool.invoke(sumTask);
            System.out.println("Total sum: " + result);
            pool.shutdown();
            pool.close();
        }


// Splitting task for parallel execution
    static class SumTask extends RecursiveTask<Integer> {
        private final ArrayList<Integer> data;
        private final int start, end;
        private static final int THRESHOLD = 50;
        SumTask(ArrayList<Integer> data,int start,int end){
            this.data = data;
            this.start = start;
            this.end = end;
        }
        @Override
        protected Integer compute() {
            int length = end - start;
            System.out.println(String.format("RecursiveTask.compute()
            called for %d elements from index %d to %d", length,
            start, end));
            if (length <= THRESHOLD) {
                // Simple computation
                System.out.println(String.format("Calculating sum of
                %d elements from index %d to %d", length, start,
                end));
                int sum = 0;
                for (int i = start; i < end; i++) {
                    sum += data.get(i);
                }
                return sum;
            } else {
                // Split task
                int mid = start + (length / 2);
                SumTask left = new SumTask(data,start,mid);
                SumTask right = new SumTask(data,mid,end);
                left.fork();
                right.fork();
                return right.join() + left.join();
            }
```

```
            }
        }
    static class ActionTask extends RecursiveAction {
        private final ArrayList<Integer> data;
        private final int start, end;
        private static final int THRESHOLD = 50;
        ActionTask(ArrayList<Integer> data,int start,
            int end){
                this.data = data;
                this.start = start;
                this.end = end;
            }
        @Override
        protected void compute() {
            int length = end - start;
            System.out.println(String.format("RecursiveAction.
            compute() called for %d elements from index %d to %d",
            length, start, end));
            if (length <= THRESHOLD) {
                // Simple processing
                for (int i = start; i < end; i++) {
                    this.data.add((int) Math.round(
                        Math.random() * 100));
                }
            } else {
                // Split task
                int mid = start + (length / 2);
                ActionTask left = new ActionTask(data,
                    start, mid);
                ActionTask right = new ActionTask(data,
                    mid, end);
                invokeAll(left, right);
            }
        }
    }
    }
}
```

Here's a breakdown of the code and its functionality:

- `SumTask` extends `RecursiveTask<Integer>` and is used for summing a portion of the array, returning the sum.

- In the `SumTask` class, the task is split when the data length exceeds a threshold, demonstrating a divide-and-conquer approach. This is similar to a head chef dividing a large recipe task among sous chefs.

- `ActionTask` extends `RecursiveAction` and is used for processing a portion of the array without returning a result.

- The `fork()` method initiates the parallel execution of a subtask, while `join()` waits for the completion of these tasks, combining their results. The `compute()` method contains the logic for either directly performing the task or further splitting it.

- Both classes split their tasks when the dataset size exceeds a threshold, demonstrating the divide-and-conquer approach.

- `ForkJoinPool` executes both tasks, illustrating how both `RecursiveTask` and `RecursiveAction` can be used in parallel processing scenarios.

This example demonstrates the practical application of the Fork/Join framework's ability to efficiently process large datasets in parallel, as discussed earlier. They exemplify how complex tasks can be decomposed and executed in a parallel manner to enhance application performance. Imagine using `SumTask` for rapidly processing large financial datasets or `ActionTask` for parallel processing in data cleaning operations in a real-time analytics application.

In the next section, we'll explore how to handle tasks with dependencies and navigate the intricacies of complex task graphs.

Beyond recursion – conquering complexities with dependencies

We've witnessed the beauty of recursive tasks in tackling smaller, independent challenges. But what about real-world scenarios where tasks have intricate dependencies like a multi-course meal where one dish relies on another to be complete? This is where `ForkJoinPool.invokeAll()` shines, a powerful tool for orchestrating parallel tasks with intricate relationships.

ForkJoinPool.invokeAll() – the maestro of intertwined tasks

Imagine a bustling kitchen with chefs working on various dishes. Some tasks, such as chopping vegetables, can be done independently. But others, such as making a sauce, depend on ingredients already being prepped. This is where the head chef, `ForkJoinPool`, steps in. With `invokeAll()`, they distribute the tasks, ensuring that dependent tasks wait for their predecessors to finish before starting.

Managing dependencies in the kitchen symphony – a recipe for efficiency

Just as a chef carefully coordinates dishes with different cooking times, parallel processing requires meticulous management of task dependencies. Let's explore this art through the lens of a kitchen, where our goal is to efficiently prepare a multi-course meal.

The following are key strategies of parallel processing:

- **Task decomposition**: Break down the workflow into smaller, manageable tasks with clear dependencies. In our kitchen symphony, we'll create tasks for preparing vegetables, making sauce, and cooking protein, each with its own prerequisites.

- **Dependency analysis**: Identify task reliance and define execution order. Tasks such as cooking protein must await prepped vegetables and sauce, ensuring a well-orchestrated meal.

- **Granularity control**: Choose the appropriate task size to balance efficiency and overhead. Too many fine-grained tasks can increase management overhead, while large tasks might limit parallelism.

- **Data sharing and synchronization**: Ensure proper access and synchronization of shared data to avoid inconsistencies. If multiple chefs use a shared ingredient, we need a system to avoid conflicts and maintain kitchen harmony.

Let's visualize dependency management with the `PrepVeggiesTask` class:

```java
import java.util.ArrayList;
import java.util.List;
import java.util.concurrent.ForkJoinPool;
import java.util.concurrent.RecursiveTask;
public class PrepVeggiesDemo {
    static interface KitchenTask {
        int getTaskId();
        String performTask();
    }
    static class PrepVeggiesTask implements KitchenTask {
        protected int taskId;
        public PrepVeggiesTask(int taskId) {
            this.taskId = taskId;
        }
        public String performTask() {
            String message = String.format(
                "[Task-%d] Prepped Veggies", this.taskId);
            System.out.println(message);
            return message;
        }
        public int getTaskId() {return this.taskId; }
    }
    static class CookVeggiesTask implements KitchenTask {
        protected int taskId;
        public CookVeggiesTask(int taskId) {
            this.taskId = taskId;
        }
```

```java
        public String performTask() {
            String message = String.format(
                "[Task-%d] Cooked Veggies", this.taskId);
            System.out.println(message);
            return message;
        }
        public int getTaskId() {return this.taskId; }
    }
    static class ChefTask extends RecursiveTask<String> {
        protected KitchenTask task;
        protected List<ChefTask> dependencies;
        public ChefTask(
            KitchenTask task,List<ChefTask> dependencies) {
                this.task = task;
                this.dependencies = dependencies;
            }
        // Method to wait for dependencies to complete
        protected void awaitDependencies() {
            if (dependencies == null || dependencies.isEmpty())
            return;
            ChefTask.invokeAll(dependencies);
        }
        @Override
        protected String compute() {
            awaitDependencies(); // Ensure all prerequisites are met
            return task.performTask(); // Carry out the specific task
        }
    }
    public static void main(String[] args) {
        // Example dataset
        int DEPENDENCY_SIZE = 10;
        ArrayList<ChefTask> dependencies = new ArrayList<ChefTask>();
        for (int i = 0; i < DEPENDENCY_SIZE; i++) {
            dependencies.add(new ChefTask(
                new PrepVeggiesTask(i), null));
        }
        ForkJoinPool pool = new ForkJoinPool();
        ChefTask cookTask = new ChefTask(
            new CookVeggiesTask(100), dependencies);
        pool.invoke(cookTask);
        pool.shutdown();
        pool.close();
    }
}
```

The provided code demonstrates the usage of the Fork/Join framework in Java to handle tasks with dependencies. It defines two interfaces: `KitchenTask` for generic tasks and `ChefTask` for tasks that return a String result.

Here are some key points:

- `PrepVeggiesTask` and `CookVeggiesTask` implement `KitchenTask`, representing specific tasks in the kitchen. The `ChefTask` class is the core of the Fork/Join implementation, containing the actual task (`task`) and its dependencies (`dependencies`).

- The `awaitDependencies()` method waits for all dependencies to complete before executing the current task. The `compute()` method is the main entry point for the Fork/Join framework, ensuring prerequisites are met and performing the actual task.

- In the main method, an example dataset is created with `PrepVeggiesTask` objects as dependencies. `ForkJoinPool` is used to manage the execution of tasks. `CookVeggiesTask` with dependencies is submitted to the pool using `pool.invoke(cookTask)`, triggering the execution of the task and its dependencies.

- `ChefTask` acts as a blueprint for tasks with dependencies.

- `awaitDependencies()` waits for prerequisites to finish.

- `PrepVeggiesTask` and `CookVeggiesTask` represent specific tasks.

- `performTask()` holds the actual task logic.

The code demonstrates how the Fork/Join framework can be used to handle tasks with dependencies, ensuring prerequisites are completed before executing a task. `ForkJoinPool` manages the execution of tasks, and the `ChefTask` class provides a structured way to define and perform tasks with dependencies.

Let's weave a real-world scenario into the mix to solidify the concept of dependency management in parallel processing.

Picture this: you're building a next-generation image rendering app that needs to handle complex 3D scenes. To efficiently manage the workload, you break down the rendering process into the following parallel tasks:

- **Task 1**: Downloading textures and model data
- **Task 2**: Building geometric primitives from the downloaded data
- **Task 3**: Applying lighting and shadows to the scene
- **Task 4**: Rendering the final image

Here's where dependencies come into play:

- Task 2 can't start until Task 1 finishes downloading the necessary data
- Task 3 needs the geometric primitives built by Task 2 before it can apply the lighting and shadows
- Finally, Task 4 depends on the completed scene from Task 3 to generate the final image

By carefully managing these dependencies and utilizing parallel processing techniques, you can significantly speed up the rendering process, delivering smooth and visually stunning 3D experiences.

This real-world example showcases how effective dependency management is crucial for harnessing the true power of parallel processing in various domains, from image rendering to scientific simulations and beyond.

Remember, just like orchestrating a kitchen symphony or rendering a complex 3D scene, mastering parallel processing lies in meticulous planning, execution, and efficient dependency management. With the right tools and techniques, you can transform your parallel processing endeavors into harmonious and high-performance symphonies of tasks.

Now, let's move on to explore the art of fine-tuning these symphonies in the next topic on performance optimization techniques!

Fine-tuning the symphony of parallelism – a journey in performance optimization

In the dynamic world of parallel programming, achieving peak performance is akin to conducting a grand orchestra. Each element plays a crucial role and fine-tuning them is essential to creating a harmonious symphony. Let's embark on a journey through the key strategies of performance optimization in Java's parallel computing.

The art of granularity control

Just as a chef balances ingredients for a perfect dish, granularity control in parallel programming is about finding the ideal task size. Smaller tasks, like having more chefs, boost parallelization but introduce dependencies and management overhead. Conversely, larger tasks simplify management but limit parallelism, like a few chefs handling everything. The key is assessing task complexity, weighing overhead against benefits, and avoiding overly fine-grained tasks that could tangle the process.

Tuning parallelism levels

Setting the right level of parallelism is like orchestrating our chefs to ensure each has just the right amount of work—neither too overwhelmed nor idly waiting. It's a delicate balance between utilizing available resources and avoiding excessive overhead from too many active threads. Consider the characteristics of your tasks and the available hardware. Remember, larger thread pools might not always benefit from work-stealing as efficiently as smaller, more focused groups.

Best practices for a smooth performance

In our parallel kitchen, the best practices are the secret recipes for success. Limiting data sharing among threads can prevent conflicts over shared resources, much like chefs working on separate stations. Opting for a smart, thread-safe data structure such as `ConcurrentHashMap` can ensure safe access to shared data. Regularly monitoring performance and being ready to adjust task sizes and thread numbers can keep your parallel applications running smoothly and efficiently.

By mastering these techniques—granularity control, tuning parallelism levels, and adhering to best practices—we can elevate our parallel computing to new heights of efficiency and performance. It's not just about running parallel tasks; it's about orchestrating them with precision and insight, ensuring each thread plays its part in this complex symphony of parallel processing.

Performance optimization lays the foundation for efficient parallelism. Now, we step into a world of refined elegance with Java's parallel streams, enabling lightning-fast data processing through concurrent execution.

Streamlining parallelism in Java with parallel streams

Fine-tuning the symphony of parallelism is akin to conducting a grand orchestra. Each element plays a crucial role and mastering them unlocks peak performance. This journey through key strategies, such as granularity control and parallelism levels, ensures harmonious execution in Java's parallel computing.

Now, we step into a world of refined elegance with Java's parallel streams. Imagine transforming a one-chef kitchen into a synchronized team, harnessing multiple cores for lightning-fast data processing. Remember that efficient parallelism lies in choosing the right tasks.

Parallel streams excel due to the following reasons:

- **Faster execution**: Especially for large datasets, they accelerate data operations remarkably
- **Handling large data**: Their strength lies in efficiently processing massive data volumes
- **Ease of use**: Switching from sequential to parallel streams is often straightforward

However, consider the following challenges:

- **Extra resource management**: Thread management incurs overhead, making smaller tasks less ideal

- **Task independence**: Parallel streams shine when tasks are independent and lack sequential dependencies

- **Caution with shared data**: Concurrent access to shared data necessitates careful synchronization to avoid race conditions

Let us now understand how to seamlessly integrate parallel streams to harness their performance benefits while addressing the potential challenges:

- **Identify suitable tasks**: Begin by pinpointing computationally expensive operations within your code that operate on independent data elements, such as image resizing, sorting large lists, or performing complex calculations. These tasks are prime candidates for parallelization.

- **Switch to parallel streams**: Effortlessly transform a sequential stream into a parallel one by simply invoking the `parallelStream()` method instead of `stream()`. This subtle change unlocks the power of multi-core processing.

   For example, consider a scenario where you need to resize a large batch of photos. The sequential approach, `photos.stream().map(photo -> resize(photo))`, processes each photo individually. By switching to `photos.parallelStream().map(photo -> resize(photo))`, you unleash the potential of multiple cores, working in concert to resize photos simultaneously, often leading to significant performance gains.

Remember that effective parallel stream integration requires careful consideration of task suitability, resource management, and data safety to ensure optimal results and avoid potential pitfalls.

Next, we'll conduct a comparative analysis, exploring different parallel processing tools and helping you choose the perfect instrument for your programming symphony.

Choosing your weapon – a parallel processing showdown in Java

Mastering the Fork/Join framework is a culinary feat in itself, but navigating the broader landscape of Java's parallel processing tools is where true expertise shines. To help you choose the perfect ingredient for your parallel processing dish, let's explore how Fork/Join stacks up against other options:

- **Fork/Join versus ThreadPoolExecutor**: Think of Fork/Join as a master chef, adept at dissecting complex tasks into bite-sized subtasks and assigning them to a dedicated team of sous chefs. It thrives on CPU-bound tasks that can be recursively split, conquering them with precision and efficiency. `ThreadPoolExecutor`, on the other hand, is a more versatile kitchen manager, handling a large volume of independent, non-divisible tasks such as prepping separate dishes for a banquet. It's ideal for simpler parallel needs where the sous chefs don't need to break down their ingredients further.

- **Fork/Join versus parallel streams**: Parallel streams are like pre-washed and chopped vegetables, ready to be tossed into the processing pan. They simplify data processing on collections by automatically parallelizing operations under the hood, using Fork/Join as their secret weapon. For straightforward data crunching, they're a quick and convenient option. However, for complex tasks with custom processing logic, Fork/Join offers the fine-grained control and flexibility of a seasoned chef, allowing you to customize the recipe for optimal results.

- **Fork/Join versus CompletableFuture**: While Fork/Join excels at dividing and conquering large tasks, `CompletableFuture` is like a multi-tasking sous chef, adept at handling asynchronous operations. It allows you to write non-blocking code and chain multiple asynchronous tasks together, ensuring your kitchen keeps running smoothly even while other dishes simmer. Think of it as preparing multiple side dishes without holding up the main course.

- **Fork/Join versus Executors.newCachedThreadPool()**: Need a temporary team of kitchen helpers for quick tasks? `Executors.newCachedThreadPool()` is like hiring temporary chefs who can jump in and out as needed. It's perfect for short-lived, asynchronous jobs such as fetching ingredients. However, for long-running, CPU-intensive tasks, Fork/Join's work-stealing algorithm shines again, ensuring each chef is optimally busy and maximizing efficiency throughout the entire cooking process.

By understanding the strengths and weaknesses of each tool, you can choose the perfect one for your parallel processing needs. Remember, Fork/Join is the master of large-scale, parallelizable tasks, while other tools cater to specific needs, such as independent jobs, simpler data processing, asynchronous workflows, or even temporary assistance.

Having explored the comparative analysis of the Fork/Join framework with other parallel processing methods in Java, we now transition to a more specialized topic. Next, we delve into unlocking the power of big data with a custom Spliterator, where we will uncover advanced techniques for optimizing parallel stream processing, focusing on custom Spliterator implementation and efficient management of computational overhead.

Unlocking the power of big data with a custom Spliterator

Java's **Splittable Iterator** (**Spliterator**) interface offers a powerful tool for dividing data into smaller pieces for parallel processing. But for large datasets, such as those found on cloud platforms such as **Amazon Web Services** (**AWS**), a custom Spliterator can be a game-changer.

For example, imagine a massive bucket of files in AWS **Simple Storage Service** (**S3**). A custom Spliterator designed specifically for this task can intelligently chunk the data into optimal sizes, considering factors such as file types and access patterns. This allows you to distribute tasks across CPU cores more effectively, leading to significant performance boosts and reduced resource utilization.

Now, imagine you have lots of files in an AWS S3 bucket and want to process them at the same time using Java Streams. Here's how you could set up a custom Spliterator for these AWS S3 objects:

```java
// Assume s3Client is an initialized AmazonS3 client
public class S3CustomSpliteratorExample {

    public static void main(String[] args) {
        String bucketName = "your-bucket-name";
        ListObjectsV2Result result = s3Client.
        listObjectsV2(bucketName);
        List<S3ObjectSummary> objects = result.getObjectSummaries();
        Spliterator<S3ObjectSummary> spliterator = new
        S3ObjectSpliterator(objects);
        StreamSupport.stream(spliterator, true)
                .forEach(S3CustomSpliteratorExample::processS3Object);
    }

    private static class S3ObjectSpliterator implements
    Spliterator<S3ObjectSummary> {
        private final List<S3ObjectSummary> s3Objects;
        private int current = 0;
        S3ObjectSpliterator(List<S3ObjectSummary> s3Objects) {
            this.s3Objects = s3Objects;
        }

        @Override
        public boolean tryAdvance(Consumer<? super S3ObjectSummary>
        action) {
            if (current < s3Objects.size()) {
                action.accept(s3Objects.get(current++));
                return true;
            }
            return false;
        }

        @Override
        public Spliterator<S3ObjectSummary> trySplit() {
            int remaining = s3Objects.size() - current;
            int splitSize = remaining / 2;
            if (splitSize <= 1) {
                return null;
            }
            List<S3ObjectSummary> splitPart = s3Objects.
            subList(current, current + splitSize);
            current += splitSize;
```

```
            return new S3ObjectSpliterator(splitPart);
        }


        @Override
        public long estimateSize() {
            return s3Objects.size() - current;
        }


        @Override
        public int characteristics() {
            return IMMUTABLE | SIZED | SUBSIZED;
        }
    }


    private static void processS3Object(S3ObjectSummary objectSummary)
{
        // Processing logic for each S3 object
    }
}
```

The Java code presented showcases how to harness the custom Spliterator to achieve efficient parallel processing of S3 objects. Let's dive into its key elements:

1. **Main method**: It sets the stage with the following:

    - Retrieves a list of S3 object summaries from a specified S3 bucket using an initialized S3 client

    - Constructs a custom S3ObjectSpliterator to divide the list for parallel processing

    - Initiates a parallel stream using the Spliterator, applying the processS3Object method to each object

2. **Custom Spliterator in action**: The S3ObjectSpliterator class implements the Spliterator<S3ObjectSummary> interface, enabling tailored data division for parallel streams. Other key methods are as follows:

    - tryAdvance: Processes the current object and advances the cursor

    - trySplit: Divides the list into smaller chunks for parallel execution, returning a new Spliterator for the divided portion

    - estimateSize: Provides an estimate of remaining objects, aiding stream optimization

    - characteristics: Specifies Spliterator traits (IMMUTABLE, SIZED, or SUBSIZED) for efficient stream operations

3. **Processing logic**: The `processS3Object` method encapsulates the specific processing steps performed on each S3 object. Implementation details are not shown, but this method could involve tasks such as downloading object content, applying transformations, or extracting metadata.

The following are the advantages of the custom Spliterator approach:

- **Fine-grained control**: A custom Spliterator allows for precise control over data splitting, enabling optimal chunk sizes for parallel processing based on task requirements and hardware capabilities

- **Optimized parallel execution**: The `trySplit` method effectively divides the workload for multi-core processors, leading to potential performance gains

- **Flexibility for diverse data handling**: A custom Spliterator can be adapted to handle different S3 object types or access patterns, tailoring processing strategies for specific use cases

In essence, this code demonstrates how a custom Spliterator empowers Java developers to take control of parallel processing for S3 objects, unlocking enhanced performance and flexibility for various data-intensive tasks within cloud environments.

Beyond a custom Spliterator, Java offers an arsenal of advanced techniques to fine-tune stream parallelism and unlock exceptional performance. Let's look at a code example showcasing three powerful strategies: custom thread pools, combining stream operations, and parallel-friendly data structures.

Let's explore these Java classes in the following code:

```java
import java.util.List;
import java.util.concurrent.ForkJoinPool;
import java.util.concurrent.ConcurrentHashMap;
import java.util.stream.Collectors;

public class StreamOptimizationDemo {
    public static void main(String[] args) {
        // Example data
        List<Integer> data = List.of(
            1, 2, 3, 4, 5, 6, 7, 8, 9, 10);
        // Custom Thread Pool for parallel streams
        ForkJoinPool customThreadPool = new ForkJoinPool(4);
        // Customizing the number of threads
        try {
            List<Integer> processedData = customThreadPool.submit(()
            ->
                data.parallelStream()
                    .filter(n -> n % 2 == 0)
// Filtering even numbers
                    .map(n -> n * n) // Squaring them
                    .collect(Collectors.toList())
```

```
// Collecting results
            ).get();
            System.out.println(
                "Processed Data: " + processedData);
        } catch (Exception e) {
            e.printStackTrace();
        } finally {
            customThreadPool.shutdown();
// Always shutdown your thread pool!
        }
// Using ConcurrentHashMap for better performance in parallel streams
        ConcurrentHashMap<Integer, Integer> map = new
        ConcurrentHashMap<>();
        data.parallelStream().forEach(n -> map.put(
            n, n * n));
        System.out.println("ConcurrentHashMap contents: " + map);
    }
}
```

In this code, we used the following techniques:

- **Custom thread pools**: We create `ForkJoinPool` with a specified number of threads (in this case, 4). This custom thread pool is used to execute our parallel stream, allowing for better resource allocation than using the common pool.

- **Combining stream operations**: The `filter` (to select even numbers) and `map` (to square the numbers) stream operations are combined into a single stream pipeline. This reduces the number of iterations over the data.

- **Parallel-friendly data structures**: We use `ConcurrentHashMap` for storing the results of a parallel stream operation. This data structure is designed for concurrent access, making it a good choice for use in parallel streams.

This class demonstrates how combining these advanced techniques can lead to more efficient and optimized parallel stream processing in Java.

A custom Spliterator offers a potent recipe for parallel processing, but is it always the tastiest dish? In the next section, we'll sprinkle in some reality checks, exploring the potential benefits and hidden costs of parallelism.

Benefits and pitfalls of parallelism

Parallel processing not only offers significant speed advantages but also comes with challenges such as thread contention and data dependency issues. This section focuses on understanding when to use parallel processing effectively. It outlines the benefits and potential problems, providing guidance on choosing between parallel and sequential processing.

The key scenarios where parallel processing excels over sequential methods are as follows:

- **Computationally intensive tasks**: Imagine crunching numbers, processing images, or analyzing vast datasets. These are the playgrounds for parallel processing.

- **Independent operations**: Parallelism thrives when tasks are independent, meaning they don't rely on each other's results. Think of filtering items in a list or resizing multiple images. Each operation can be handled concurrently by a separate thread, boosting efficiency without causing tangled dependencies.

- **Input/Output (I/O) bound operations**: Tasks waiting for data from a disk or network are prime candidates for parallel processing. While one thread waits for data, others can tackle other independent tasks, maximizing resource utilization and keeping your code humming along.

- **Real-time applications**: Whether it's rendering dynamic visuals or handling user interactions, responsiveness is crucial in real-time applications. Parallel processing can be your secret sauce, ensuring smooth, lag-free experiences by splitting the workload and keeping the **user interface (UI)** responsive even under heavy load.

Beyond these specific scenarios, the potential performance gains of parallel processing are vast. From accelerating video encoding to powering real-time simulations, its ability to unleash the power of multiple cores can dramatically improve the efficiency and responsiveness of your applications.

We've witnessed the exhilarating potential of parallel processing, but now comes the crucial question: how much faster is it? How can we quantify the performance gains of parallelism processing?

The most common metric for measuring parallel processing efficiency is speedup. It simply compares the execution time of a task running sequentially with its parallel execution time. The formula is straightforward:

Speedup = Sequential Execution Time / Parallel Execution Time

A speedup of 2 means the parallel version took half the time of the sequential version.

However, parallel processing isn't just about raw speed; it's also about resource utilization and efficiency. Here are some additional metrics to consider:

- **Efficiency**: The percentage of CPU time utilized by the parallel program. Ideally, you'd like to see efficiency close to 100%, indicating all cores are working hard.

- **Amdahl's Law**: A 1960s principle by Gene Amdahl, which sets limits on parallel processing. Amdahl's Law says that adding processors won't magically speed up everything. Focus on bottlenecks first, then parallelize wisely. Why? Accelerating part of a task only helps if the rest is fast too. So, as tasks become more parallel, adding more processors gives less and less benefit. Optimize the slowest parts first! Even highly parallel tasks have *unparallelizable bits* that cap the overall speedup.

- **Scalability**: How well does the parallel program perform as the number of cores increases? Ideally, we want to see a near-linear speedup with additional cores.

Here are some notable tools for performance tuning in cloud environments and Java frameworks:

- **Profilers**: Identify hotspots and bottlenecks in your code:

  - **Cloud**:

    - **Amazon CodeGuru Profiler**: Identifies performance bottlenecks and optimization opportunities in AWS environments

    - **Azure Application Insights**: Provides profiling insights for .NET applications running in Azure

    - **Google Cloud Profiler**: Analyzes the performance of Java and Go applications on the **Google Cloud Platform (GCP)**

  - **Java frameworks**:

    - **JProfiler**: Commercial profiler for detailed analysis of CPU, memory, and thread usage

    - **YourKit Java Profiler**: Another commercial option with comprehensive profiling capabilities

    - **Java VisualVM**: Free tool included in the JDK, offering basic profiling and monitoring features

    - **Java Flight Recorder (JFR)**: Built-in tool for low-overhead profiling and diagnostics, especially useful in production environments

- **Benchmarks**: Compare the performance of different implementations of the same task:

  - **Cloud**:

    - **AWS Lambda power tuning**: Optimizes memory and concurrency settings for Lambda functions

    - **Azure performance benchmarks**: Provides reference scores for various VM types and workloads in Azure

    - **Google Cloud benchmarks**: Offers performance data for different compute options on GCP

- **Java frameworks**:

  - **Java Microbenchmark Harness (JMH)**: Framework for creating reliable and accurate microbenchmarks

  - **Caliper**: Another Microbenchmark framework from Google

  - **SPECjvm2008**: Standardized benchmark suite for measuring Java application performance

- **Monitoring tools**: Continuously track and assess the performance and health of diverse resources such as CPU, disk, and network usage, and application performance metrics:

  - **Cloud**:

    - **Amazon CloudWatch**: Monitors various metrics across AWS services, including CPU, memory, disk, and network usage

    - **Azure Monitor**: Provides comprehensive monitoring for Azure resources, including application performance metrics

    - **Google Cloud Monitoring**: Offers monitoring and logging capabilities for GCP resources

  - **Java frameworks**:

    - **Java Management Extensions (JMX)**: Built-in API for exposing management and monitoring information from Java applications

    - **Micrometer**: Framework for collecting and exporting metrics to different monitoring systems (e.g., Prometheus and Graphite)

    - **Spring Boot Actuator**: Provides production-ready endpoints for monitoring Spring Boot applications.

By mastering these tools and metrics, you can transform from a blindfolded speed demon to a data-driven maestro, confidently wielding the power of parallel processing while ensuring optimal performance and efficiency.

In the next section, we'll tackle the other side of the coin: the potential pitfalls of parallelism. We'll delve into thread contention, race conditions, and other challenges you might encounter.

Challenges and solutions in parallel processing

Parallel processing accelerates computation but comes with challenges such as thread contention, race conditions, and debugging complexities. Understanding and addressing these issues is crucial for efficient parallel computing. Let us dive into gaining an insight into each of these issues:

- **Thread contention**: This occurs when multiple threads compete for the same resources, leading to performance issues such as increased waiting times, resource starvation, and deadlocks.

- **Race conditions**: These happen when multiple threads access shared data unpredictably, causing problems such as data corruption and unreliable program behavior.

- **Debugging complexities**: Debugging in a multithreaded environment is challenging due to non-deterministic behavior and hidden dependencies, such as shared state dependency and order of execution dependency. These dependencies often arise from the interactions between threads that are not explicit in the code but can affect the program's behavior.

While these challenges may seem daunting, they're not insurmountable. Let's dive into practical strategies for mitigating these pitfalls:

- **Avoiding thread contention**:

  - **Minimize shared resources**: Analyze your code and identify opportunities to reduce the number of shared resources accessed by multiple threads. This can involve data partitioning, private copies of frequently accessed data, or alternative synchronization strategies.

  - **Choose appropriate data structures**: Opt for thread-safe data structures such as `ConcurrentHashMap` or `ConcurrentLinkedQueue` when dealing with shared data, preventing concurrent access issues and data corruption.

  - **Employ lock-free algorithms**: Consider lock-free algorithms such as **compare-and-swap** (**CAS**) operations, which avoid overhead associated with traditional locks and can improve performance while mitigating contention.

- **Conquering race conditions**:

  - **Embrace immutability**: Whenever possible, design your data structures and objects to be immutable. This eliminates the need for synchronization and prevents accidental data corruption by concurrent modifications.

  - **Utilize synchronized blocks**: Carefully use synchronized blocks when accessing shared state, ensuring only one thread can operate on the data at a time. However, excessive synchronization can introduce bottlenecks, so use it judiciously.

  - **Leverage atomic operations**: For specific operations such as incrementing a counter, consider atomic operations such as `AtomicInteger`, which guarantee thread-safe updates to underlying values.

- **Mastering parallel debugging**:

  - **Use visual debuggers with thread views**: Debuggers such as Eclipse or IntelliJ IDEA offer specialized views for visualizing thread execution timelines, identifying deadlocks, and pinpointing race conditions

- **Leverage logging with timestamps:** Strategically add timestamps to your logs in multithreaded code, helping you reconstruct the sequence of events and identify the thread responsible for issues

- **Employ assertion checks:** Place assertional checks at critical points in your code to detect unexpected data values or execution paths that might indicate race conditions

- **Consider automated testing tools:** Tools such as JUnit with parallel execution capabilities can help you uncover concurrency-related issues early on in the development process

Here are a few real-world examples of how to avoid these issues in AWS:

- **Amazon SQS – Parallel processing for message queue:**

  - **Use case:** Implementing parallel processing for message queue handling with **Amazon Simple Queue Service (SQS)** using its batch operations

  - **Scenario:** A system needs to process a high volume of incoming messages efficiently

  - **Implementation:** Instead of processing messages one by one, the system uses Amazon SQS's batch operations to process multiple messages in parallel.

  - **Advantage:** This approach minimizes thread contention, as multiple messages are read and written in batches rather than competing for individual message handling

- **Amazon DynamoDB – Atomic updates and conditional writes:**

  - **Use case:** Utilizing DynamoDB's atomic updates and conditional writes for safe parallel data access and modification.

  - **Scenario:** An online store tracks product inventory in DynamoDB and needs to update inventory levels safely when multiple purchases occur simultaneously.

  - **Implementation:** When processing a purchase, the system uses DynamoDB's atomic updates to adjust inventory levels. Conditional writings ensure that updates happen only if the inventory level is sufficient, preventing race conditions.

  - **Advantage:** This ensures inventory levels are accurately maintained even with concurrent purchase transactions.

- **AWS Lambda – Stateless functions and resource management:**

  - **Use case:** Designing AWS Lambda functions to be stateless and avoiding shared resources for simpler and safer concurrent executions.

  - **Scenario:** A web application uses Lambda functions to handle user requests, such as retrieving user data or processing transactions.

- **Implementation**: Each Lambda function is designed to be stateless, meaning it doesn't rely on or alter shared resources. Any required data is passed to the function in its request.

- **Advantage**: This stateless design simplifies Lambda execution and reduces the risk of data inconsistencies or conflicts when the same function is invoked concurrently for different users.

In each of these cases, the goal is to leverage AWS' built-in features to handle concurrency effectively, ensuring that applications remain robust, scalable, and error-free. By embracing these best practices and practical solutions, you can navigate the complexities of parallel processing with confidence. Remember, mastering concurrency requires a careful balance between speed, efficiency, and reliability.

In the next section, we'll explore the trade-offs of parallel processing, helping you make informed decisions about when to harness its power and when to stick with proven sequential approaches.

Evaluating parallelism in software design – balancing performance and complexity

Implementing parallel processing in software design involves critical trade-offs between the potential for increased performance and the added complexity it brings. A careful assessment is essential to determine whether parallelization is justified.

Here are the considerations for parallelization:

- **Task suitability**: Evaluate whether the task is suitable for parallelization and whether the expected performance gains justify the added complexity

- **Resource availability**: Assess the hardware capabilities, such as CPU cores and memory, needed for effective parallel execution

- **Development constraints**: Consider available time, budget, and expertise for developing and maintaining a parallelized system

- **Expertise requirements**: Ensure your team has the skills required for parallel programming

The approach to parallel processing should begin with simple, modular designs for an easier transition to parallelism. Benchmarking is vital to gauge potential performance improvements. Opt for incremental refactoring, supported by comprehensive testing at each step, to ensure smooth integration of parallel processes.

From all this discussion, we conclude that parallel processing can substantially enhance performance, but successful implementation demands a balanced approach, considering task suitability, resource availability, and the development team's expertise. It's a potent tool that, when used judiciously and designed with clarity, can lead to efficient and maintainable code. Remember, while parallel processing is powerful, it's not a universal solution and should be employed strategically.

Summary

This chapter was your invitation to this fascinating world of parallel processing, where we explored the tools at your disposal. First up was the Fork/Join framework. Your head chef, adept at breaking down daunting tasks into bite-sized sub-recipes, ensured everyone had a role to play. But efficiency is key, and that's where the work-stealing algorithm kicked in. Think of it as chefs who glanced over each other's shoulders, jumped in to help if anyone fell behind, and kept the kitchen humming like a well-oiled machine.

However, not all tasks are created equal. That's where `RecursiveTask` and `RecursiveAction` stepped in. They were like chefs specializing in different courses, one meticulously chopped vegetables while the other stirred a simmering sauce, each focused on their own piece of the culinary puzzle.

Now, let's talk about efficiency. Parallel streams were like pre-washed and chopped ingredients, ready to be tossed into the processing pan. We saw how they simplify data processing on collections, using the Fork/Join framework as their secret weapon to boost speed, especially for those dealing with mountains of data.

However, choosing the right tool is crucial. That's why we dived into a parallel processing showdown, pitting Fork/Join against other methods such as `ThreadPoolExecutor` and `CompletableFuture`. This helped you understand their strengths and weaknesses and enabled you to make informed decisions.

However, complexity lurks in the shadows. So, we also tackled the art of handling tasks with dependencies, learned how to break them down, and kept data synchronized. This ensured your culinary masterpiece didn't turn into a chaotic scramble.

And who doesn't love a bit of optimization? So, we explored strategies to fine-tune your parallel processing and learned how to balance task sizes and parallelism levels for the most efficient performance, like a chef adjusting the heat and seasoning to perfection.

Finally, we delved into the advanced realm of a custom Spliterator, giving you the power to tailor parallel stream processing for specific needs.

As every dish comes with its own trade-offs, we discussed the balance between performance gains and complexity, guiding you in making informed software design decisions that leave you feeling satisfied, not burnt out.

We've orchestrated a symphony of parallel processing in this chapter, but what happens when your culinary creations clash and pots start boiling over? That's where *Chapter 4* steps in, where we will dive deep into the Java concurrency utilities and testing, your essential toolkit for handling the delicate dance of multithreading.

Questions

1. What is the primary purpose of the Fork/Join Framework in Java?

    A. To provide a GUI interface for Java applications

    B. To enhance parallel processing by recursively splitting and executing tasks

    C. To simplify database connectivity in Java applications

    D. To manage network connections in Java applications

2. How do `RecursiveTask` and `RecursiveAction` differ in the Fork/Join Framework?

    A. `RecursiveTask` returns a value, while `RecursiveAction` does not

    B. `RecursiveAction` returns a value, while `RecursiveTask` does not

    C. Both return values but `RecursiveAction` does so asynchronously

    D. There is no difference; they are interchangeable

3. What role does the work-stealing algorithm play in the Fork/Join Framework?

    A. It encrypts data for secure processing

    B. It allows idle threads to take over tasks from busy threads

    C. It prioritizes task execution based on complexity

    D. It reduces the memory footprint of the application

4. Which of the following is the best practice for optimizing parallel processing performance in Java?

    A. Increasing the use of shared data

    B. Balancing task granularity and parallelism level

    C. Avoiding the use of thread-safe data structures

    D. Consistently using the highest possible level of parallelism

5. What factors should be considered when implementing parallel processing in software design?

    A. Color schemes and UI design

    B. The task's nature, resource availability, and team expertise

    C. The brand of hardware being used

    D. The programming language's popularity

4

Java Concurrency Utilities and Testing in the Cloud Era

Remember the bustling kitchen from the last chapter, where chefs collaborated to create culinary magic? Now, imagine a cloud kitchen, where orders fly in from all corners, demanding parallel processing and perfect timing. That's where Java concurrency comes in, the secret sauce for building high-performance cloud applications.

This chapter is your guide to becoming a master chef of Java concurrency. We'll explore the Executor framework, your trusty sous chef for managing threads efficiently. We'll dive into Java's concurrent collections, ensuring data integrity even when multiple cooks are stirring the pot.

But a kitchen thrives on coordination! We'll learn synchronization tools such as `CountDownLatch`, `Semaphore`, and `CyclicBarrier`, guaranteeing ingredients arrive at the right time and chefs don't clash over shared equipment. We'll even unlock the secrets of Java's locking mechanisms, mastering the art of sharing resources without culinary chaos.

Finally, we'll equip you with testing and debugging strategies, the equivalent of a meticulous quality check before serving your dishes to the world. By the end, you'll be a Java concurrency ninja, crafting cloud applications that run smoothly and efficiently, and leave your users raving for more.

Technical requirements

You will need **Visual Studio Code (VS Code)** installed. Here is the URL to download it: `https://code.visualstudio.com/download`.

VS Code offers a lightweight and customizable alternative to the other options on this list. It's a great choice for developers who prefer a less resource-intensive **integrated development environment (IDE)** and want the flexibility to install extensions tailored to their specific needs. However, it may not have all the features out of the box compared to the more established Java IDEs.

You will need to install Maven. To do so, follow these steps:

1. **Download Maven**:

   - Go to the Apache Maven website: `https://maven.apache.org/download.cgi`

   - Select the **Binary zip archive** if you are on Windows or the **Binary tar.gz archive** if you are on Linux or macOS.

2. **Extract the archive**:

   - Unzip or untar the downloaded file to the directory where you want to install Maven (e.g., `C:\Program Files\Apache\Maven` on Windows or `/opt/apache/maven` on Linux).

3. **Set environment variables**:

   - **Windows**:

     - `MAVEN_HOME`: Create an environment variable named `MAVEN_HOME` and set its value to the directory where you extracted Maven (e.g., `C:\Program Files\Apache\Maven\apache-maven-3.8.5`).

     - `PATH`: Update your `PATH` environment variable to include the Maven bin directory (e.g., `%MAVEN_HOME%\bin`).

   - **Linux/macOS**:

     - Open the terminal and add the following line to your `~/.bashrc or ~/.bash_profile file: export PATH=/opt/apache-maven-3.8.5/bin:$PATH`.

4. **Verify installation**:

   - Open a command prompt or terminal and type `mvn -version`. If installed correctly, you'll see the Maven version, Java version, and other details.

Uploading your JAR file to AWS Lambda

Here are the prerequisites:

- **AWS account**: You'll need an AWS account with permission to create a Lambda function.

- **JAR file**: Your Java project is compiled and packaged into a JAR file (using tools such as Maven or Gradle).

Log in to the AWS console:

1. **Go to AWS Lambda**: Navigate to the AWS Lambda service within your AWS console.

2. **Create function**: Click **Create Function**. Choose **Author from Scratch**, give your function a name, and select the Java runtime.

3. **Upload code**: In the **Code source** section, choose **Upload from: Upload .zip or .jar file**, and then click **Upload**. Select your JAR file.

4. **Handler**: Enter the fully qualified name of your handler class (e.g., `com.example.MyHandler`). A Java AWS Lambda handler class is a Java class that defines the entry point for your Lambda function's execution, containing a method named `handleRequest` to process incoming events and provide an appropriate response. For detailed information, see the following documentation:

   • **Java**: `https://docs.aws.amazon.com/lambda/latest/dg/java-handler.html`

5. **Save**: Click **Save** to create your Lambda function.

Here are some important things to consider:

• **Dependencies**: If your project has external dependencies, you'll either need to package them into your JAR (sometimes called an *uber-jar* or *fat jar*) or utilize Lambda layers for those dependencies.

• **IAM role**: Your Lambda function needs an IAM role with appropriate permissions to interact with other AWS services if it will do so.

Further, the code in this chapter can be found on GitHub:

`https://github.com/PacktPublishing/Java-Concurrency-and-Parallelism`

Introduction to Java concurrency tools – empowering cloud computing

In the ever-expanding realm of cloud computing, building applications that can juggle multiple tasks simultaneously is no longer a luxury, but a necessity. This is where **Java concurrency utilities (JCU)** emerge as a developer's secret weapon, offering a robust toolkit to unlock the true potential of concurrent programming in the cloud. Here are the useful features of JCU:

• **Unleashing scalability**: Imagine a web application effortlessly handling a sudden surge in user traffic. This responsiveness and ability to seamlessly scale up is a key benefit of JCU. By leveraging features such as thread pools, applications can dynamically allocate resources based on demand, preventing bottlenecks and ensuring smooth performance even under heavy load.

- **Speed is king**: In today's fast-paced world, latency is the enemy of a positive user experience. JCU helps combat this by optimizing communication and minimizing wait times. Techniques such as non-blocking I/O and asynchronous operations ensure requests are processed swiftly, leading to quicker response times and happier users.

- **Every resource counts**: Cloud environments operate on a pay-as-you-go model, making efficient resource utilization crucial. JCU acts as a wise steward, carefully managing threads and resources to avoid wastage. Features such as concurrent collections, designed for concurrent access, reduce locking overhead and ensure efficient data handling, ultimately keeping cloud costs under control.

- **Resilience in the face of adversity**: No system is immune to occasional hiccups. In the cloud, these can manifest as temporary failures or glitches. Thankfully, JCU's asynchronous operations and thread safety act as a shield, enabling applications to recover quickly from setbacks and maintain functionality with minimal disruption.

- **Seamless integration**: Modern cloud development often involves integrating with various cloud-specific services and libraries. JCU's standards-compliant design ensures smooth integration, providing a unified approach to managing concurrency across different cloud platforms and technologies.

- **The road ahead**: While JCU offers immense power, navigating the cloud environment requires careful consideration. Developers need to monitor and fine-tune JCU configurations to ensure optimal performance, just like carefully optimizing server configurations. Distributed cloud deployments introduce the challenge of managing concurrency across regions, which JCU tools such as `ConcurrentHashMap` readily address, but others might require additional configuration for cross-region communication and synchronization.

- **Security first**: As with any powerful tool, security is paramount. JCU offers features such as atomic variables and proper locking mechanisms to help prevent concurrency vulnerabilities such as race conditions, but it's crucial to adopt secure coding practices to fully fortify cloud applications against potential threats.

In conclusion, JCU are not just tools, but an empowering force for developers seeking to build cloud applications that are not only efficient and scalable but also resilient. By understanding and harnessing their power, along with navigating the considerations with care, developers can create digital solutions that thrive in the ever-evolving cloud landscape.

Real-world example – building a scalable application on AWS

Imagine an e-commerce platform experiencing surges in image uploads during product launches or promotions. Traditional, non-concurrent approaches can struggle with such spikes, leading to slow processing, high costs, and frustrated customers. This example demonstrates how JCU and AWS Lambda can be combined to create a highly scalable and cost-effective image processing pipeline.

Let's look at this scenario – our e-commerce platform needs to process uploaded product images by resizing them for various display sizes, optimizing them for web delivery, and storing them with relevant metadata for efficient retrieval. This process must handle sudden bursts in image uploads without compromising performance or incurring excessive costs.

The following Java code demonstrates how to use JCU within an AWS Lambda function to perform image processing tasks in parallel. This example includes using `ExecutorService` for executing tasks such as image resizing and optimization, `CompletableFuture` for asynchronous operations, such as calling external APIs or fetching data from DynamoDB, and illustrates a conceptual approach for non-blocking I/O operations with Amazon S3 integration.

For Maven users, add the `aws-java-sdk` dependency to `pom.xml`:

```
<dependencies>
        <dependency>
            <groupId>com.amazonaws</groupId>
            <artifactId>aws-java-sdk</artifactId>
            <version>1.12.118</version>
    <!-- Check https://mvnrepository.com/artifact/com.amazonaws/aws-java-
sdk for the latest version -->
        </dependency>
    </dependencies>
```

Here is the code snippet:

```
public class ImageProcessorLambda implements RequestHandler<S3Event,
String> {

    private final ExecutorService executorService = Executors.
newFixedThreadPool(10);
    private final AmazonS3 s3Client = AmazonS3ClientBuilder.
standard().build();

    @Override
    public String handleRequest(S3Event event, Context context) {
        event.getRecords().forEach(record -> {
            String bucketName = record.getS3().getBucket().getName();
            String key = record.getS3().getObject().getKey();

            // Asynchronously resize and optimize image
            CompletableFuture.runAsync(() -> {
                // Placeholder for image resizing and optimization
                    logic
                System.out.println("Resizing and optimizing image: " +
                key);
                // Simulate image processing
```

```
        try {
            Thread.sleep(500);
// Simulate processing delay
        } catch (InterruptedException e) {
            Thread.currentThread().interrupt();
        }
        // Upload the processed image to a different bucket or
            prefix
        s3Client.putObject(new PutObjectRequest(
            "processed-bucket", key,
            "processed-image-content"));
    }, executorService);


    // Asynchronously call external APIs or fetch user
        preferences from DynamoDB
    CompletableFuture.supplyAsync(() -> {
        // Placeholder for external API call or fetching user
            preferences
        System.out.println("Fetching additional data for
        image: " + key);
        return "additional-data";
// Simulated return value
    }, executorService).thenAccept(additionalData -> {
// Process additional data (e.g., tagging based on content)
        System.out.println("Processing additional data: " +
        additionalData);
    });
    });


    // Shutdown the executor to allow the Lambda function to
complete
    // Note: In a real-world scenario, consider carefully when to
shut down the executor,
    // as it may be more efficient to keep it alive across
multiple invocations if possible
    executorService.shutdown();
    return "Image processing initiated";
    }
}
```

Here is the code explanation:

- ExecutorService: This manages a pool of threads for concurrent tasks. Here, it is used to resize and optimize images asynchronously.

- `CompletableFuture`: This enables asynchronous programming. This example uses it for making non-blocking calls to external APIs or services such as DynamoDB and processing their results.

- **Amazon S3 integration**: `AmazonS3ClientBuilder` is used to create an S3 client, which is then used to upload processed images.

- **Lambda handler**: This implements `RequestHandler<S3Event, String>` to process incoming S3 events, indicating it's triggered by S3 events (e.g., new image uploads).

This example omits actual image processing, API calls, and AWS SDK setup details for brevity.

This example showcases how JCU, combined with the serverless architecture of AWS Lambda, empowers developers to build highly scalable, cost-effective, and efficient cloud-based applications. By leveraging JCU's concurrency features and integrating them seamlessly with AWS services, developers can create robust solutions that thrive in the dynamic and demanding cloud environment.

Taming the threads – conquering the cloud with the Executor framework

Remember those single-threaded applications, struggling to keep up with the ever-changing demands of the cloud? Well, forget them! The **Executor framework** is here to unleash your inner cloud architect, empowering you to build applications that adapt and thrive in this dynamic environment.

Think of it like this: your cloud application is a bustling city, constantly handling requests and tasks. The Executor framework is your trusty traffic manager, ensuring smooth operation even during peak hours.

The key players of the Executor framework are as follows:

- `ExecutorService`: The adaptable city planner, dynamically adjusting the number of available *lanes* (threads) based on real-time traffic (demand). No more idle threads or bottlenecked tasks!

- `ScheduledExecutorService`: The punctual timekeeper, meticulously scheduling events, reminders, and tasks with precision. Whether it's daily backups or quarterly reports, everything runs like clockwork.

- `ThreadPoolExecutor`: The meticulous jeweler, carefully crafts thread pools with just the right size and configuration. They balance the city's needs with resource efficiency, ensuring every thread shines like a gem.

- **Work queues**: The city's storerooms, each with unique strategies for organizing tasks before execution. Choose the right strategy (such as first in first out or priority queues) to keep tasks flowing smoothly and avoid resource overload.

The Executor framework doesn't just manage resources; it prioritizes them too. Imagine a sudden surge in visitors (requests). The framework ensures critical tasks are handled first, even when resources are stretched thin, keeping your city (application) running smoothly.

The symphony of cloud integration and adaptation

Our city, though grand, does not stand alone. It is but a part of a greater kingdom – the cloud. By integrating the Executor framework with the cloud's myriad services and APIs, our city can stretch beyond its walls, tapping into the vast reservoirs of the cloud to dynamically adjust its resources, much like drawing water from the river during a drought or opening the gates during a flood.

Adaptive execution strategies are the city's scouts, constantly surveying the landscape and adjusting the city's strategies based on the ever-changing conditions of the cloud. Whether it's a surge in visitors or an unexpected storm, the city adapts, ensuring optimal performance and resource utilization.

The chronicles of best practices

As our tale comes to a close, the importance of monitoring and metrics emerges as the sage's final piece of advice. Keeping a vigilant eye on the city's operations ensures that decisions are made not in the dark, but with the full light of knowledge, guiding the city to scale gracefully and efficiently.

So, our journey through the realms of the Executor framework for cloud-based applications concludes. By embracing dynamic scalability, mastering resource management, and integrating seamlessly with the cloud, developers can forge applications that not only withstand the test of time but thrive in the ever-evolving landscape of cloud computing. The tale of the Executor framework is a testament to the power of adaptation, efficiency, and strategic foresight in the era of cloud computing.

Real-world examples of thread pooling and task scheduling in cloud architectures

Moving beyond theory, let's dive into real-world scenarios where Java's concurrency tools shine in cloud architectures. These examples showcase how to optimize resource usage and ensure application responsiveness under varying loads.

Example 1 – keeping data fresh with scheduled tasks

Imagine a cloud-based application that needs to regularly crunch data from various sources. Scheduled tasks are your secret weapon, ensuring data is always up to date, even during peak hours.

Objective: Process data from multiple sources periodically, scaling with data volume.

Environment: A distributed system gathering data from APIs for analysis.

Here is the code snippet:

```
public class DataAggregator {
    private final ScheduledExecutorService scheduler = Executors.
    newScheduledThreadPool(5);
    public DataAggregator() {
        scheduleDataAggregation();
    }
    private void scheduleDataAggregation() {
        Runnable dataAggregationTask = () -> {
            System.out.println(
                "Aggregating data from sources...");
            // Implement data aggregation logic here
        };
        // Run every hour, adjust based on your needs
        scheduler.scheduleAtFixedRate(
            dataAggregationTask, 0, 1, TimeUnit.HOURS);
    }
}
```

The key points from the preceding example are as follows:

- **Scheduled task execution**: The `scheduleAtFixedRate` method ensures regular data updates, even under varying loads.
- **Resource efficiency**: A dedicated executor with a configurable thread pool size allows for efficient resource management, scaling up during peak processing.

Example 2 – adapting to the cloud's dynamics

Cloud resources are like the weather – ever-changing. This example shows how to customize thread pools for optimal performance and resource utilization in AWS, handling diverse workloads and fluctuating resource availability.

Objective: Adapt a thread pool to handle varying computational demands in AWS, ensuring efficient resource use and cloud resource adaptability.

Environment: An application processing both lightweight and intensive tasks, deployed in an AWS environment with dynamic resources.

Here is the code snippet:

```
public class AWSCloudResourceManager {
    private ThreadPoolExecutor threadPoolExecutor;
    public AWSCloudResourceManager() {
        // Initial thread pool configuration based on baseline
```

```
resource availability
        int corePoolSize = 5;
// Core number of threads for basic operational capacity
        int maximumPoolSize = 20;
// Maximum threads to handle peak loads
        long keepAliveTime = 60;
// Time (seconds) an idle thread waits before terminating
        TimeUnit unit = TimeUnit.SECONDS;


        // WorkQueue selection: ArrayBlockingQueue for a fixed-size
queue to manage task backlog
        ArrayBlockingQueue<Runnable> workQueue = new
        ArrayBlockingQueue<>(100);
        // Customizing ThreadPoolExecutor to align with cloud resource
          dynamics
        threadPoolExecutor = new ThreadPoolExecutor(
            corePoolSize,
            maximumPoolSize,
            keepAliveTime,
            unit,
            workQueue,
            new ThreadPoolExecutor.CallerRunsPolicy()
// Handling tasks when the system is saturated
        );
    }
    // Method to adjust ThreadPoolExecutor parameters based on real-
time cloud resource availability
    public void adjustThreadPoolParameters(int newCorePoolSize, int
    newMaxPoolSize) {
        threadPoolExecutor.setCorePoolSize(
            newCorePoolSize);
        threadPoolExecutor.setMaximumPoolSize(
            newMaxPoolSize);
        System.out.println("ThreadPool parameters adjusted:
        CorePoolSize = " + newCorePoolSize + ", MaxPoolSize = " +
        newMaxPoolSize);
    }
    // Simulate processing tasks with varying computational demands
    public void processTasks() {
        for (int i = 0; i < 500; i++) {
            final int taskId = i;
            threadPoolExecutor.execute(() -> {
                System.out.println(
                    "Processing task " + taskId);
                // Task processing logic here
```

```
        });
    }
}

    public static void main(String[] args) {
        AWSCloudResourceManager manager = new
        AWSCloudResourceManager();

        // Simulate initial task processing
        manager.processTasks();

        // Adjust thread pool settings based on simulated change in
resource availability
        manager.adjustThreadPoolParameters(10, 30);
// Example adjustment for increased resources
    }
}
```

The key points from the preceding example are as follows:

> **Dynamic thread pool customization**: The AWSCloudResourceManager class initializes
> ThreadPoolExecutor with a configurable core and maximum pool sizes. This setup
> allows the application to start with a conservative resource usage model, scaling up as demand
> increases or more AWS resources become available.

- **Adaptable resource management**: By providing the adjustThreadPoolParameters
  method, the application can dynamically adapt its thread pool configuration in response to
  AWS resource availability changes. This might be triggered by metrics from AWS CloudWatch
  or other monitoring tools, enabling real-time scaling decisions.

- **Work queue strategy**: The selection of ArrayBlockingQueue for the executor's work queue
  provides a clear strategy for managing task overflow. By limiting the queue size, the system can
  apply backpressure when under heavy load, preventing resource exhaustion.

- **Handling diverse workloads**: This approach allows the application to efficiently process a
  mixture of task types – ranging from quick, lightweight tasks to more prolonged, compute-
  intensive operations. The CallerRunsPolicy rejection policy ensures that tasks are not
  lost during peak loads but rather executed on the calling thread, adding a layer of robustness.

These examples demonstrate how Java's concurrency tools empower cloud-based applications to
thrive in dynamic environments. By embracing dynamic scaling, resource management, and cloud
integration, you can build applications that are both responsive and cost-effective, regardless of the
ever-changing cloud landscape.

Utilizing Java's concurrent collections in distributed systems and microservices architectures

In the intricate world of distributed systems and microservices architectures, akin to a bustling city where data zips across the network like cars on a freeway, managing shared resources becomes a vital endeavor. Java's concurrent collections step into this urban sprawl, offering efficient pathways and junctions for data to flow unhindered, ensuring that every piece of information reaches its destination promptly and accurately. Let's embark on a journey through two pivotal structures in this landscape: `ConcurrentHashMap` and `ConcurrentLinkedQueue` and explore how they enable us to build applications that are not only scalable and reliable but also high performing.

Navigating through data with ConcurrentHashMap

Let us first understand the landscape of `ConcurrentHashMap`.

Scenario: Picture a scenario in a sprawling metropolis where every citizen (microservice) needs quick access to a shared repository of knowledge (data cache). Traditional methods might cause traffic jams – delays in data access and potential mishaps in data consistency.

Solution: `ConcurrentHashMap` acts as a high-speed metro system for data, offering a thread-safe way to manage this shared repository. It enables concurrent read and write operations without the overhead of full-scale synchronization, akin to having an efficient, automated traffic system that keeps data flowing smoothly at rush hour.

Here is an example of the usage of `ConcurrentHashMap`:

```
ConcurrentHashMap<String, String> cache = new ConcurrentHashMap<>();
cache.put("userId123", "userData");
String userData = cache.get("userId123");}
```

This simple snippet demonstrates how a user's data can be cached and retrieved with `ConcurrentHashMap`, ensuring fast access and thread safety without the complexity of manual synchronization.

Processing events with ConcurrentLinkedQueue

Now, let us explore the landscape of `ConcurrentLinkedQueue`.

Scenario: Imagine our city bustling with events – concerts, parades, and public announcements. There needs to be a system to manage these events efficiently, ensuring they're organized and processed in a timely manner.

Solution: `ConcurrentLinkedQueue` serves as the city's event planner, a non-blocking, thread-safe queue that efficiently handles the flow of events. It's like having a dedicated lane on the freeway for emergency vehicles; events are processed swiftly, ensuring the city's life pulse remains vibrant and uninterrupted.

Here is an example of the usage of `ConcurrentLinkedQueue`:

```
ConcurrentLinkedQueue<String> eventQueue = new
ConcurrentLinkedQueue<>();
eventQueue.offer("New User Signup Event");
String event = eventQueue.poll();
```

In this example, events such as user signups are added to and processed from the queue, showcasing how `ConcurrentLinkedQueue` supports concurrent operations without locking, making event handling seamless and efficient.

Best practices for using Java's concurrent collections

Here are the best practices for our consideration:

- **Choose the right collection**: Just like selecting the optimal route for your commute, choosing the right concurrent collection for your needs is crucial. `ConcurrentHashMap` is ideal for caches or frequent read/write operations, while `ConcurrentLinkedQueue` excels in FIFO event processing scenarios.

- **Understand collection behavior**: Familiarize yourself with the nuances of each collection, such as iteration safety with `CopyOnWriteArrayList` or the non-blocking nature of `ConcurrentLinkedQueue`, to fully leverage their capabilities.

- **Monitor performance**: Keep an eye on the performance of these collections, especially in high-load scenarios. Tools such as JMX or Prometheus can help identify bottlenecks or contention points, allowing for timely optimizations.

By integrating Java's concurrent collections into your distributed systems and microservices, you empower your applications to handle the complexities of concurrency with grace, ensuring data is managed efficiently and reliably amidst the bustling activity of your digital ecosystem.

Advanced locking strategies for tackling cloud concurrency

This section delves into sophisticated locking strategies within Java, spotlighting mechanisms that extend well beyond basic synchronization techniques. These advanced methods provide developers with enhanced control and flexibility, crucial for addressing concurrency challenges in environments marked by high concurrency or intricate resource management needs.

Revisiting lock mechanisms with a cloud perspective

Here's a breakdown of how each advanced locking strategy can benefit cloud applications:

- **Reentrant locks for cloud resources**: `ReentrantLock` surpasses traditional intrinsic locks by offering detailed control, including the ability to specify a timeout for lock attempts. This prevents threads from being indefinitely blocked, a vital feature for cloud applications dealing with shared resources such as cloud storage or database connections. For example, managing access to a shared cloud service can leverage `ReentrantLock` to ensure that if one task is waiting too long for a resource, other tasks can continue, enhancing overall application responsiveness.

- **Optimizing cloud data access with read/write locks**: `ReadWriteLock` is pivotal in scenarios where cloud applications experience a high volume of read operations but fewer write operations, such as caching layers or configuration data stores. Utilizing `ReadWriteLock` can significantly improve performance by allowing concurrent reads, while still ensuring data integrity during writes.

- **Stamped locks for dynamic cloud environments**: `StampedLock`, introduced in Java 8, is particularly suited for cloud applications due to its versatility in handling read and write access. It supports optimistic reading, which can reduce lock contention in read-heavy environments such as real-time data analytics or monitoring systems. The ability to upgrade from a read to a write lock is especially useful in cloud environments where data states can change frequently.

- **Utilizing condition objects for cloud task coordination**: Condition objects, when used with `ReentrantLock`, offer a refined mechanism for managing inter-thread communication, crucial for orchestrating complex workflows in cloud applications. This approach is more advanced and flexible compared to the traditional wait-notify mechanism, facilitating efficient resource utilization and synchronization among distributed tasks.

Consider a scenario managing comments in a cloud-based application, showcasing how to apply different locking mechanisms for optimizing both read-heavy and write-heavy operations.

Here is a code snippet:

```
public class BlogManager {
    private final ReadWriteLock readWriteLock = new
    ReentrantReadWriteLock();
    private final StampedLock stampedLock = new StampedLock();
    private List<Map<String, Object>> comments = new ArrayList<>();

    // Method to read comments using ReadWriteLock for concurrent
access
    public List<Map<String, Object>> getComments() {
        readWriteLock.readLock().lock();
        try {
            return Collections.unmodifiableList(comments);
```

```
    } finally {
        readWriteLock.readLock().unlock();
    }
}

// Method to add a comment with StampedLock for efficient locking
public void addComment(String author, String content, long
timestamp) {
    long stamp = stampedLock.writeLock();
    try {
        Map<String, Object> comment = new HashMap<>();
        comment.put("author", author);
        comment.put("content", content);
        comment.put("timestamp", timestamp);
        comments.add(comment);
    } finally {
        stampedLock.unlock(stamp);
    }
}
}
```

The key points from the preceding code example are as follows:

- **Optimized reading**: Using ReadWriteLock ensures that multiple threads can concurrently read comments without blocking each other, maximizing efficiency in high-read scenarios typical in cloud applications.

- **Efficient writing**: StampedLock is used for adding comments, providing a mechanism to ensure that writes are performed with exclusive access, yet efficiently managed to minimize blocking.

Understanding and leveraging these advanced Java locking strategies empowers developers to address cloud-specific concurrency challenges effectively. By judiciously applying these techniques, cloud applications can achieve improved performance, scalability, and resilience, ensuring robust management of shared resources in complex, distributed cloud environments. Each locking mechanism serves a distinct purpose, allowing for tailored solutions based on the application's requirements and the concurrency model it employs.

Advanced concurrency management for cloud workflows

Cloud architectures introduce unique challenges in workflow management, necessitating precise coordination across multiple services and efficient resource allocation. This section advances the discussion from *Chapter 2, Introduction to Java's Concurrency Foundations: Threads, Processes, and Beyond*, introducing sophisticated Java synchronizers suited for orchestrating complex cloud workflows and ensuring seamless inter-service communication.

Sophisticated Java synchronizers for cloud applications

This section explores advanced Java synchronizers that go beyond basic functionality, empowering you to orchestrate complex service startups with grace and efficiency.

Enhanced CountDownLatch for service initialization

Beyond basic synchronization, an advanced **CountDownLatch** can facilitate the phased startup of cloud services, integrating health checks and dynamic dependencies.

Let's delve into an enhanced example of using `CountDownLatch` for initializing cloud services, incorporating dynamic checks and dependencies resolution. This example illustrates how an advanced `CountDownLatch` mechanism can be employed to manage the complex startup sequence of cloud services, ensuring that all initialization tasks are completed, considering service dependencies and health checks:

```
public class CloudServiceInitializer {
    private static final int TOTAL_SERVICES = 3;
    private final CountDownLatch latch = new CountDownLatch(
    TOTAL_SERVICES);

    public CloudServiceInitializer() {
        // Initialization tasks for three separate services
        for (int i = 0; i < TOTAL_SERVICES; i++) {
            new Thread(new ServiceInitializer(
                i, latch)).start();
        }
    }
    public void awaitServicesInitialization() throws
    InterruptedException {
        // Wait for all services to be initialized
        latch.await();
        System.out.println("All services initialized. System is ready
        to accept requests.");
    }
    static class ServiceInitializer implements Runnable {
        private final int serviceId;
        private final CountDownLatch latch;
        ServiceInitializer(
            int serviceId, CountDownLatch latch) {
                this.serviceId = serviceId;
                this.latch = latch;
        }

        @Override
```

```
    public void run() {
        try {
            // Simulate service initialization with varying time
            delays
            System.out.println(
                "Initializing service " + serviceId);
            Thread.sleep((long) (
                Math.random() * 1000) + 500);
            System.out.println("Service " + serviceId + "
            initialized.");
        } catch (InterruptedException e) {
            Thread.currentThread().interrupt();
        } finally {
            // Signal that this service has been initialized
            latch.countDown();
        }
    }
}

public static void main(String[] args) {
    CloudServiceInitializer initializer = new
    CloudServiceInitializer();
    try {
        initializer.awaitServicesInitialization();
    } catch (InterruptedException e) {
        Thread.currentThread().interrupt();
        System.out.println("Service initialization was
        interrupted.");
    }
}
}}
```

The key points from the preceding code example are as follows:

- **Initialization logic**: The `CloudServiceInitializer` class encapsulates the logic for initializing a predefined number of services, defined by `TOTAL_SERVICES`. It creates and starts a separate thread for each service initialization task, passing a shared `CountDownLatch` to each.

- `ServiceInitializer`: Each instance of `ServiceInitializer` represents a task to initialize a particular service. It simulates the initialization process with a random sleep duration. Upon completion, it decrements the latch's count using `countDown()`, signaling that it has finished its initialization task.

- **Synchronization on service readiness**: The awaitServicesInitialization method in CloudServiceInitializer waits for the count of CountDownLatch to reach zero, indicating that all services have been initialized. This method blocks the main thread until all services report readiness, after which it prints a message indicating that the system is ready to accept requests.

- **Dynamic service initialization**: This approach provides flexibility in managing cloud service dependencies. Services are initialized in parallel, with CountDownLatch ensuring that the main application flow proceeds only after all services are up and running. This model is particularly useful in cloud environments where services may have interdependencies or require health checks before they can be deemed ready.

This enhanced CountDownLatch usage showcases how Java concurrency utilities can be effectively applied to manage complex initialization sequences in cloud applications, ensuring robust startup behavior and dynamic dependency management.

Semaphore for controlled resource access

In cloud environments, **Semaphore** can be fine-tuned to manage access to shared cloud resources such as databases or third-party APIs, preventing overloading while maintaining optimal throughput. This mechanism is critical in environments where resource constraints are dynamically managed based on current load and **service-level agreements (SLAs)**.

Here's an example of how Semaphore can be used to coordinate access to a shared data resource in a cloud environment:

```java
public class DataAccessCoordinator {
    private final Semaphore semaphore;

    public DataAccessCoordinator(int permits) {
        this.semaphore = new Semaphore(permits);
    }

    public void accessData() {
        try {
            semaphore.acquire();
            // Access shared data resource
            System.out.println("Data accessed by " + Thread.
            currentThread().getName());
            // Simulate data access
            Thread.sleep(100);
        } catch (InterruptedException e) {
            Thread.currentThread().interrupt();
        } finally {
            semaphore.release();
```

```
        }
    }

    public static void main(String[] args) {
        DataAccessCoordinator coordinator = new
        DataAccessCoordinator(5);
        // Simulate multiple services accessing data concurrently
        for (int i = 0; i < 10; i++) {
            new Thread(coordinator::accessData,
                "Service-" + i).start();
        }
    }
}
```

Here is the code explanation:

- `Semaphore`: It uses a **Semaphore** object with limited permits (configurable via constructor) to control access

- `acquire()`: Threads trying to access data call `acquire()`, blocking if no permits are available

- **Shared data access**: Once acquired, a permit allows the thread to access shared data (simulated by `System.out.println` and sleep)

- `release()`: After accessing data, `release()` is called to return the permit and allow other threads to acquire it

- **Main method**: This demonstrates usage by creating a coordinator with 5 permits, then starting 10 threads that concurrently call `accessData`

CyclicBarrier for batch processing

Imagine a complex data pipeline in the cloud, where processing happens in distinct stages across distributed services. Ensuring each stage is completed successfully before moving on is crucial. This is where **CyclicBarrier** shines as a powerful tool for coordinating batch-processing workflows:

```
public class BatchProcessingWorkflow {
    private final CyclicBarrier barrier;
    private final int batchSize = 5;
// Number of parts in each batch
    public BatchProcessingWorkflow() {
        // Action to take when all threads reach the barrier
        Runnable barrierAction = () -> System.out.println(
            "Batch stage completed. Proceeding to next stage.");
        this.barrier = new CyclicBarrier(batchSize, barrierAction);
    }
```

```
public void processBatchPart(int partId) {
    try {
        System.out.println(
            "Processing part " + partId);
        // Simulating time taken to process part of the batch
        Thread.sleep((long) (Math.random() * 1000));
        System.out.println("Part " + partId + " processed. Waiting
        at barrier.");

        // Wait for other parts to reach this point
        barrier.await();
        // After all parts reach the barrier, proceed with the
        next stage
    } catch (Exception e) {
        Thread.currentThread().interrupt();
    }
}

public static void main(String[] args) {
    BatchProcessingWorkflow workflow = new
    BatchProcessingWorkflow();
    // Simulating concurrent processing of batch parts
    for (int i = 0; i < workflow.batchSize; i++) {
        final int partId = i;
        new Thread(() -> workflow.processBatchPart(
            partId)).start();
    }
}
}
```

The key points from the preceding code example are as follows:

- `CyclicBarrier`: Utilizes `CyclicBarrier` to synchronize batch processing stages. The barrier is set with a specific number of permits (`batchSize`) and an optional action to perform when all threads reach the barrier.

- **Processing method**:

  - **Processing simulation**: Each thread simulates processing a part of the batch by printing a message and sleeping for a random duration

  - **Barrier synchronization**: After processing, threads call `barrier.await()`, blocking until the specified number of threads (`batchSize`) reaches this point, ensuring all parts of the batch are processed before moving on

- **Shared data access**: While this example doesn't directly manipulate shared data, it simulates processing and synchronization points. In real scenarios, threads would operate on shared resources here.

- **Barrier action**: A *runnable action* defined during `CyclicBarrier` initialization executes once all participating threads reach the barrier. It marks the completion of a batch stage and allows for collective post-processing or setup before the next stage begins.

- **Main method**:

  - **Workflow initialization**: It instantiates `BatchProcessingWorkflow` with a `CyclicBarrier` configured for 5 permits (matching `batchSize`).

  - **Concurrent execution**: It starts 10 threads to simulate concurrent processing of batch parts. Since the barrier is set for 5 permits, it demonstrates two rounds of batch processing, waiting for 5 parts to complete before proceeding in each round.

This code structure is ideal for scenarios requiring precise coordination between threads, like in distributed systems or complex data processing pipelines, where each processing stage must be completed across all services before moving to the next stage.

Utilizing tools for diagnosing concurrency problems

In the world of Java development, especially when navigating the complexities of cloud-based applications, understanding and diagnosing concurrency issues becomes a critical skill. Like detectives at a crime scene, developers often need to piece together evidence to solve the mysteries of application slowdowns, freezes, or unexpected behavior. This is where thread dumps and lock monitors come into play.

Thread dumps – the developer's snapshot

Imagine you're walking through a bustling marketplace – each stall and shopper representing threads within a **Java virtual machine** (**JVM**). Suddenly, everything freezes. A thread dump is like taking a panoramic photo of this scene, capturing every detail: who's talking to whom, who's waiting in line, and who's just browsing. It's a moment-in-time snapshot that reveals the state of all threads running in the JVM, including their current actions, who they're waiting for, and who's blocking their path.

Here are the features of thread dumps:

- **Capturing the moment**: Generating these insightful snapshots can be done in various ways, each like choosing the right lens for your camera

- **JDK command-line tools**: `jstack`, a tool as handy as a Swiss army knife, allows developers to generate a thread dump from the command line

- **IDEs:** Modern IDEs, such as IntelliJ IDEA or Eclipse, come equipped with built-in tools or plugins for generating and analyzing thread dumps

- **JVM options**: For those who prefer setting traps to catch the moment automatically, configuring the JVM to generate thread dumps under specific conditions is like installing a high-tech security camera system in the marketplace

Real-world cloud adventures

Consider a cloud-based Java application, akin to a sprawling marketplace spread across multiple cloud regions. This application begins to experience intermittent slowdowns, much like congestion happening at unpredictable intervals. The development team suspects deadlocks or thread contention but needs evidence.

The investigation process involves the following:

- **Monitoring and alerting**: First, set up surveillance using cloud-native tools or third-party solutions
- **Generating thread dumps**: Upon an alert, akin to a congestion notification, they use cloud-native tools such as CloudWatch with AWS Lambda, Azure Monitor with Azure Functions, or Stackdriver logging with Google Cloud Monitoring to take snapshots within the affected cloud *regions* (containers)
- **Analyzing the evidence**: With snapshots in hand, the team analyzes them to identify any threads stuck in a deadlock, to see where the congestion started

Lock monitors – the guardians of synchronization

Lock monitors are like sentries guarding access to resources within your application. Tools such as Java VisualVM and JConsole act as the central command center, providing real-time insights into thread lock dynamics, memory usage, and CPU usage.

Imagine your microservice architecture experiencing latency spikes like a flash mob suddenly flooding the marketplace. With Java VisualVM, you can connect to the affected service's JVM and see threads waiting in line, blocked by a single lock. This real-time observation helps you identify bottlenecks and take immediate action, like dispatching security to manage the crowd.

The takeaway after exploring thread dumps and lock monitors is that they maintain order and performance. By utilizing thread dumps and lock monitors, you can transform the chaotic scenes of concurrency issues into orderly queues. This ensures each thread completes its tasks efficiently, keeping your cloud applications running smoothly and delivering a positive user experience.

Remember, these tools are just a starting point. Combine them with your understanding of your application's architecture and behavior for even more effective troubleshooting!

The quest for clarity – advanced profiling techniques

The vast landscapes of cloud-native applications, with their intricate networks of microservices, can pose challenges for traditional profiling methods. These methods often struggle to navigate the distributed nature and complex interactions within these environments. Enter advanced profiling techniques, acting as powerful tools to shed light on performance bottlenecks and optimize your cloud applications. Here are three powerful techniques to demystify your cloud journeys:

- **Distributed tracing – illuminating the request journey**: Think of distributed tracing as charting the stars. While traditional profiling shines a light on individual nodes, tracing follows requests as they hop between microservices, revealing hidden latency bottlenecks and intricate service interactions. Imagine the following:

  - **Pinpointing slow service calls**: Identify which service is causing delays and focus optimization efforts

  - **Visualizing request flow**: Understand the intricate dance of microservices and identify potential bottlenecks

- **Service-level aggregation – zooming out for the big picture**: Imagine profiling data as scattered islands. Service-level aggregation gathers them into a cohesive view, showing how each service contributes to overall performance. It's like looking at the forest, not just the trees:

  - **Spot service performance outliers**: Quickly identify services impacting overall application responsiveness

  - **Prioritize optimization efforts**: Focus resources on services with the most room for improvement

- **Automated anomaly detection – predicting performance storms**: Leveraging machine learning, automated anomaly detection acts as a weather forecaster for your application. It scans for subtle shifts in performance patterns, alerting you to potential issues before they cause major disruptions:

  - **Catch performance regressions early**: Proactively address issues before they impact users.

  - **Reduce time spent troubleshooting**: Focus your efforts on confirmed problems, not chasing ghosts.

These techniques are just the starting point. Choosing the right tool for your specific needs and workflow is crucial.

Weaving the web – integrating profiling tools into CI/CD pipelines

As your cloud application evolves, continuous performance optimization is key. Embedding profiling tools into your CI/CD pipeline is akin to giving your application a heart that beats in rhythm with performance best practices.

Think of your tools as weapons in your performance optimization arsenal, and consider the following:

- **Seamless integration**: Select tools that integrate smoothly into your existing CI/CD workflow
- **Automation capability**: Opt for tools that support automated data collection and analysis
- **Actionable insights**: Ensure the tools provide clear, actionable insights to guide optimization efforts

Some popular options include the following:

- **Distributed tracing tools**: Jaeger and Zipkin
- **Service-level profiling tools**: JProfiler and Dynatrace
- **CI/CD integration tools**: Jenkins and GitLab CI

In addition to these tools, consider tools such as Grafana for visualizing performance data, and leverage machine learning-powered insights from tools such as Dynatrics and New Relic.

Continuously refine your tools and practices based on experience and evolving needs.

By weaving performance into the fabric of your CI/CD pipeline, you can ensure your cloud applications operate at their peak, delivering consistent and exceptional performance for your users.

In the following sections, we'll delve deeper into specific techniques such as service mesh integration and APM solutions, further enriching your performance optimization toolbox.

Service mesh and APM – your cloud performance powerhouse

Imagine your cloud application as a bustling marketplace, with microservices such as vendors conducting transactions. Without a conductor, things get chaotic. Service mesh, such as Istio and Linkerd, ensures each microservice plays its part flawlessly:

- **Transparent observability**: See how data flows between services, identify bottlenecks, and debug issues, all without modifying your code
- **Traffic management**: Route requests efficiently, avoiding overloads and ensuring smooth performance even during peak traffic

- **Consistent policy enforcement**: Set rules (e.g., retry policies, rate limits) globally for all services, simplifying management and guaranteeing predictable behavior

Now, imagine a skilled musician analyzing the marketplace soundscape. That's what APM solutions such as Dynatrace, New Relic, and Elastic APM do:

- **Observability beyond monitoring**: Go beyond basic metrics to correlate logs, traces, and metrics for a holistic view of application health and performance

- **AI-powered insights**: Leverage machine learning to predict issues, diagnose problems faster, and suggest optimizations, keeping your application performing at its best

- **Business impact analysis**: Understand how performance affects user satisfaction and business outcomes, enabling data-driven decisions

By combining service mesh and APM, you gain a comprehensive performance powerhouse for your cloud applications.

Incorporating concurrency frameworks

In the grand tapestry of Java application development, where the threads of concurrency and distributed systems intertwine, frameworks such as Akka and Vert.x emerge as the artisans, sculpting scalable, resilient, and responsive systems from the raw fabric of code.

Akka – building resilient real-time systems with actors

Imagine a bustling marketplace, where merchants and customers work independently yet collaborate seamlessly. This analogy captures the essence of **Akka**, a concurrency framework empowering you to build scalable, resilient, and responsive real-time systems in Java.

Actors rule the roost in Akka's domain. Actors are sovereign entities, each tasked with their own responsibilities, communicating through immutable messages. This design sidesteps the quagmires of shared-memory concurrency, rendering the system more comprehensible and less prone to errors.

Here's what makes Akka stand out:

- **Actor-based design**: Each actor handles its own tasks independently, simplifying concurrent programming and reducing the risk of errors.

- **Location transparency**: Actors can reside anywhere within your cluster, allowing you to scale your application dynamically across nodes.

- **Built-in resilience**: Akka embraces the *let it crash* philosophy. If an actor fails, it's automatically restarted, ensuring your system remains highly available.

Akka shines in scenarios where you need to process data streams in real time. Imagine receiving data from various sources such as sensors or social media feeds. Using Akka actors, you can efficiently process each data point independently, achieving high throughput and low latency.

In order to run an Akka project with Maven, you'll need to set up your pom.xml file to include dependencies for Akka actors and any other Akka modules you plan to use.

Include the akka-actor-typed library in your pom.xml file under <dependencies> to use Akka Typed actors:

```
<properties>
    <akka.version>2.6.19</akka.version>
</properties>
<dependency>
    <groupId>com.typesafe.akka</groupId>
    <artifactId>akka-actor-typed_2.13</artifactId>
    <version>${akka.version}</version>
</dependency>
```

Akka uses SLF4J for logging. You must add an SLF4J implementation, such as Logback, as a dependency:

```
<dependency>
    <groupId>com.typesafe.akka</groupId>
    <artifactId>akka-slf4j_2.13</artifactId>
    <version>${akka.version}</version>
</dependency>
<dependency>
    <groupId>ch.qos.logback</groupId>
    <artifactId>logback-classic</artifactId>
    <version>1.2.11</version>
</dependency>
```

Here is the simplified code to demonstrate how Akka is used for a data processing project:

```
import akka.actor.typed.Behavior;
import akka.actor.typed.javadsl.*;

public class DataProcessor extends AbstractBehavior<DataProcessor.
DataCommand> {
    interface DataCommand {}
    static final class ProcessData implements DataCommand {
        final String content;
        ProcessData(String content) {
```

```java
        this.content = content;
    }
}

static final class DataResult implements DataCommand {
    final String result;
    DataResult(String result) {
        this.result = result;
    }
}

static Behavior<DataCommand> create() {
    return Behaviors.setup(DataProcessor::new);
}

private DataProcessor(
    ActorContext<DataCommand> context) {
        super(context);
    }

@Override
public Receive<DataCommand> createReceive() {
    return newReceiveBuilder()
            .onMessage(ProcessData.class,
                this::onProcessData)
            .onMessage(DataResult.class,
                this::onDataResult)
            .build();
}

private Behavior<DataCommand> onProcessData(
    ProcessData data) {
        try {
            getContext().getLog().info(
                "Processing data: {}", data.content);
        // Data processing logic here
            DataResult result = new DataResult(
                "Processed: " + data.content);
            return this;
        } catch (Exception e) {
            getContext().getLog().error(
                "Error processing data: {}",
                data.content, e);
```

```
                    return Behaviors.stopped();
            }
        }


    private Behavior<DataCommand> onDataResult(
        DataResult result) {
        // Handle DataResult if needed
            return this;
        }
    }
```

This code snippet demonstrates how Akka actors can be used for simple data processing. Here's a breakdown of how it works:

- **Actor definition:**

  - The `DataProcessor` class extends `AbstractBehavior<DataProcessor.DataCommand>`, which is a base class provided by Akka for defining actors

  - The `DataCommand` interface serves as the base type for the messages that the `DataProcessor` actor can receive

- **Message handling:**

  - The `createReceive()` method defines the behavior of the actor when it receives messages

  - It uses the `newReceiveBuilder()` to create a `Receive` object that specifies how the actor should handle different message types

- **Processing data:**

  - When the actor receives a `ProcessData` message, the `onProcessData()` method is invoked

  - This method contains the logic for processing the data received in the message

- **Error handling:**

  - The `onProcessData()` method includes error handling using a try-catch block

  - If an exception occurs during data processing, the actor's behavior is changed to `Behaviors.stopped()`, which stops the actor

Akka's actor model provides a way to structure the application around individual units of computation (actors) that can process messages concurrently and independently. In the context of processing real-time data streams, Akka actors offer benefits such as concurrency, isolation, asynchronous communication, and scalability.

This is a simplified example. Real-world scenarios involve more complex data structures, processing logic, and potential interactions with other actors.

In the next section, we'll explore Vert.x, another powerful framework for building reactive applications in Java. We'll also delve into advanced testing and debugging techniques crucial for mastering concurrency in cloud environments.

Vert.x – embracing the reactive paradigm for web applications

Imagine a vibrant city humming with activity, its residents and systems constantly interacting. **Vert.x** embodies this dynamic spirit, enabling you to build reactive, responsive, and scalable web applications in Java, JavaScript, Kotlin, and more.

The key highlights of Vert.x are as follows:

- **Event-driven magic**: Unlike traditional approaches, Vert.x revolves around a non-blocking event loop, handling multiple requests simultaneously, making it ideal for I/O-intensive tasks.

- **Polyglot prowess**: Ditch language limitations! Vert.x embraces diverse tongues, from Java and JavaScript to Python and Ruby, empowering you to choose the tool that best suits your project and team.

- **Reactive revolution**: Vert.x champions the reactive programming paradigm, fostering applications that are resilient, elastic, and responsive to user interactions and system changes.

- **Microservices made easy**: Vert.x shines in the microservices ecosystem. Its lightweight, modular architecture and event-driven nature make it a perfect fit for building independent, yet interconnected, microservices that seamlessly collaborate.

Let's dive into a simplified example: creating an HTTP server. This server will greet every request with a cheerful *Hello, World!*, showcasing Vert.x's straightforward approach to web development:

1. **Setting up your project**: for Maven users, this means adding the Vert.x core dependency to your pom.xml file:

```
<dependency>
    <groupId>io.vertx</groupId>
    <artifactId>vertx-core</artifactId>
    <version>4.1.5</version>
</dependency>
<dependency>
    <groupId>io.vertx</groupId>
    <artifactId>vertx-web</artifactId>
    <version>4.1.5</version>
</dependency>
```

2. **Create a Java class**: The class should extend `AbstractVerticle`, the fundamental unit of Vert.x execution:

```java
import io.vertx.core.AbstractVerticle;
import io.vertx.core.Vertx;
import io.vertx.core.http.HttpServer;

public class VertxHttpServerExample extends AbstractVerticle {

    @Override
    public void start() {
        HttpServer server = vertx.createHttpServer();
        server.requestHandler(request -> {
            String path = request.path();
            if ("/hello".equals(path)) {
                request.response().putHeader(
                    "content-type", "text/plain").end(
                        "Hello, Vert.x!");
            } else {
                request.response().setStatusCode(
                    404).end("Not Found");
            }
        });
        server.listen(8080, result -> {
            if (result.succeeded()) {
                System.out.println(
                    "Server started on port 8080");
            } else {
                System.err.println("Failed to start server: " +
                result.cause());
            }
        });
    }

    public static void main(String[] args) {
        Vertx vertx = Vertx.vertx();
        vertx.deployVerticle(
            new VertxHttpServerExample());
    }
}
```

In this example, we create a `VertxHttpServerExample` class that extends `AbstractVerticle`, which is the base class for Vert.x verticles:

- In the `start()` method, we create an instance of `HttpServer` using `vertx.createHttpServer()`.

- We set up a request handler using `server.requestHandler()` to handle incoming HTTP requests. In this example, we check the request path and respond with `"Hello, Vert.x!"` for the `"/hello"` path and a `"Not Found"` response for any other path.

- We start the server using `server.listen()`, specifying the port number (`8080` in this case) and a handler to handle the result of the server startup.

- In the `main()` method, we create an instance of `Vertx` and deploy our `VertxHttpServerExample` verticle using `vertx.deployVerticle()`.

To run this example, compile the Java file and run the main class. Once the server is started, you can access it in your web browser or using a tool such as cURL: `curl http://localhost:8080/hello`, which will output: *Hello, Vert.x!*.

This simple example highlights Vert.x's ability to quickly build web applications. Its event-driven approach and polyglot nature make it a versatile tool for modern web development, empowering you to create flexible, scalable, and responsive solutions.

Both Akka and Vert.x offer unique strengths for building concurrent and distributed applications. While Akka excels in real-time processing with actors, Vert.x shines in web development with its event-driven and polyglot nature. Explore these frameworks and discover which aligns best with your specific needs and preferences.

In the following sections, we'll delve deeper into advanced testing and debugging techniques for ensuring the robustness of your cloud-based Java applications.

Mastering concurrency in cloud-based Java applications – testing and debugging tips

Building robust, scalable Java applications in cloud environments demands expertise in handling concurrency challenges. Here are key strategies and tools to elevate your testing and debugging game.

The key testing strategies are as follows:

- **Unit testing with concurrency**: Use frameworks such as JUnit to test individual units with concurrent scenarios. Mocking frameworks help simulate interactions for thorough testing.

- **Integration testing for microservices**: Tools such as Testcontainers and WireMock help test how interconnected components handle concurrent loads in distributed architectures.

- **Stress and load testing**: Tools such as Gatling and JMeter push your applications to their limits, revealing bottlenecks and scalability issues under high concurrency.

- **Chaos engineering for resilience**: Introduce controlled chaos with tools such as Netflix's Chaos Monkey to test how your application handles failures and extreme conditions, fostering resilience.

Here are the best practices for robust concurrency:

- **Embrace immutability**: Design with immutable objects whenever possible to avoid complexity and ensure thread safety

- **Use explicit locking**: Go for explicit locks over synchronized blocks for finer control over shared resources and to prevent deadlocks

- **Leverage modern Java concurrency tools**: Utilize the rich set of utilities in the `java.util.concurrent` package for effective thread, task, and synchronization management

- **Stay up to date**: Continuously learn about the latest advancements in Java concurrency and cloud computing to adapt and improve your practices

By combining these strategies, you can build cloud-based Java applications that are not only powerful but also resilient, scalable, and ready to handle the demands of modern computing.

Summary

This chapter provided a deep dive into the advanced facets of Java concurrency, focusing on the Executor framework and Java's concurrent collections. This chapter is instrumental for developers aiming to optimize thread execution and maintain data integrity within concurrent applications, especially in cloud-based environments. The journey began with the Executor framework, which highlighted its role in efficient thread management and task delegation, akin to a head chef orchestrating a kitchen's operations. Concurrent collections were explored after that, which offered insights into managing data access amidst concurrent operations effectively.

Key synchronization tools such as `CountDownLatch`, `Semaphore`, and `CyclicBarrier` were detailed, and their importance in ensuring coordinated execution across different parts of an application was demonstrated. The chapter further delved into Java's locking mechanisms, which provided strategies to safeguard shared resources and prevent concurrency-related issues. The narrative extended to cover service mesh and APM for optimizing application performance, alongside frameworks such as Akka and Vert.x for building reactive and resilient systems. It concluded with a focus on testing and debugging, which equipped developers with essential tools and methodologies for identifying and resolving concurrency challenges and ensuring high-performing, scalable, and robust Java applications in cloud environments. Through practical examples and expert advice, this chapter armed readers with the knowledge to master advanced concurrency concepts and apply them successfully in their cloud computing endeavors.

This groundwork sets the stage for delving into **Java concurrency patterns** in the next chapter, promising deeper insights into asynchronous programming and thread pool management for crafting efficient, robust cloud solutions.

Questions

1. What is the primary purpose of the Executor framework in Java?

    A. To schedule future tasks for execution

    B. To manage a fixed number of threads within an application

    C. To efficiently manage thread execution and resource allocation

    D. To lock resources for synchronized access

2. Which Java utility is best suited for handling scenarios with high read operations and fewer write operations to ensure data integrity during writes?

    A. `ConcurrentHashMap`

    B. `CopyOnWriteArrayList`

    C. `ReadWriteLock`

    D. `StampedLock`

3. What advantage does `CompletableFuture` provide in Java concurrency?

    A. Reduces the need for callbacks by blocking the thread until completion

    B. Enables asynchronous programming and non-blocking operations

    C. Simplifies the management of multiple threads

    D. Allows for manual locking and unlocking of resources

4. In the context of cloud computing, why are Java's concurrent collections important?

    A. They provide a mechanism for manual synchronization of threads

    B. They enable efficient data handling and reduce locking overhead in concurrent access scenarios

    C. They are necessary for creating new threads and processes

    D. They replace traditional collections for all use cases

5. How do advanced locking mechanisms such as `ReentrantLock` and `StampedLock` improve application performance in the cloud?

    A. By allowing unlimited concurrent read operations

    B. By completely removing the need for synchronization

    C. By offering more control over lock management and reducing lock contention

    D. By automatically managing thread pools without developer input

Mastering Concurrency Patterns in Cloud Computing

Mastering concurrency is crucial for unlocking the full potential of cloud computing. This chapter equips you with the knowledge and skills required to leverage concurrency patterns, the cornerstones of building high-performance, resilient, and scalable cloud applications.

These patterns are more than just theory. They empower you to harness the distributed nature of cloud resources, ensuring smooth operation under high loads and a seamless user experience. Leader-Follower, Circuit Breaker, and Bulkhead are indeed fundamental design patterns that serve as essential building blocks for robust cloud systems. They provide a strong foundation for understanding how to achieve high availability, fault tolerance, and scalability. We'll explore these core patterns, which are designed to address challenges such as network latency and failures. While there are many other patterns beyond these three, these chosen patterns serve as a solid starting point for mastering concurrency in cloud computing. They provide a basis for understanding the principles and techniques that can be applied to a wide range of cloud architectures and scenarios.

We'll then delve into patterns for asynchronous operations and distributed communication, including Producer-Consumer, Scatter-Gather, and Disruptor. The true power lies in combining these patterns strategically. We'll explore techniques for integrating and blending patterns to achieve synergistic effects, boosting both performance and resilience.

By the end of this chapter, you'll be equipped to design and implement cloud applications that excel at handling concurrent requests, are resilient to failures, and effortlessly scale to meet growing demands. We'll conclude with practical implementation strategies to solidify your learning and encourage further exploration.

Technical requirements

Package and run a Java class as an AWS Lambda function.

First, prepare your Java class:

1. Ensure your class implements the `RequestHandler<Input, Output>` interface from the `com.amazonaws:aws-lambda-java-core` library. This defines the handler method that processes events.

2. Include any necessary dependencies in your `pom.xml` file (if you're using Maven):

    ```
    <dependency>
        <groupId>com.amazonaws</groupId>
        <artifactId>aws-lambda-java-core</artifactId>
        <version>1.2.x</version>
    </dependency>
    ```

Be sure to replace `1.2.x` with the latest compatible version of the `aws-lambda-java-core` library.

Then, package your code:

Create a JAR file containing your compiled Java class and all its dependencies. You can use a tool such as Maven or a simple command such as `jar cvf myLambdaFunction.jar target/classes/*.class` (assuming compiled classes are in target/classes).

Create a Lambda function in AWS:

1. Go to the AWS Lambda console and click **Create function**.

2. Choose **Author from scratch** and select **Java 11** or a compatible runtime for your code.

3. Provide a name for your function and choose **Upload** for the code source.

4. Upload your JAR file in the **Code entry type** section.

5. Configure your function's memory allocation, timeout, and other settings as needed.

6. Click **Create function**.

Test your function:

1. In the Lambda console, navigate to your newly created function.

2. Click on **Test** and provide a sample event payload (if applicable).

3. Click on **Invoke** to run your function with the provided test event.

4. The Lambda console will display the output or error message returned by your function's handler method.

For a more comprehensive guide with screenshots and additional details, you can refer to the official AWS documentation on deploying Java Lambda functions: `https://docs.aws.amazon.com/lambda/latest/dg/java-package.html`

This documentation provides step-by-step instructions on packaging your code, creating a deployment package, and configuring your Lambda function in the AWS console. It also covers additional topics such as environment variables, logging, and handling errors.

The code in this chapter can be found on GitHub:

`https://github.com/PacktPublishing/Java-Concurrency-and-Parallelism`

Core patterns for robust cloud foundations

In this section, we delve into the foundational design patterns that are essential for building resilient, scalable, and efficient cloud-based applications. These patterns provide the architectural groundwork necessary to address common challenges in cloud computing, including system failures, resource contention, and service dependencies. Specifically, we will explore the Leader-Follower pattern, the Circuit Breaker pattern, and the Bulkhead pattern, each offering unique strategies to enhance fault tolerance, system reliability, and service isolation in the dynamic environment of cloud computing.

The Leader-Follower pattern

The **Leader-Follower** pattern is a concurrency design pattern that's particularly suited to distributed systems where tasks are dynamically allocated to multiple worker units. This pattern helps manage resources and tasks efficiently by organizing the worker units into a leader and multiple followers. The leader is responsible for monitoring and delegating work, while the followers wait to become leaders or to execute tasks assigned to them. This role-switching mechanism ensures that at any given time, one unit is designated to handle task distribution and management, optimizing resource utilization, and improving system scalability.

In distributed systems, efficient task management is key. The Leader-Follower pattern addresses this in the following ways:

- **Maximizing resource usage**: The pattern minimizes idle time by always assigning tasks to available workers.

- **Streamlining distribution**: A single leader handles task allocation, simplifying the process and reducing overhead.

- **Enabling easy scaling**: You can seamlessly add more follower threads to handle increased workloads without significantly altering the system's logic.

- **Promoting fault tolerance**: If the leader fails, a follower can take its place, ensuring system continuity.

- **Enhancing uptime and availability**: The Leader-Follower pattern improves system uptime and availability by efficiently distributing and processing tasks. Dynamic task allocation to available followers minimizes the impact of individual worker failures. If a follower becomes unresponsive, the leader can quickly reassign the task, reducing downtime. Moreover, promoting a follower to a leader role in case of leader failure enhances the system's resilience and availability. This fault-tolerant characteristic contributes to higher levels of uptime and availability in distributed systems.

To illustrate the Leader-Follower pattern in Java, we focus on its use for task delegation and coordination through a simplified code example. This pattern involves a central Leader that assigns tasks to a pool of Followers, effectively managing task execution.

The following is a simplified code snippet (key elements; for the full code, please refer to the GitHub repository accompanying this title):

```
public interface Task {
    void execute();
}

public class TaskQueue {
    private final BlockingQueue<Task> tasks;
    // ... addTask(), getTask()
}

public class LeaderThread implements Runnable {
    // ...
    @Override
    public void run() {
        while (true) {
            // ... Get a task from TaskQueue
            // ... Find an available Follower and assign the task
        }
    }
}


public class FollowerThread implements Runnable {
    // ...
    public boolean isAvailable() { ... }
}
```

Here is the code explanation:

- `Task` interface: This defines the contract for the work units. Any class implementing this interface must have an `execute()` method that performs the actual work.

- `TaskQueue`: This class manages a queue of tasks using `BlockingQueue` for thread safety. `addTask()` allows the addition of tasks to the queue, and `getTask()` retrieves tasks for processing.

- `LeaderThread`: This thread continuously retrieves tasks from the queue using `getTask()`. It then iterates through the list of followers and assigns the task to the first available Follower.

- `FollowerThread`: This thread processes tasks and signals its availability to the leader. The `isAvailable()` method allows the leader to check if a follower is ready for new work.

This overview encapsulates the Leader-Follower pattern's core logic. For a detailed exploration and the complete code, visit the GitHub repository accompanying this book. There, you'll find extended functionalities and customization options, enabling you to tailor the implementation to your specific needs, such as electing a new leader or prioritizing urgent tasks.

Remember, this example serves as a foundation. You're encouraged to expand upon it, integrating features such as dynamic leader election, task prioritization, and progress monitoring to build a robust task management system suited to your application's requirements.

Next, in *The Leader-Follower pattern in action*, we'll see how this pattern empowers different real-world applications.

The Leader-Follower pattern in action

The Leader-Follower pattern offers flexibility and adaptability for various distributed systems scenarios, particularly in cloud computing environments. Here are a few key use cases where it excels:

- **Scaling a cloud-based image processing service**: Imagine a service receiving numerous image manipulation requests. The leader thread monitors incoming requests, delegating them to available follower threads (worker servers). This distributes the workload, reduces bottlenecks, and improves response times.

- **Real-time data stream processing**: In applications handling continuous streams of data (e.g., sensor readings and financial transactions), a leader thread can receive incoming data and distribute it among follower threads for analysis and processing. This parallelization enables real-time insights by maximizing resource utilization.

- **Distributed job scheduling**: For systems with various computational tasks (e.g., scientific simulations and machine learning models), the Leader-Follower pattern promotes efficient distribution of these jobs across a cluster of machines. The leader coordinates task assignments based on resource availability, accelerating complex executions.

- **Work queue management**: In applications with unpredictable bursts of activity (e.g., e-commerce order processing), a leader thread can manage a central work queue and delegate tasks to follower threads as they become available. This design promotes responsiveness and optimizes resource usage during peak activity.

The Leader-Follower pattern's core advantage lies in its ability to distribute workloads across multiple threads or processes. This distribution increases efficiency and scalability and is highly beneficial in cloud-based environments where resources can be scaled dynamically.

Picture our distributed system as a complex machine. The Leader-Follower pattern helps it run smoothly. But, like with any machine, parts can malfunction. The Circuit Breaker acts like a safety switch, preventing a single faulty component from bringing down the entire system. Let's see how this protective mechanism operates.

The Circuit Breaker pattern – building resilience in cloud applications

Think of the Circuit Breaker pattern like its electrical counterpart—it prevents cascading failures in your distributed system. In cloud applications, where services rely on remote components, the **Circuit Breaker** pattern safeguards against the ripple effects of failing dependencies.

How it works? The Circuit Breaker monitors failures when calling a remote service. Once a failure threshold is crossed, the circuit *trips*. Tripping means calls to the remote service are blocked for a set amount of time. This timeout allows the remote service a chance to recover. During the timeout, your application can gracefully handle the error or use a fallback strategy. After the timeout, the circuit transitions to *half-open*, testing the service's health with a limited number of requests. If those succeed, normal operation resumes; if they fail, the circuit reopens, and the timeout cycle begins again.

Let's look at the following diagram:

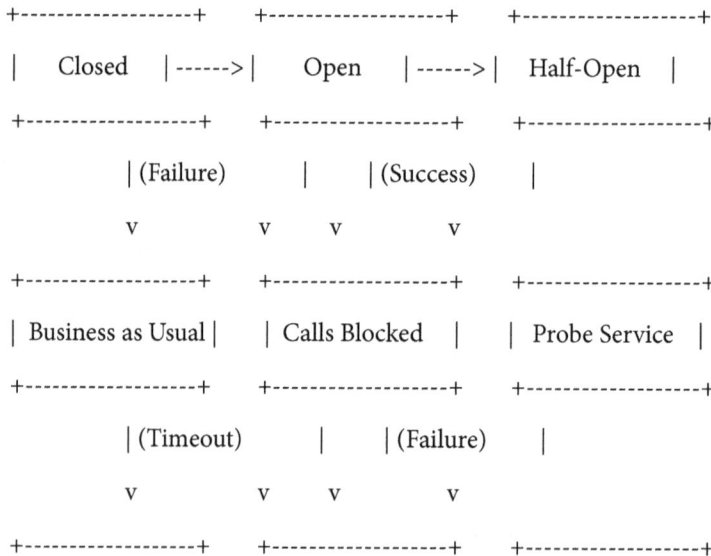

```
+-------------------+      +-------------------+      +-------------------+
|     Closed    | ------>|     Open      | ------>|   Half-Open   |
+-------------------+      +-------------------+      +-------------------+
        | (Failure)              |      | (Success)      |
        v                v     v           v
+-------------------+      +-------------------+      +-------------------+
| Business as Usual |      | Calls Blocked  |      | Probe Service  |
+-------------------+      +-------------------+      +-------------------+
        | (Timeout)              |      | (Failure)      |
        v                v     v           v
+-------------------+      +-------------------+      +-------------------+
```

Figure 5.1: States of the Circuit Breaker

The Circuit Breaker has three states:

- **Closed**: This is the initial state. Calls to the service are allowed to flow through (business as usual).
- **Open**: This state is reached if the error threshold is hit (consecutive failures). Calls to the service are blocked, preventing further failures and giving the service time to recover.
- **Half-Open**: A single call is allowed through to probe the health of the service. If the call is successful, the circuit transitions back to *Closed*. However, if the call fails, the circuit transitions back to *Open*.

There are the following transition events:

- **Closed -> Open**: This transition occurs when the error threshold is reached
- **Open -> Closed**: This transition occurs after a timeout period in the *Open* state (assuming the service has had enough time to recover)
- **Open -> Half-Open**: This transition can be triggered manually or automatically after a configurable time in the *Open* state
- **Half-Open -> Closed**: This transition occurs if the probe call in the *Half-Open* state is successful
- **Half-Open -> Open**: This transition occurs if the probe call in the *Half-Open* state fails

Next, we'll demonstrate the Circuit Breaker pattern in Java, focusing on safeguarding an e-commerce application's order service from failures in its service dependencies. The pattern acts as a state machine with *Closed*, *Open*, and *Half-Open* states, along with implementing a fallback strategy for handling operations when failures occur.

First, we create the `CircuitBreakerDemo` class:

```
public class CircuitBreakerDemo {
    private enum State {
        CLOSED, OPEN, HALF_OPEN
    }

    private final int maxFailures;
    private final Duration openDuration;
    private final Duration retryDuration;
    private final Supplier<Boolean> service;
    private State state;
    private AtomicInteger failureCount;
    private Instant lastFailureTime;
    public CircuitBreakerDemo(int maxFailures, Duration
        openDuration, Duration retryDuration,
        Supplier<Boolean> service) {
            this.maxFailures = maxFailures;
```

```
                this.openDuration = openDuration;
                this.retryDuration = retryDuration;
                this.service = service;
                this.state = State.CLOSED;
                this.failureCount = new AtomicInteger(0);
            }
        }
```

The CircuitBreakerDemo class defines an enum State to represent the three states: CLOSED, OPEN, and HALF_OPEN. The class has fields to store the maximum number of failures allowed (maxFailures), the duration for which the circuit breaker remains open (openDuration), the duration between consecutive probe calls in the HALF_OPEN state (retryDuration), and a Supplier representing the service being monitored. The constructor() initializes the state to CLOSED and sets the provided configuration values.

Next, we create the call() method and state transitions:

```
public boolean call() {
    switch (state) {
        case CLOSED:
            return callService();
        case OPEN:
            if (lastFailureTime.plus(
                openDuration).isBefore(Instant.now())) {
                    state = State.HALF_OPEN;
                }
            return false;
        case HALF_OPEN:
            boolean result = callService();
            if (result) {
                state = State.CLOSED;
                failureCount.set(0);
            } else {
                state = State.OPEN;
                lastFailureTime = Instant.now();
            }
            return result;
        default:
            throw new IllegalStateException(
                "Unexpected state: " + state);
    }
}
```

This code performs the following actions:

- The `call()` method is the entry point for making requests to the service.

- In the CLOSED state, it calls the `callService()` method and returns the result.

- In the OPEN state, it blocks requests and transitions to the HALF_OPEN state after the `openDuration` has elapsed.

- In the HALF_OPEN state, it sends a probe request by calling `callService()`. If the probe succeeds, it transitions to CLOSED; otherwise, it transitions back to OPEN.

Lastly, we have a service call and failure handling:

```
private boolean callService() {
    try {
        boolean result = service.get();
        if (!result) {
            handleFailure();
        } else {
            failureCount.set(0);
        }
        return result;
    } catch (Exception e) {
        handleFailure();
        return false;
    }
}

private void handleFailure() {
    int currentFailures = failureCount.incrementAndGet();
    if (currentFailures >= maxFailures) {
        state = State.OPEN;
        lastFailureTime = Instant.now();
    }
}
```

This code performs the following functions:

- The `callService()` method invokes the service's `get()` method and returns the result.

- If the service call fails (returns false or throws an exception), the `handleFailure()` method is called.

- The `handleFailure()` method increments the failure count (`failureCount`).

- If the `failure count` reaches the maximum allowed (`maxFailures`), the state is transitioned to OPEN, and the `lastFailureTime` is updated.

Remember, this is a simplified illustration of the Circuit Breaker pattern. For the full implementation, including detailed state management and customizable thresholds, please check out the accompanying GitHub repository. Also, consider using robust libraries such as Resilience4j for production-ready solutions, and remember to tailor failure thresholds, timeouts, and fallback behaviors to match your specific application's needs.

The key takeaway is to understand the pattern's underlying logic: how it transitions between states, handles failures gracefully with fallbacks, and ultimately shields your services from cascading breakdowns.

Unleashing resilience – Circuit Breaker use cases in the cloud

The Circuit Breaker pattern can be used in the following situations:

- **Online retail overload**: Circuit breakers protect dependent services (e.g., payment processing) during high-traffic events. They enable graceful degradation, provide time for service recovery, and help automate the restoration of service.
- **Real-time data processing**: Circuit breakers safeguard analytics systems if data sources become slow or unresponsive, preventing overload.
- **Distributed job scheduling**: In job scheduling systems, circuit breakers prevent jobs from overwhelming failing resources, promoting overall system health.

To maximize resilience, proactively integrate circuit breakers into your distributed cloud application's design. Strategically position them at service boundaries, implement robust fallback mechanisms (e.g., caching and queuing), and couple them with monitoring tools to track circuit states and fine-tune configurations. Remember to weigh the added complexity against the resilience gains for your specific application.

The Bulkhead pattern – enhancing cloud application fault tolerance

The **Bulkhead** pattern, drawing inspiration from the maritime industry, involves compartmentalizing sections of a ship's hull to prevent it from sinking if one part fills with water. Similarly, in software architecture, the Bulkhead pattern isolates elements of an application into separate sections (bulkheads) to prevent failures in one part from cascading throughout the entire system. This pattern is particularly useful in distributed systems and microservices architectures, where different components handle various functionalities.

The Bulkhead pattern safeguards your applications by dividing them into isolated compartments. This does the following:

- **Prevents cascading failures**: If one component fails, others remain unaffected
- **Optimizes resources**: Each compartment gets its own resources, preventing one area from hogging them all
- **Boosts resilience**: Critical parts of your application stay functional even during problems
- **Simplifies scaling**: Scale individual components independently as needed

Let's look at practical examples and dive into how to implement the Bulkhead pattern in Java microservices and your projects.

Imagine an e-commerce application with a recommendation engine. This engine might be resource-intensive. We want to protect other services (order processing and search) from being starved of resources if the recommendation feature experiences high traffic.

Here is a code snippet using Resilience4j:

```java
import io.github.resilience4j.bulkhead.Bulkhead;
import io.github.resilience4j.bulkhead.BulkheadConfig;
// ... Other imports
public class OrderService {
    Bulkhead bulkhead = Bulkhead.of(
        "recommendationServiceBulkhead",
        BulkheadConfig.custom().maxConcurrentCalls(
            10).build());

    // Existing order processing logic...
    public void processOrder(Order order) {
        // ... order processing ...
        Supplier<List<Product>> recommendationCall = Bulkhead
                .decorateSupplier(bulkhead, () ->
recommendationEngine.getRecommendations(order.getItems())));

        try {
            List<Product> recommendations = recommendationCall.get();
            // Display recommendations
        } catch (BulkheadFullException e) {
            // Handle scenario where recommendation service is
            unavailable (show defaults)
        }
    }
}}
```

Here is an explanation of the code:

- **Bulkhead creation**: We create a Bulkhead named `recommendationServiceBulkhead`, limiting the number of concurrent calls to 10.

- **Wrapping the call**: We decorate the call to the recommendation engine with the bulkhead.

- **Handling exceptions**: If the bulkhead is full, a `BulkheadFullException` is thrown. Implement a fallback (e.g., display default products) to handle this gracefully.

The Bulkhead pattern safeguards your application by isolating resources; in this example, we limit the recommendation service to only 10 concurrent calls. This strategy ensures that order processing remains unaffected even if the recommendation engine is overloaded. For enhanced visibility, integrate the bulkhead with a metrics system to track how often the limit is reached. Remember that Resilience4j offers a Bulkhead implementation, but you can also explore alternative libraries or design your own.

This code snippet demonstrates the Bulkhead pattern in action, showcasing how to isolate services within a single application. Now, let's explore some essential use cases of this pattern in cloud environments that can significantly enhance your system's resilience.

Essential Bulkhead pattern use cases in cloud environments

Let's focus on some highly practical use cases of the Bulkhead pattern in cloud environments that you would find immediately valuable:

- **Multi-tenant applications**: Isolate tenants within a shared cloud application. This ensures that one tenant's heavy usage won't starve resources for others, guaranteeing fairness and consistent performance. Consider a multi-tenant e-commerce application. Each tenant (store) has its own product catalog, customer data, and order processing tasks. Using the Bulkhead pattern, each store would have a dedicated database connection pool for its product and customer data, separate message queues would be used for processing orders for each store, and there could be thread pools dedicated to handling order processing tasks for specific stores. This ensures that a surge in activity from one store won't affect the performance of other stores in the application.

- **Mixed workload environments**: Separate critical services from less-critical ones (e.g., production batch jobs versus real-time user requests). Bulkheads ensure that lower-priority workloads don't cannibalize resources needed by critical services.

- **Unpredictable traffic**: Protect systems against sudden traffic spikes to specific components. Bulkheads isolate the impact, preventing a surge in one area from causing a total collapse.

- **Microservice architectures**: A core principle in microservices! Bulkheads limit cascading failures. If one microservice fails, bulkheads help to prevent that failure from rippling through the entire application.

When implementing the Bulkhead pattern, pay close attention to these key considerations: decide the granularity of isolation (service level, endpoint level, etc.) and meticulously configure bulkhead sizes (max calls and queues) based on thorough workload analysis. Always design robust fallback strategies (such as caching or default responses) for when bulkheads reach capacity. The Bulkhead pattern complements the cloud's advantages—use it to dynamically scale isolated compartments and add a vital layer of resilience in your distributed cloud applications, where network reliance can increase the chances of failure.

Java concurrency patterns for asynchronous operations and distributed communications

In this section, we'll explore three crucial patterns that transform applications: the Producer-Consumer pattern for efficient data exchange, the Scatter-Gather pattern for distributed systems, and the Disruptor pattern for high-performance messaging. We'll analyze each pattern and provide Java implementations, use cases, and their benefits in real-world cloud architectures emphasizing asynchronous operations and distributed communications.

The Producer-Consumer pattern – streamlining data flow

The **Producer-Consumer** pattern is a fundamental design pattern that addresses the mismatch between the rate of data generation and data processing. It decouples the producers, which generate tasks or data, from the consumers, which process those tasks or data, often asynchronously using a shared queue as a buffer. This pattern offers several benefits, particularly in cloud and distributed architectures, but it also introduces the need to handle the producer-consumer mismatch problem effectively.

The producer-consumer mismatch occurs when the rate of data production differs from the rate of data consumption. This mismatch can lead to two potential issues:

- **Overproduction**: If the producers generate data faster than the consumers can process it, the shared queue can become overwhelmed, leading to increased memory usage, potential out-of-memory errors, and overall system instability.

- **Underproduction**: If the producers generate data slower than the consumers can process it, the consumers may become idle, leading to underutilized resources and reduced system throughput.

To address the producer-consumer mismatch problem, several strategies can be employed:

- **Backpressure**: Implementing backpressure mechanisms allows consumers to signal to producers when they are overwhelmed, prompting producers to slow down or pause data generation temporarily. This helps prevent the shared queue from becoming overloaded and ensures a balanced flow of data.

- **Queue size management**: Configuring the shared queue with an appropriate size limit can prevent unbounded memory growth in the case of overproduction. When the queue reaches its maximum size, producers can be blocked or data can be dropped, depending on the specific requirements of the system.

- **Dynamic scaling**: In cloud and distributed environments, dynamically scaling the number of producers or consumers based on the observed load can help maintain a balanced data flow. Additional producers can be launched when data generation is high, and more consumers can be added when data processing lags behind.

- **Load shedding**: In extreme cases, when the system is overloaded and cannot keep up with the incoming data, load shedding techniques can be employed to selectively drop or discard lower-priority data or tasks, ensuring that the most critical data is processed first.

- **Monitoring and alerting**: Implementing monitoring and alerting mechanisms can provide visibility into the data flow rates and queue lengths, allowing timely intervention or automatic scaling when imbalances are detected.

By effectively managing the producer-consumer mismatch problem, the Producer-Consumer pattern can offer several advantages, such as decoupling, workload balancing, asynchronous flow, and improved performance through concurrency. It is the cornerstone of building robust and scalable applications where efficient data flow management is crucial, particularly in cloud and distributed architectures where components may not be immediately available, and workloads can vary dynamically.

The Producer-Consumer pattern in Java – a real-world example

Let's explore a practical example of how the Producer-Consumer pattern can be applied in a cloud-based image processing system, where the goal is to generate thumbnails for uploaded images asynchronously:

```
public class ThumbnailGenerator implements RequestHandler<SQSEvent,
Void> {
    private static final AmazonS3 s3Client = AmazonS3ClientBuilder.
    defaultClient();
    private static final String bucketName = "your-bucket-name";
    private static final String thumbnailBucket = "your-thumbnail-
    bucket-name";
    @Override
    public Void handleRequest(SQSEvent event, Context context) {
        String imageKey = extractImageKey(event);
// Assume this method extracts the image key from the //SQSEvent
        try (ByteArrayOutputStream outputStream = new
        ByteArrayOutputStream()) {
            // Download from S3
            S3Object s3Object = s3Client.getObject(
                bucketName, imageKey);
```

```
            InputStream objectData = s3Object.getObjectContent();
            // Load image
            BufferedImage image = ImageIO.read(objectData);
            // Resize (Maintain aspect ratio example)
            int targetWidth = 100;
            int targetHeight = (int) (
                image.getHeight() * targetWidth / (
                    double) image.getWidth());
            BufferedImage resized = getScaledImage(image,
                targetWidth, targetHeight);

            // Save as JPEG
            ImageIO.write(resized, "jpg", outputStream);
            byte[] thumbnailBytes = outputStream.toByteArray();
            // Upload thumbnail to S3
            s3Client.putObject(thumbnailBucket,
                imageKey + "-thumbnail.jpg",
                new ByteArrayInputStream(thumbnailBytes));
        } catch (IOException e) {
            // Handle image processing errors
            e.printStackTrace();
        }
        return null;
    }
    // Helper method for resizing
    private BufferedImage getScaledImage(BufferedImage src,
        int w, int h) {
            BufferedImage result = new BufferedImage(w, h,
                src.getType());
            Graphics2D g2d = result.createGraphics();
            g2d.drawImage(src, 0, 0, w, h, null);
            g2d.dispose();
            return result;
        }
    private String extractImageKey(SQSEvent event) {
        // Implementation to extract the image key from the SQSEvent
        return "image-key";
    }
}
```

This code demonstrates the Producer-Consumer pattern in the context of a cloud-based thumbnail generation system. Let's break down how the pattern works in this example:

- **Producer**:

  - The *producer* uploads images to an S3 bucket and sends messages to an *SQS queue*

  - Each message contains information about the uploaded image, such as the image key

- **Consumer**:

  - The `ThumbnailGenerator` class acts as the consumer and handles SQS events

  - When an `SQS event` is triggered, the `handleRequest()` method is invoked

- **Message consumption**:

  - The `handleRequest()` method receives an `SQSEvent` object representing the message from the SQS queue

  - The `extractImageKey()` method extracts the `image key` from the `SQS event`

- **Image processing**:

  - The `consumer` retrieves the image from the S3 bucket using the `image key`

  - The `image` is loaded, resized while maintaining its aspect ratio, and saved as a JPEG

  - The resized image bytes are stored in a `ByteArrayOutputStream`

- **Thumbnail upload**:

  - The generated thumbnail bytes are uploaded to a separate S3 bucket

  - The thumbnail is stored with a key that includes the original image key and a *thumbnail.jpg* suffix

- **Asynchronous processing**:

  - The `handleRequest()` method returns `null`, indicating no response is sent back to the producer

  - This allows the `consumer` to process messages asynchronously, without blocking the producer

This code demonstrates how the Producer-Consumer pattern enables asynchronous processing of image thumbnails in a cloud environment. The producer uploads images and sends messages, while the consumer processes the messages, generates thumbnails, and uploads them to a separate S3 bucket. This decoupling allows scalable and efficient image processing.

Next, we will delve into the practical use cases of the Producer-Consumer pattern within cloud architectures.

The Producer-Consumer pattern – a foundation for efficient, scalable cloud systems

Here is a list of high-value use cases of the Producer-Consumer pattern within cloud environments:

- **Task offloading and distribution**: Decouple a computationally intensive process (image processing, video transcoding, etc.) from the main application. This allows scaling worker components independently to handle varying loads without impacting the primary application's responsiveness.

- **Microservice communication**: In microservice architectures, the Producer-Consumer pattern facilitates asynchronous communication between services. Services can produce messages without needing immediate responses, enhancing modularity and resilience.

- **Event-driven processing**: Design highly reactive cloud systems. Sensors, log streams, and user actions can trigger events, leading producers to generate messages that trigger downstream processing in a scalable way.

- **Data pipelines**: Build multi-stage data processing workflows. Each stage can act as a consumer and a producer, enabling complex data transformations that operate asynchronously.

The Producer-Consumer pattern offers significant benefits in cloud environments. It enables flexible scaling by allowing independent scaling of producers and consumers, ideal for handling unpredictable traffic. The pattern enhances system resilience with its queueing mechanism, preventing failures from cascading in the event of temporary component unavailability. It also encourages clean modular design through loose coupling, as components communicate indirectly. Finally, it promotes efficient resource usage by ensuring consumers process tasks only when they have capacity, optimizing resource allocation in dynamic cloud environments.

The Scatter-Gather pattern: distributed processing powerhouse

The **Scatter-Gather** pattern optimizes parallel processing in distributed systems by dividing a large task into smaller subtasks (scatter phase). These subtasks are then processed concurrently across multiple nodes. Finally, the results are collected and combined (gather phase) to produce the final output.

The core concept involves the following:

- **Scatter**: A coordinator splits a task into independent subtasks

- **Parallel processing**: Subtasks are distributed for concurrent execution

- **Gather**: The coordinator collects partial results

- **Aggregation**: Results are combined into the final output

Its key benefits are as follows:

- **Improved performance**: Parallel processing significantly reduces execution time

- **Scalability**: Easily add more processing nodes to handle larger workloads

- **Flexibility**: Subtasks can run on nodes with specific capabilities

- **Fault tolerance**: Potential for reassigning subtasks if a node fails

This pattern is ideal for distributed systems and cloud environments where tasks can be parallelized for faster execution and dynamic resource allocation.

Next, we will explore how to apply Scatter-Gather in a specific use case!

Implementing Scatter-Gather in Java with ExecutorService

Here's a compact Java example that illustrates the Scatter-Gather pattern, tailored for an AWS environment. This example conceptually demonstrates how you might use AWS Lambda functions (as the scatter phase) to perform parallel processing of tasks and then gather the results. It uses AWS SDK for Java to interact with AWS services such as Lambda and S3 for simplicity in code demonstration. Please note that this example assumes you have a basic setup done in AWS, such as Lambda functions and S3 buckets in place.

```java
// ... imports
public class ScatterGatherAWS {
    // ... constants
    public static void main(String[] args) {
        // ... task setup

        // Scatter phase
        ExecutorService executor = Executors.newFixedThreadPool(tasks.
        size());
        List<Future<InvokeResult>> futures = executor.submit(tasks.
        stream()
                .map(task -> (Callable<InvokeResult>
                    ) () -> invokeLambda(task))
                .collect(Collectors.toList()));
        executor.shutdown();

        // Gather phase
        List<String> results = futures.stream()
            .map(f -> {
                try {
                    return f.get();
                } catch (Exception e) {
                    // Handle error
```

```
                       return null;
                       // Example - Replace with actual error handling
                   }
               })
               .filter(Objects::nonNull)
               .map(this::processLambdaResult)
               .collect(Collectors.toList());

        // ... store aggregated results

    }


    // Helper methods for brevity
    private static InvokeResult invokeLambda(String task) {
        // ... configure InvokeRequest with task data
        return lambdaClient.invoke(invokeRequest);
    }

    private static String processLambdaResult(InvokeResult result) {
        // ... extract and process the result payload
        return new String(result.getPayload().array(),
            StandardCharsets.UTF_8);
    }
}
```

This code demonstrates the Scatter-Gather pattern using AWS services for distributed task processing:

- **Scatter phase**:

  - A fixed-size thread pool (`ExecutorService`) is created to match the number of tasks

  - Each task is submitted to the pool. Within each task, we have the following:

    - An InvokeRequest is prepared for an AWS Lambda function, carrying the task data

    - The Lambda function is invoked (`lambdaClient.invoke(...)`)

- **Gather phase**:

  - A list of Future< InvokeResult> holds references to the pending Lambda execution results

  - The code iterates over the futures list and retrieves the `InvokeResult` for each task using future.get()

  - Lambda results are processed (assuming the payload is a string) and collected into a list

- **Aggregation (optional)**:

  - The collected results are joined into a single string

  - The aggregated result is stored in an S3 bucket

This code exemplifies the Scatter-Gather pattern by distributing tasks to AWS Lambda functions for parallel execution (scatter), awaiting their completion, and then aggregating the results (gather). The use of AWS Lambda highlights the pattern's compatibility with cloud-native technologies. For a production-ready implementation, it's crucial to incorporate robust error handling, timeout mechanisms, and proper resource management to ensure system resilience.

Next, we will delve into the practical use cases.

Practical applications of Scatter-Gather in cloud environments

Here's a breakdown of practical applications where the Scatter-Gather pattern excels within cloud environments:

- **High-performance computation**:

  - **Scientific simulations**: Break down complex simulations into smaller, independent sub-calculations that can be distributed across a cluster of machines or serverless functions for parallel execution.

  - **Financial modeling**: Apply Monte Carlo simulations or complex risk models in parallel to a large dataset, significantly reducing computation time.

  - **Machine learning (model training)**: Distribute the training of machine learning models across multiple GPUs or instances. Each worker trains on a subset of the data, and results are aggregated to update the global model.

- **Large-scale data processing**:

  - **Batch processing**: Divide large datasets into smaller chunks for parallel processing. This is useful for tasks such as **Extract, Transform, Load (ETL)** pipelines in data warehouses.

  - **MapReduce-style operations**: Implement custom MapReduce-like frameworks in the cloud. Split a large input, have workers process in parallel (map), and gather results to be combined (reduce).

  - **Web crawling**: Distribute web page crawling tasks across multiple nodes (avoiding overwhelming individual websites), then combine results into a searchable index.

- **Real-time or event-driven workflows**:

  - **Fan-out processing**: An event (e.g., an IoT device reading) triggers multiple parallel actions. These could include sending notifications, updating databases, or initiating calculations. Results are then potentially aggregated.

  - **Microservices request-response**: A client request sent to an API Gateway might require calling multiple backend microservices in parallel, potentially with each service responsible for a different data source. Gather responses to provide a comprehensive response to the client.

The Scatter-Gather pattern is a powerful tool in your cloud development toolkit. Consider it when you need to accelerate computationally intensive tasks, process massive datasets, or architect responsive event-driven systems. Experiment with this pattern and witness the efficiency gains it brings to your cloud applications.

The Disruptor pattern – streamlined messaging for low-latency applications

The **Disruptor** pattern is a high-performance messaging and event processing framework designed to achieve exceptionally low latency. Its key elements are as follows:

- **Ring buffer**: A pre-allocated circular data structure where producers place events and consumers retrieve them. This prevents dynamic memory allocation and garbage collection overheads.

- **Lock-Free design**: The Disruptor pattern employs sequence numbers and atomic operations to eliminate the need for traditional locking, boosting concurrency and reducing latency.

- **Batching**: Events are processed in batches for increased efficiency, minimizing context switching and cache misses.

- **Multi-producer/consumer**: The pattern supports multiple producers and consumers working concurrently, crucial for scalable distributed systems.

Let's look at *Figure 5.2*:

A[Producer] --> B {Claim slot}

B --> C {Check availability}

C --> D {Wait (Optional)}

C --> E {Reserve slot (sequence number)}

E --> F {Publish event}

F --> G {Update sequence number}

G --> H {Notify consumers}

H --> I [Consumer]

I --> J {Check sequence}

J --> K {Process events (up to sequence)}

K --> L {Update consumer sequence}

L --> I

Figure 5.2: Disruptor pattern flowchart (left-right)

Here is an explanation of the Disruptor pattern flowchart:

1. The producer initiates the process by claiming a slot in the ring buffer (A --> B).
2. The Disruptor checks if a slot is available (B --> C).
3. If a slot is unavailable, the producer might wait (C --> D).
4. If a slot is available, the producer reserves a slot using a sequence number (C --> E).
5. The event data is published to the reserved slot (E --> F).
6. The sequence number is updated atomically (F --> G).
7. Consumers are notified about the updated sequence (G --> H).
8. A consumer wakes up and checks the latest sequence (H --> I, J).
9. The consumer processes events in a batch up to the available sequence (J --> K).
10. The consumer's sequence number is updated (K --> L).
11. The process loops back for the consumer to check for new events (L --> I)

The Disruptor pattern delivers remarkable performance benefits. It's known for its ability to process millions of events per second, achieving ultra-low latency with processing times in the microsecond range. This exceptional performance makes it ideal for use cases such as financial trading systems, real-time analytics platforms, and high-volume event processing scenarios such as IoT or log analysis. The Disruptor pattern outperforms traditional queue-based approaches when speed and low latency are critical requirements.

Now we will explore a practical implementation to see how the Disruptor pattern is used in specific cloud-based applications.

Disruptor in cloud environments – real-time stock market data processing

Let's explore how the Disruptor pattern is used in cloud-based applications. We'll use a simplified example to illustrate the key concepts, understanding that production-ready implementations will involve greater detail.

Imagine a system that needs to ingest a continuous stream of stock price updates and perform real-time calculations (e.g., moving averages and technical indicators). These calculations must be lightning-fast to enable rapid trading decisions. How does the Disruptor fit in? Here is a simple Java example.

First, to use the Disruptor library in your Java project with Maven, you need to add the following dependency to your pom.xml file:

```xml
<dependency>
    <groupId>com.lmax</groupId>
    <artifactId>disruptor</artifactId>
    <version>3.4.6</version>
</dependency>
```

Next, we create an event class, StockPriceEvent, and a MovingAverageCalculator class:

```java
import com.lmax.disruptor.*;
import com.lmax.disruptor.dsl.Disruptor;
import com.lmax.disruptor.dsl.ProducerType;

// Event Class
class StockPriceEvent {
    String symbol;
    long timestamp;
    double price;

    // Getters and setters (optional)
}

// Sample Calculation Consumer (Moving Average)
class MovingAverageCalculator implements EventHandler<StockPriceEvent>
{
    private double average; // Maintain moving average state

    @Override
    public void onEvent(StockPriceEvent event, long
        sequence, boolean endOfBatch) throws Exception {
            average = (average * (
                sequence + 1) + event.getPrice()) / (
                    sequence + 2);
        // Perform additional calculations or store the average
            System.out.println("Moving average for " + event.symbol +
            ": " + average);
    }
}
```

In the above code snippet, the `StockPriceEvent` class represents the event that will be processed by the `Disruptor`. It contains fields for the stock symbol, timestamp, and price.

The `MovingAverageCalculator` class implements the `EventHandler` interface and acts as a consumer for the `StockPriceEvent`. It calculates the moving average of the stock prices as events are processed.

Finally, we create the `DisruptorExample` class:

```java
public class DisruptorExample {
    public static void main(String[] args) {
        // Disruptor configuration
        int bufferSize = 1024; // Adjust based on expected event
        volume
        Executor executor = Executors.newCachedThreadPool();
        // Replace with your thread pool
        ProducerType producerType = ProducerType.MULTI;
        // Allow multiple producers
        WaitStrategy waitStrategy = new BlockingWaitStrategy();
        // Blocking wait for full buffers


        // Create Disruptor
        Disruptor<StockPriceEvent> disruptor = new
        Disruptor<>(StockPriceEvent::new, bufferSize,
        executor,producerType, waitStrategy);


        // Add consumer (MovingAverageCalculator)
        disruptor.handleEventsWith(new MovingAverageCalculator());


        // Start Disruptor
        disruptor.start();


        // Simulate producers publishing events (replace with your
        actual data source)
        for (int i = 0; i < 100; i++) {
            StockPriceEvent event = new StockPriceEvent();
            event.symbol = "AAPL";
            event.timestamp = System.currentTimeMillis();
            event.price = 100.0 + Math.random() * 10;
    // Simulate random price fluctuations


            disruptor.publishEvent((eventWriter) -> eventWriter.
            onData(event)); // Publish event using lambda
        }
```

```
        // Shutdown Disruptor (optional)
        disruptor.shutdown();
    }
}
```

This code demonstrates the Disruptor pattern for low-latency processing of stock price updates with a moving average calculation as a consumer. Let's break down the key steps:

- **Disruptor configuration**:

  - `bufferSize`: Defines the size of the pre-allocated ring buffer where events (stock price updates) are stored. This prevents memory allocation overhead during runtime.

  - `executor`: A thread pool responsible for executing event handlers (consumers) concurrently.

  - `producerType`: Set to `ProducerType.MULTI` to allow multiple sources (producers) to publish stock price updates concurrently.

  - `waitStrategy`: A `BlockingWaitStrategy` is used here. This strategy causes producers to wait if the ring buffer is full, ensuring no data loss.

- **Disruptor creation**:

  - `Disruptor<StockPriceEvent>`: An instance of the `Disruptor` class is created, specifying the event type (`StockPriceEvent`). This Disruptor object manages the entire event processing pipeline.

- **Adding consumer**:

  - `disruptor.handleEventsWith(new MovingAverageCalculator())`: This line adds the `MovingAverageCalculator` class as an event handler (consumer) to the Disruptor. The consumer will be invoked for each published stock price update event.

- **Disruptor startup**:

  - `disruptor.start()`: Starts the Disruptor, initializing the ring buffer and consumer threads.

- **Simulating producers**:

  - The `for` loop simulates 100 stock price updates for the symbol `"AAPL"` with random prices.

  - `disruptor.publishEvent(...)`: This line publishes each event to the `Disruptor` using a lambda function. The lambda calls `eventWriter.onData(event)` to populate the event data in the ring buffer.

- **Overall flow**:

  - `Producers` (simulated in this example) publish stock price update events to the Disruptor's ring buffer.

  - The `Disruptor` assigns sequence numbers to events and makes them available to consumers.

  - The `MovingAverageCalculator` consumer concurrently processes these events, updating the moving average based on each stock price.

  - The Disruptor's lock-free design ensures efficient event handling and prevents bottlenecks caused by traditional locking mechanisms.

Remember that this is a simple illustration. Production code would include error handling, multiple consumers for different calculations, and integration with cloud-specific services for data input.

Now, let's delve into some practical use cases where the Disruptor pattern can significantly enhance the performance of cloud applications.

High-performance cloud applications – essential Disruptor pattern use cases

The top use cases where the Disruptor pattern shines within cloud environments:

- **High-throughput, low-latency processing**:

  - **Financial trading**: Execute trades at lightning speed and make rapid decisions based on real-time market data. The Disruptor's low latency processing is paramount in this domain.

  - **Real-time analytics**: Process massive streams of data (website clicks, sensor readings, etc.) to gain insights and trigger actions in near real time.

  - **High-frequency event logging**: Ingest and process vast amounts of log data for security monitoring, analysis, or troubleshooting in large-scale systems.

- **Microservice architectures**:

  - **Inter-service communication**: Use the Disruptor as a high-performance message bus. Producers and consumers can be decoupled, enhancing modularity and scalability.

  - **Event-driven workflows**: Orchestrate complex workflows where different microservices react to events in a responsive and efficient manner.

- **Cloud-specific use cases**:

  - **IoT event processing**: Handle the deluge of data from IoT devices. The Disruptor can quickly process sensor readings or device state changes to trigger alerts or updates.

  - **Serverless event processing**: Integrate with serverless functions (e.g., AWS Lambda), where the Disruptor can coordinate event processing with ultra-low overhead.

While the Disruptor pattern offers exceptional performance benefits, it's essential to be mindful of its potential complexities. Careful tuning of parameters such as ring buffer size and consumer batch sizes is often necessary to achieve optimal results. In a cloud environment, consider integrating with cloud-native services to enhance the system's resilience through features such as replication or persistence of the ring buffer. Properly understanding and addressing potential bottlenecks is crucial to fully harness the Disruptor's power and ensure your cloud-based system remains highly efficient and robust.

The Disruptor pattern versus the Producer-Consumer pattern – a comparative analysis

Let's compare the Disruptor pattern and the Producer-Consumer pattern, highlighting their key differences:

- **Design purpose**:

  - **Producer-Consumer**: A general-purpose pattern for decoupling the production and consumption of data or events

  - **Disruptor**: A specialized high-performance variant optimized for low-latency and high-throughput scenarios

- **Data structure**:

  - **Producer-Consumer**: Uses a shared queue or buffer, which can be bounded or unbounded

  - **Disruptor**: Employs a pre-allocated ring buffer with a fixed size to minimize memory allocation and garbage collection overhead

- **Locking mechanism**:

  - **Producer-Consumer**: Often relies on traditional locking mechanisms, such as locks or semaphores, for synchronization

  - **Disruptor**: Utilizes a lock-free design using sequence numbers and atomic operations, reducing contention and enabling higher concurrency

- **Batching**:

  - **Producer-Consumer**: Typically processes events or data one at a time, with no inherent support for batching

  - **Disruptor**: Supports batching of events, allowing consumers to process events in batches for improved efficiency

- **Performance**:

    - **Producer-Consumer**: Performance depends on the implementation and chosen synchronization mechanisms, and may suffer from lock contention and increased latency

    - **Disruptor**: Optimized for high performance and low latency, thanks to its lock-free design, pre-allocated ring buffer, and batching capabilities

The choice between the two patterns depends on the system's requirements. The Disruptor pattern is suitable for low-latency and high-throughput scenarios, while the Producer-Consumer pattern is more general-purpose and simpler to implement.

As we move into the next section, keep in mind that combining these core patterns opens up possibilities for even more sophisticated and robust cloud solutions. Let's explore how they can work together to push the boundaries of performance and resilience!

Combining concurrency patterns for enhanced resilience and performance

By strategically blending these patterns, you can achieve new levels of cloud system efficiency and robustness. Harness the power of combined concurrency patterns to build cloud systems that are both exceptionally performant and resilient, unlocking the hidden potential of your cloud architecture.

Integrating the Circuit Breaker and Producer-Consumer patterns

Combining the Circuit Breaker and Producer-Consumer patterns significantly boosts resilience and data flow efficiency in asynchronous cloud applications. The Circuit Breaker safeguards against failures, while the Producer-Consumer pattern optimizes data processing. Here's how to integrate them effectively:

- **Decouple with Circuit Breakers**: Place a Circuit Breaker between producers and consumers to prevent consumer overload during failures or slowdowns. This allows the system to recover gracefully.

- **Adaptive load management**: Use the Circuit Breaker's state to dynamically adjust the producer's task generation rate. Reduce the rate when the Circuit Breaker trips to maintain throughput while ensuring reliability.

- **Prioritize data**: Use multiple queues with individual Circuit Breakers to protect each queue. This ensures that high-priority tasks are processed even during system stress.

- **Self-healing feedback loop**: Have the Circuit Breaker's state trigger resource allocation, error correction, or alternative task routing, enabling autonomous system recovery.

- **Implement graceful degradation**: Employ fallback mechanisms in consumers to maintain service (even in a reduced form) when Circuit Breakers trip.

To demonstrate how this integration enhances fault tolerance, let's examine a code demo for resilient order processing using the Circuit Breaker and Producer-Consumer patterns.

Resilient order processing – Circuit Breaker and Producer-Consumer demo

In an e-commerce platform, use a queue to buffer orders (the Producer-Consumer pattern). Wrap external service calls (e.g., payment processing) within circuit breakers for resilience. If a service fails, the Circuit Breaker pattern prevents cascading failures and can trigger fallback strategies.

Here is an example code snippet:

```
// Pseudo-code for a Consumer processing orders with a Circuit Breaker
for the Payment Service
public class OrderConsumer implements Runnable {
    private OrderQueue queue;
    private CircuitBreaker paymentCircuitBreaker;

    public OrderConsumer(OrderQueue queue, CircuitBreaker
    paymentCircuitBreaker) {
        this.queue = queue;
        this.paymentCircuitBreaker = paymentCircuitBreaker;
    }

    @Override
    public void run() {
        while (true) {
            Order order = queue.getNextOrder();
            if (paymentCircuitBreaker.isClosed()) {
                try {
                    processPayment(order);
                } catch (ServiceException e) {
                    paymentCircuitBreaker.trip();
                    handlePaymentFailure(order);
                }
            } else {
                // Handle the case when the circuit breaker is open
                retryOrderLater(order);
            }
        }
    }
}
```

This code demonstrates the integration of the Circuit Breaker and Producer-Consumer patterns to enhance the resilience of an order processing system. Let's look at the code in detail:

- **Producer-Consumer**: `OrderQueue` acts as a buffer between order generation and processing. `OrderConsumer` pulls orders from this queue for asynchronous processing.

- **Circuit Breaker**: `paymentCircuitBreaker` protects an external payment service. If the payment service is experiencing issues, the circuit breaker prevents cascading failures.

- **Failure handling**: When a `ServiceException` occurs during `processPayment`, the circuit breaker is tripped (`paymentCircuitBreaker.trip()`), temporarily halting further calls to the payment service.

- **Graceful degradation**: If the circuit breaker is open, the `retryOrderLater` method signals that the order should be processed at a later time, allowing the dependent service to recover.

Overall, this code snippet highlights how these patterns work together to improve system robustness and maintain functionality even during partial failures.

Integrating Bulkhead with Scatter-Gather for enhanced fault tolerance

Combine the Bulkhead pattern with Scatter-Gather to build more resilient and efficient microservice architectures in the cloud. Bulkhead's focus on isolation helps manage failures and optimize resource usage within the Scatter-Gather framework. Here's how:

- **Isolated scatter components**: Employ the Bulkhead pattern to isolate scatter components. This prevents failures or heavy loads in one component from affecting others.

- **Dedicated gather resources**: Allocate distinct resources to the gather component using Bulkhead principles. This ensures efficient result aggregation, even under heavy load on the scatter services.

- **Dynamic resource allocation**: Bulkhead enables dynamic adjustment of resources for each scatter service based on its needs, optimizing overall system usage.

- **Fault tolerance and redundancy**: Bulkhead isolation ensures that the entire system doesn't fail if one scatter service goes down. Create redundant scatter service instances with separate resource pools for high fault tolerance.

To illustrate the benefits of this integration, let's consider a real-world use case: a weather forecasting service.

Weather data processing with Bulkhead and Scatter-Gather

Imagine a weather forecasting service that gathers data from multiple weather stations spread across a vast geographical region. The system needs to process this data efficiently and reliably to generate accurate weather forecasts. Here's how we can use the combined power of Bulkhead and Scatter-Gather patterns:

```
// Interface for weather data processing (replace with actual logic)
interface WeatherDataProcessor {
    ProcessedWeatherData processWeatherData(
        List<WeatherStationReading> readings);
}


// Bulkhead class to encapsulate processing logic for a region
class Bulkhead {
    private final String region;
    private final List<WeatherDataProcessor> processors;

    public Bulkhead(String region,
        List<WeatherDataProcessor> processors) {
            this.region = region;
            this.processors = processors;
        }


    public ProcessedWeatherData processRegionalData(
        List<WeatherStationReading> readings) {
    // Process data from all stations in the region
        List<ProcessedWeatherData> partialResults = new ArrayList<>();
        for (WeatherDataProcessor processor : processors) {
            partialResults.add(
                processor.processWeatherData(readings));
        }
    // Aggregate partial results (replace with specific logic)
        return mergeRegionalData(partialResults);
    }
}


// Coordinator class to manage Scatter-Gather and bulkheads
class WeatherDataCoordinator {
    private final Map<String, Bulkhead> bulkheads;

    public WeatherDataCoordinator(
        Map<String, Bulkhead> bulkheads) {
            this.bulkheads = bulkheads;
        }
```

```
public ProcessedWeatherData processAllData(
    List<WeatherStationReading> readings) {
// Scatter data to appropriate bulkheads based on region
        Map<String, List<WeatherStationReading>> regionalData =
        groupDataByRegion(readings);
        Map<String, ProcessedWeatherData> regionalResults = new
        HashMap<>();
        for (String region : regionalData.keySet()) {
            regionalResults.put(region, bulkheads.get(
                region).processRegionalData(
                    regionalData.get(region)));
        }
// Gather and aggregate results from all regions (replace with
specific logic)
    return mergeGlobalData(regionalResults);
}
}
```

This code demonstrates the integration of the Bulkhead and Scatter-Gather patterns for weather data processing. Here is the explanation:

- **Scatter-Gather**: `WeatherDataCoordinator` orchestrates parallel processing. It scatters weather readings to regional bulkhead instances and gathers the results for final aggregation.

- **Bulkhead**: Each bulkhead represents a region, isolating the processing logic. It encapsulates multiple `WeatherDataProcessor` instances, potentially allowing further parallelization within a region.

- **Resilience**: Bulkheads prevent failures in one region from affecting others. If a region's processing experiences issues, other regions can continue working.

This is a simple example. Real-world implementations would involve error handling, communication mechanisms between coordinator and bulkheads, and specific logic for processing weather data and merging results.

This integration not only enhances the resilience of distributed systems by isolating failures but also optimizes resource utilization across parallel processing tasks, making it an ideal strategy for complex, cloud-based environments.

Blending concurrency patterns – a recipe for high-performance cloud applications

Blending different concurrency patterns in cloud applications can significantly enhance both performance and resilience. By carefully integrating patterns that complement each other's strengths, developers can create more robust, scalable, and efficient systems. In this section, we'll explore strategies for the synergistic integration of concurrency patterns, highlighting scenarios where such blends are particularly effective.

Blending the Circuit Breaker and Bulkhead patterns

In a microservices architecture, where each service may depend on several other services, combining the Circuit Breaker and Bulkhead patterns can prevent failures from cascading across services and overwhelming the system.

Integration strategy: Use the Circuit Breaker pattern to protect against failures in dependent services. In parallel, apply the Bulkhead pattern to limit the impact of any single service's failure on the overall system. This approach ensures that if a service does become overloaded or fails, it doesn't take down unrelated parts of the application.

Combining Scatter-Gather with the Actor model

Building on our previous discussion of the Actor model in *Chapter 4, Java Concurrency Utilities and Testing in the Cloud Era*, let's see how it complements the Scatter-Gather pattern for distributed data processing tasks requiring result aggregation.

Integration strategy: Use the Actor model to implement the scatter component, distributing tasks among a group of actor instances. Each actor processes a portion of the data independently. Then, employ a gather actor to aggregate the results. This setup benefits from the Actor model's inherent message-passing concurrency, ensuring that each task is handled efficiently and in isolation.

Merging Producer-Consumer with the Disruptor pattern

In high-throughput systems where processing speed is critical, such as real-time analytics or trading platforms, the Producer-Consumer pattern can be enhanced with the Disruptor pattern for lower latency and higher performance.

Integration strategy: Implement the Producer-Consumer infrastructure using the Disruptor pattern's ring buffer to pass data between producers and consumers. This blend takes advantage of the Disruptor pattern's high-performance, lock-free queues to minimize latency and maximize throughput, all while maintaining the clear separation of concerns and scalability of the Producer-Consumer pattern.

Synergizing event sourcing with CQRS

Both event sourcing and **Command Query Responsibility Segregation (CQRS)** are software architectural patterns. They address different aspects of system design:

- **Event sourcing**: Focuses fundamentally on how the state of an application is represented, persisted, and derived. It emphasizes an immutable history of events as the source of truth.

- **CQRS**: Focuses on separating the actions that change an application's state (commands) from those actions that retrieve information without changing the state (queries). This separation can improve scalability and performance.

While they are distinct, event sourcing and CQRS are often used together in a complementary way: event sourcing provides a natural source of events for CQRS, and CQRS allows the independent optimization of read and write models within an event-sourced system.

Integration strategy: Use event sourcing to capture changes to the application state as a sequence of events. Combine this with CQRS to separate the models for reading and writing data. This blend allows highly efficient, scalable read models optimized for query operations while maintaining an immutable log of state changes for system integrity and replayability.

To maximize the benefits of pattern integration, choose patterns with complementary objectives, such as those focused on fault tolerance and scalability. Combine patterns that promote isolation (such as Bulkhead) with those offering efficient resource management (such as Disruptor) to achieve both resilience and performance. Utilize patterns that decouple components (such as Event Sourcing and CQRS) to make a simpler system architecture that's easier to scale and maintain over time. This strategic blending of concurrency patterns helps you address the complexities of cloud applications, resulting in systems that are more resilient, scalable, and easier to manage.

Summary

Think of this chapter as your journey into the heart of cloud application design. We started by building a strong foundation—exploring patterns such as Leader-Follower, Circuit Breaker, and Bulkhead to create systems that can withstand the storms of cloud environments. Think of these patterns as your architectural armor!

Next, we ventured into the realm of asynchronous operations and distributed communication. Patterns such as the Producer-Consumer, Scatter/Gather, and Disruptor became your tools for streamlining data flow and boosting performance. Imagine them as powerful engines propelling your cloud applications forward.

Finally, we uncovered the secret to truly exceptional cloud systems: the strategic combination of patterns. You learned how to integrate Circuit Breaker and Bulkhead for enhanced resilience, enabling you to create applications that can adapt and recover gracefully. This is like giving your cloud systems superpowers!

With your newfound mastery of concurrency patterns, you're well equipped to tackle complex challenges. *Chapter 6, Java in the Realm of Big Data*, throws you a new curveball: processing massive datasets. Let's see how Java and these patterns come together to conquer this challenge.

Questions

1. What is the main purpose of the Circuit Breaker pattern in a distributed system?

    A. To enhance data encryption

    B. To prevent a high number of requests from overwhelming a service

    C. To prevent failures in one service from affecting other services

    D. To schedule tasks for execution at a later time

2. When implementing the Disruptor pattern, which of the following is crucial for achieving high performance and low latency?

    A. Using a large number of threads to increase concurrency

    B. Employing a lock-free ring buffer to minimize contention

    C. Prioritizing tasks based on their complexity

    D. Increasing the size of the message payload

3. In the context of microservices, what is the primary advantage of implementing the Bulkhead pattern?

    A. It allows a single point of operation for all services.

    B. It encrypts messages exchanged between services.

    C. It isolates services to prevent failures in one from cascading to others.

    D. It aggregates data from multiple sources into a single response.

4. Which concurrency pattern is particularly effective for operations that require results to be aggregated from multiple sources in a distributed system?

    A. Leader Election pattern

    B. Scatter-Gather pattern

    C. Bulkhead pattern

    D. Actor model

5. Integrating the Circuit Breaker and Producer-Consumer patterns in cloud applications primarily enhances the system's:

    A. Memory efficiency

    B. Computational complexity

    C. Security posture

    D. Resilience and data flow management

Part 2:
Java's Concurrency in
Specialized Domains

The second part explores Java's concurrency capabilities across specialized domains, demonstrating how these features tackle complex challenges in big data, machine learning, microservices, and serverless computing.

This part includes the following chapters:

- *Chapter 6, Java and Big Data – a Collaborative Odyssey*

- *Chapter 7, Concurrency in Java for Machine Learning*

- *Chapter 8, Microservices in the Cloud and Java's Concurrency*

- *Chapter 9, Serverless Computing and Java's Concurrent Capabilities*

6

Java and Big Data – a Collaborative Odyssey

Embark on a transformative journey as we harness the power of Java to navigate the vast landscape of big data. In this chapter, we'll explore how Java's proficiency in distributed computing, coupled with its robust ecosystem of tools and frameworks, empowers you to tackle the complexities of processing, storing, and extracting insights from massive datasets. As we delve into the world of big data, we'll showcase how Apache Hadoop and Apache Spark seamlessly integrate with Java to overcome the limitations of conventional methods.

Throughout this chapter, you'll gain hands-on experience in building scalable data processing pipelines, using Java alongside the Hadoop and Spark frameworks. We'll explore Hadoop's core components, such as **Hadoop Distributed File System (HDFS)** and MapReduce, and dive deep into Apache Spark, focusing on its primary abstractions, including **Resilient Distributed Datasets (RDDs)** and DataFrames.

We'll place a strong emphasis on the DataFrame API, which has become the de facto standard for data processing in Spark. You'll discover how DataFrames provide a more efficient, optimized, and user-friendly way to work with structured and semi-structured data. We'll cover essential concepts such as transformations, actions, and SQL-like querying using DataFrames, enabling you to perform complex data manipulations and aggregations with ease.

To ensure a comprehensive understanding of Spark's capabilities, we'll explore advanced topics such as the Catalyst optimizer, the execution **Directed Acyclic Graph (DAG)**, caching and persistence techniques, and strategies to handle data skew and minimize data shuffling. We'll also introduce you to the equivalent managed services offered by major cloud platforms, enabling you to harness the power of big data within the cloud environment.

As we progress, you'll have the opportunity to apply your newfound skills to real-world big data challenges, such as log analysis, recommendation systems, and fraud detection. We'll provide detailed code examples and explanations, emphasizing the use of DataFrames and demonstrating how to leverage Spark's powerful APIs to solve complex data processing tasks.

By the end of this chapter, you'll be equipped with the knowledge and tools to conquer the realm of big data using Java. You'll understand the core characteristics of big data, the limitations of traditional approaches, and how Java's concurrency features and big data frameworks enable you to overcome these challenges. Moreover, you'll have gained practical experience in building real-world applications that leverage the power of Java and big data technologies, with a focus on utilizing the DataFrame API for optimal performance and productivity.

Technical requirements

- **Set up the Hadoop/Spark environment**: Setting up a small Hadoop and Spark environment can be a crucial step for hands-on practice and deepening your understanding of big data processing. Here's a simplified guide to get you started in creating your own *sandbox* environment:

- **Prerequisites**:

  - **System requirements**: 64-bit OS, at least 8 GB of RAM, and a multi-core processor

  - **Java Installation**: Install the **Java Development Kit (JDK)** 8 or 11

- **Install Hadoop**:

  - **Download Hadoop**: Go to the https://hadoop.apache.org/releases.html page and download the binary suitable for your OS.

  - **Extract and configure**: Unpack the download and configure Hadoop by editing the core-site.xml, hdfs-site.xml, and mapred-site.xml in the etc/hadoop directory. Refer to the official documentation for detailed configuration steps (https://hadoop.apache.org/docs/stable/).

  - **Environment variables**: Add Hadoop's bin and sbin to your PATH, and set JAVA_HOME to your JDK path.

  - **Initialize and start HDFS**: Format the HDFS filesystem with the hdfs namenode format, and then start HDFS and **Yet Another Resource Negotiator (YARN)** with start-dfs.sh and start-yarn.sh.

- **Install Spark**:

  - **Download Spark**: Visit the page (https://spark.apache.org/downloads.html) and download a pre-built binary of Spark for Hadoop.

  - **Extract Spark**: Unpack the Spark download to a chosen directory

  - **Configure Spark**: Edit conf/spark-env.sh to set JAVA_HOME and HADOOP_CONF_DIR as required

  - **Run Spark**: Start a Spark shell with ./bin/spark-shell or submit a job using ./bin/spark-submit

- **Testing:**

  - **Hadoop test**: Run a Hadoop example (e.g., calculating Pi with `hadoop jar share/hadoop/mapreduce/hadoop-mapreduce-examples-*.jar pi 4 100`)

  - **Spark test**: Execute an example Spark job, such as `./bin/run-example SparkPi`

This streamlined guide provides the essentials to get a Hadoop and Spark environment running. Detailed configurations might vary, so refer to the official documentation for in-depth instructions.

Download Visual Studio Code (VS Code) from `https://code.visualstudio.com/download`. VS Code offers a lightweight and customizable alternative to the other options on this list. It's a great choice for developers who prefer a less resource-intensive IDE and want the flexibility to install extensions tailored to their specific needs. However, it may not have all the features out of the box compared to the more established Java IDEs.

The code in this chapter can be found on GitHub:

`https://github.com/PacktPublishing/Java-Concurrency-and-Parallelism`

The big data landscape – the evolution and need for concurrent processing

Within this torrent of data lies a wealth of potential – insights that can drive better decision-making, unlock innovation, and transform entire industries. To seize this opportunity, however, we need specialized tools and a new approach to data processing. Let's begin by understanding the defining characteristics of big data and why it demands a shift in our thinking.

Navigating the big data landscape

Big data isn't merely about the sheer amount of information. It's a phenomenon driven by the explosion in the volume, speed, and diversity of data being generated every second.

The core characteristics of big data are volume, velocity, and variety:

- **Volume**: The massive scale of datasets, often reaching petabytes (millions of GB) or even exabytes (billions of GB).

- **Velocity**: The unprecedented speed at which data is created and needs to be collected – think social media feeds, sensor streams, and financial transactions.

- **Variety**: Data no longer fits neatly into structured rows and columns. We now have images, videos, text, sensor data, and so on.

Imagine a self-driving car navigating the streets. Its cameras, lidar sensors, and onboard computers constantly collect a torrent of data to map the environment, recognize objects, and make real-time driving decisions. This relentless stream of information can easily amount to terabytes of data each day – that's more storage than many personal laptops even hold.

Now, think of a massive online retailer. Every time you search for a product, click on an item, or add something to your cart, your actions are tracked. Multiply this by millions of shoppers daily, and you can see how an e-commerce platform captures a colossal dataset of user behavior.

Finally, picture the constant flow of posts, tweets, photos, and videos flooding social networks every second. This vast and ever-changing collection of text, images, and videos embodies the diversity and speed inherent in big data.

The tools and techniques that served us well in the past simply can't keep pace with the explosive growth and complexity of big data. Here's how traditional data processing struggles:

- **Scalability roadblocks**: Relational databases, optimized for structured data, buckle under massive datasets. Adding more data often translates to sluggish performance and skyrocketing hardware and maintenance costs.

- **Data diversity dilemma**: Traditional systems expect data in neat rows and columns, while big data embraces unstructured and semi-structured formats such as text, images, and sensor data.

- **Batch processing bottlenecks**: Conventional methods rely on batch processing, analyzing data in large chunks, which is slow and inefficient for real-time insights that are crucial in the big data world.

- **Centralized architecture woes**: Centralized storage solutions become overloaded and bottleneck when handling vast amounts of data flowing from multiple sources.

The limitations of relational databases become even clearer when considering specific aspects of big data:

- **Volume**: Distributing relational databases is difficult, and single nodes can't handle the sheer volume of big data.

- **Velocity**: **Atomicity, consistency, isolation, and durability** (ACID) transactions in relational databases slow down writes, making them unsuitable for the high velocity of incoming big data. Batching writes offers a partial solution, but it locks tables for other operations.

- **Variety**: Storing unstructured data (images, binary, etc.) is cumbersome due to size limitations and other challenges. While some semi-structured support exists (XML/JSON), it depends on the database implementation and doesn't fit well with the relational model.

These limitations underscore the immense potential hidden within big data but also reveal the inadequacy of traditional methods. To unlock this potential, we need a new paradigm – one built on distributed systems and the power of Java. Frameworks such as Hadoop and Spark represent this paradigm shift, offering the tools and techniques to effectively navigate the big data deluge.

Concurrency to the rescue

At its heart, concurrency is about managing multiple tasks that seem to happen at the same time. In big data, this translates to breaking down large datasets into smaller, more manageable chunks for processing.

Imagine you have a monumental task – sorting through a vast, dusty library filled with thousands of unorganized books. Doing this alone would take months! Thankfully, you're not limited to working solo. Java provides you with a team of helpers – threads and processes – to tackle this challenge:

- **Divide and conquer**: Think of threads and processes as your librarian assistants. Threads are lightweight helpers, perfect for tackling smaller tasks within the library, such as sorting bookshelves or searching through specific sections. Processes are your heavy-lifters, capable of taking on major sections of the library independently.

- **The coordination challenge**: Since your assistants work simultaneously, imagine the potential chaos without careful planning. Books could end up in the wrong places or go missing entirely! This is where synchronization comes in. It's like having a master catalog to track where books belong, ensuring that everything stays consistent even amid the whirlwind of activity.

- **Maximizing resource utilization**: Your library of computing power isn't just about how many helpers you have; it's also about using them wisely. Efficient resource utilization means spreading a workload evenly. Picture making sure each bookshelf in our metaphorical library gets attention and that assistants don't sit idle while others are overloaded.

Let's bring this story to life! Say you need to analyze that massive log dataset. A concurrent approach is like splitting the library into sections and assigning teams of assistants:

- **Filtering**: Assistants sift through log files for relevant entries, much like sorting through bookshelves to find those on specific topics

- **Transforming**: Other assistants clean and format the data for consistency – it's like standardizing book titles for a catalog

- **Aggregating**: Finally, some assistants compile statistics and insights from the data, just as you might summarize books on a particular subject

By dividing the work and coordinating these efforts, this huge task becomes not only manageable but amazingly fast!

Now that we've harnessed the power of concurrency and parallelism, let's explore how frameworks such as Hadoop leverage these principles to build a robust foundation for distributed big data processing.

Hadoop – the foundation for distributed data processing

As a Java developer, you're in the perfect position to harness this power. Hadoop is built with Java, offering a rich set of tools and APIs to craft scalable big data solutions. Let's dive into the core components of Hadoop's HDFS and MapReduce. Here's a detailed explanation of each component.

Hadoop distributed file system

Hadoop distributed file system or **HDFS** is the primary storage system used by Hadoop applications. It is designed to store massive amounts of data across multiple commodity hardware nodes, providing scalability and fault tolerance. The key characteristics of HDFS include the following:

- **Scaling out, not up**: HDFS splits large files into smaller blocks (typically, 128 MB) and distributes them across multiple nodes in a cluster. This allows for parallel processing and enables a system to handle files that are larger than the capacity of a single node.

- **Resilience through replication**: HDFS ensures data durability and fault tolerance by replicating each block across multiple nodes (the default replication factor is 3). If a node fails, the data can still be accessed from the replicated copies on other nodes.

- **Scalability**: HDFS is designed to scale horizontally by adding more nodes to the cluster. As the data size grows, the system can accommodate the increased storage requirements by simply adding more commodity hardware.

- **Namenode and datanodes**: HDFS follows a master-slave architecture. The *Namenode* acts as the master, managing the filesystem namespace and regulating client access to files. *Datanodes* are slave nodes that store the actual data blocks and serve read/write requests from clients.

MapReduce – the processing framework

MapReduce is a distributed processing framework that allows developers to write programs that process large datasets in parallel across a cluster of nodes. It consists of the following:

- **Simplified parallelism**: MapReduce simplifies the complexities of distributed processing. At its core, it consists of two primary phases:

  - **Map**: Input data is divided, and *mapper* tasks process these smaller chunks simultaneously

  - **Reduce**: Results from the mappers are aggregated by *reducer* tasks, producing the final output

- **A data-centric approach**: MapReduce moves code to where data resides on the cluster, rather than the traditional approach of moving data to code. This optimizes data flow and makes processing highly efficient.

HDFS and MapReduce form the core of Hadoop's distributed computing ecosystem. HDFS provides the distributed storage infrastructure, while MapReduce enables the distributed processing of large datasets. Developers can write MapReduce jobs in Java to process data stored in HDFS, leveraging the power of parallel computing to achieve scalable and fault-tolerant data processing.

In the next section, we will explore how Java and Hadoop work hand in hand, and we will also provide a basic MapReduce code example to demonstrate the data processing logic.

Java and Hadoop – a perfect match

Apache Hadoop revolutionized big data storage and processing. At its core lies a strong connection with Java, the widely used programming language known for its versatility and robustness. This section explores how Java and Hadoop work together, providing the necessary tools for effective Hadoop application development.

Why Java? A perfect match for Hadoop development

Several factors make Java the go-to language for Hadoop development:

- **Java as the foundation of Hadoop**:

  - Hadoop is written in Java, making it the native language for Hadoop development

  - Java's object-oriented programming paradigm aligns perfectly with Hadoop's distributed computing model

  - Java's platform independence allows Hadoop applications to run seamlessly across different hardware and operating systems

- **Seamless integration with the Hadoop ecosystem**:

  - The Hadoop ecosystem encompasses a wide range of tools and frameworks, many of which are built on top of Java

  - Key components such as HDFS and MapReduce heavily rely on Java for their functionality

  - Java's compatibility ensures smooth integration between Hadoop and other Java-based big data tools, such as *Apache Hive*, *Apache HBase*, and *Apache Spark*

- **Rich API support for Hadoop development**:

  - Hadoop provides comprehensive Java APIs that enable developers to interact with its core components effectively

  - The Java API for HDFS allows programmatic access and manipulation of data stored in the distributed filesystem

- MapReduce, the heart of Hadoop's data processing engine, exposes Java APIs to write and manage MapReduce jobs

- These APIs empower developers to leverage Hadoop's functionalities and build powerful data-processing applications

As the Hadoop ecosystem continues to evolve, Java remains the foundation upon which new tools and frameworks are built, cementing its position as the perfect match for Hadoop development.

Now that we understand the strengths of Java in the Hadoop ecosystem, let's delve into the heart of Hadoop data processing. In the next section, we'll explore how to write MapReduce jobs using Java, with a basic code example to solidify these concepts.

MapReduce in action

The following example demonstrates how MapReduce in Java can be used to analyze website clickstream data and identify user browsing patterns.

We have a large dataset containing clickstream logs, where each log entry records details such as the following:

- A user ID

- A timestamp

- A visited web page URL

We will analyze user clickstream data to understand user browsing behavior and identify popular navigation patterns, which incorporates a custom grouping logic within the reducer to group user sessions based on a time window (e.g., 15 minutes), and then we will analyze web page visit sequences within each session.

Here is the Mapper code snippet:

```
public static class Map extends Mapper<LongWritable, Text, Text, Text>
{
    @Override
    public void map(LongWritable key, Text value,
        Context context) throws IOException,
        InterruptedException {
    // Split the log line based on delimiters (e.g., comma)
            String[] logData = value.toString().split(",");

    // Extract user ID, timestamp, and webpage URL
            String userId = logData[0];
            long timestamp = Long.parseLong(logData[1]);
            String url = logData[2];
```

```
        // Combine user ID and timestamp (key for grouping by session)
        String sessionKey = userId + "-" + String.valueOf(
            timestamp / (15 * 60 * 1000));

        // Emit key-value pair: (sessionKey, URL)
        context.write(new Text(sessionKey),
            new Text(url));
    }
}
```

This code defines a `Mapper` class, a core component in MapReduce responsible for processing individual input data records. The key points in this class are as follows:

- **Input/output types**: The `Mapper<LongWritable, Text, Text, Text>` class declaration specifies its input and output key-value pairs:

  - **Input**: `LongWritable` key for line numbers and, `Text` value for text lines

  - **Output**: `Text` key for session keys and `Text` value for URLs

- **The map function**:

  - **Input data handling**: The map function receives a `key-value pair`, representing a line from the input data. The line is split into an array using a comma (,) delimiter, assuming a comma-separated log format.

  - **Data extraction**: Extracts relevant information from the logline:

    - `userId`: The user ID from the first element of the array

    - `timestamp`: The long value parsed from the second element

    - `url`: The web page URL from the third element

  - **Key generation for grouping**:

    - Creates a unique session key by combining the user ID and a downsampled timestamp

    - Divides the raw timestamp by `15 * 60 * 1000` (15 minutes) to group events within 15-minute intervals, usually for session-based analysis

- **Key-value emission**: Emits a new `key-value pair` for downstream processing:

  - **Key**: The text representing the generated session key

  - **Value**: The text representing the extracted URL

- **Purpose and context**: This mapper functions within a larger MapReduce job, designed for session-based analysis of user activity logs. It groups events belonging to each user into 15-minute sessions. The emitted key-value pairs (session key and URL) will undergo shuffling and sorting before being processed by the reducers. These reducers will perform further aggregation or analysis based on the session keys.

Here is the Reducer code snippet:

```java
public static class Reduce extends Reducer<Text, Text,
    Text, Text> {

    @Override
    public void reduce(Text key,
        Iterable<Text> values,
        Context context) throws IOException,
        InterruptedException {
            StringBuilder sessionSequence = new StringBuilder();

        // Iterate through visited URLs within the same session
(defined by key)
            for (Text url : values) {
                sessionSequence.append(url.toString()
                ).append(" -> ");
            }

        // Remove the trailing " -> " from the sequence
        sessionSequence.setLength(
            sessionSequence.length() - 4);

        // Emit key-value pair: (sessionKey, sequence of visited
URLs)
            context.write(key, new Text(
                sessionSequence.toString()));
        }
    }
```

This code defines a `Reducer` class, responsible for aggregating and summarizing data grouped by a common key after the map phase. The key points in this class are as follows:

- **Input/output types**: The Reducer's `reduce()` function operates on key-value pairs:

  - **Input**: The text key representing a session key, `Iterable<Text> values`, containing a collection of URLs associated with that session

  - **Output**: The text key for the session key and the text value for the constructed URL sequence

- **The reduce function**:

  - **Initialization**: Creates a StringBuilder named `sessionSequence` to accumulate the URL sequence for the current session.

  - **URL concatenation**: Iterates through the collection of URLs (values) associated with the given session key. Appends each URL to `sessionSequence`, followed by `- >` to maintain order.

  - **Trailing space adjustment**: Removes the redundant `- >` at the end of the constructed sequence for cleaner output.

  - **Key-value emission**: Emits a new key-value pair:

    - **Key**: The input session key (unchanged)

    - **Value**: The text representation of the constructed URL sequence, representing the ordered sequence of URLs visited within that session

- **Purpose and context**: This reducer works alongside the mapper code to facilitate session-based analysis of user activity logs. It aggregates URLs associated with each session key, effectively reconstructing the order of web page visits within those user sessions. The final output of this MapReduce job takes the form of key-value pairs. These keys represent unique user-session combinations, while the values hold the corresponding sequences of visited URLs. This valuable output enables various analyses, such as understanding user navigation patterns, identifying common paths taken during sessions, and uncovering frequently visited page transitions.

For Hadoop developers, *writing MapReduce jobs in Java is essential*. Java's object-oriented features and the Hadoop API empower developers to distribute complex data processing tasks across a cluster. The `Mapper` and `Reducer` classes, the heart of a MapReduce job, handle the core logic. Java's rich ecosystem and tooling support streamline the writing and debugging of these jobs. As you progress, mastering efficient MapReduce development in Java unlocks the full potential of big data processing with Hadoop.

Beyond the basics – advanced Hadoop concepts for Java developers and architects

While understanding the core concepts of Hadoop, such as HDFS and MapReduce, is essential, there are several advanced Hadoop components and technologies that Java developers and architects should be familiar with. In this section, we'll explore YARN and HBase, two important components of the Hadoop ecosystem, focusing on their practical applications and how they can be leveraged in real-world projects.

Yet another resource negotiator

Yet Another Resource Negotiator (**YARN**) is a resource management and job scheduling framework in Hadoop. It separates resource management and processing components, allowing multiple data processing engines to run on Hadoop. Its key concepts are as follows:

- **ResourceManager**: Manages the global assignment of resources to applications

- **NodeManager**: Monitors and manages resources on individual nodes in a cluster

- **ApplicationMaster**: Negotiates resources and manages the life cycle of an application

Its benefits for Java developers and architects are as follows:

- YARN enables running various data processing frameworks, such as Spark and Flink, on the same Hadoop cluster, providing flexibility and efficiency

- It allows better resource utilization and multitenancy, enabling multiple applications to share the same cluster resources

- Java developers can leverage YARN's APIs to develop and deploy custom applications on Hadoop

HBase

HBase is a column-oriented, NoSQL database built on top of Hadoop. It provides real-time, random read/write access to large datasets. Its key concepts are as follows:

- **Table:** Consists of rows and columns, similar to a traditional database table

- **Row key**: Uniquely identifies a row in an HBase table

- **Column family**: Groups related columns together for better data locality and performance

Its benefits for Java developers and architects are as follows:

- HBase is ideal for applications that require low-latency and random access to large datasets, such as real-time web applications or sensor data storage

- It integrates seamlessly with Hadoop and allows you to run MapReduce jobs on HBase data

- Java developers can use the HBase Java API to interact with HBase tables, perform **Create, Read, Update, Delete** (**CRUD**) operations, and execute scans and filters

- HBase supports high write throughput and scales horizontally, making it suitable to handle large-scale, write-heavy workloads

Integration with the Java ecosystem

Hadoop integrates well with various Java-based tools and frameworks commonly used in enterprise environments. Some notable integrations are as follows:

- **Apache Hive**: A data warehousing and SQL-like querying framework built on top of Hadoop. Java developers can use Hive to query and analyze large datasets using familiar SQL syntax.

- **Apache Kafka**: A distributed streaming platform that integrates with Hadoop for real-time data ingestion and processing. Java developers can use Kafka's Java API to publish and consume data streams.

- **Apache Oozie**: A workflow scheduler for Hadoop jobs. Java developers can define and manage complex workflows using Oozie's XML-based configuration or Java API.

For Java developers and architects, Hadoop's power extends beyond core components. Advanced features such as YARN (resource management) and HBase (real-time data access) enable flexible, scalable big data solutions. Seamless integration with other Java-based tools, such as Hive and Kafka, expands Hadoop's capabilities.

One real-life system where Hadoop's capabilities have been expanded through integration with Hive and Kafka is LinkedIn's data processing and analytics infrastructure. LinkedIn has built a robust data handling infrastructure, leveraging Hadoop for large-scale data storage and processing, complemented by Kafka for real-time data streaming and Hive for SQL-like data querying and analysis. Kafka channels vast streams of user activity data into the Hadoop ecosystem, where it's stored and processed. Hive then enables detailed data analysis and insight generation. This integrated system supports LinkedIn's diverse analytical needs, from operational optimization to personalized recommendations, showcasing the synergy between Hadoop, Hive, and Kafka in managing and analyzing big data.

Mastering these concepts empowers architects to build robust big data applications for modern businesses. As processing needs evolve, frameworks such as Spark offer even faster in-memory computations, complementing Hadoop for complex data pipelines.

Understanding DataFrames and RDDs in Apache Spark

Apache Spark provides two primary abstractions to work with distributed data – **Resilient Distributed Datasets (RDDs)** and **DataFrames**. Each offers unique features and benefits tailored to different types of data processing tasks.

RDDs

RDDs are the fundamental data structure of Spark, providing an immutable distributed collection of objects that allows data to be processed in parallel across a distributed environment. Each dataset in RDD is divided into logical partitions, which can be computed on different nodes of the cluster.

RDDs are well-suited for low-level transformations and actions that require fine-grained control over physical data distribution and transformations, such as custom partitioning schemes or when performing complex algorithms that involve iterative data processing over a network.

RDDs support two types of operations – transformations, which create a new RDD from an existing one, and actions, which return a value to the driver program after running a computation on the dataset.

DataFrames

Introduced as an abstraction on top of RDDs, DataFrames are a distributed collection of data organized into named columns, similar to a table in a relational database but with richer optimizations under the hood.

Here are the advantages of DataFrames:

- **Optimized execution**: Spark SQL's Catalyst optimizer compiles DataFrame operations into highly efficient physical execution plans. This optimization allows for faster processing compared to RDDs, which do not benefit from such optimization.

- **Ease of use**: The DataFrame API provides a more declarative programming style, making complex data manipulations and aggregations easier to express and understand.

- **Interoperability**: DataFrames support various data formats and sources, including Parquet, CSV, JSON, and JDBC, making data integration and processing simpler and more robust.

DataFrames are ideal for handling structured and semi-structured data. They are preferred for data exploration, transformation, and aggregation tasks, especially when ease of use and performance optimization are priorities.

Emphasizing DataFrames over RDDs

Since the introduction of Spark 2.0, DataFrames have been recommended as the standard abstraction for data processing tasks due to their significant advantages in terms of optimization and usability. While RDDs remain useful for specific scenarios requiring detailed control over data operations, DataFrames provide a powerful, flexible, and efficient way to work with large-scale data.

RDDs are the foundation of Spark's distributed data processing capabilities. This section dives into how to create and manipulate RDDs to efficiently analyze large-scale datasets across a cluster.

RDDs can be created in several ways, including the following:

- Parallelizing an existing collection:

```
List<Integer> data = Arrays.asList(1, 2, 3, 4, 5);
JavaRDD<Integer> rdd = sc.parallelize(data);
```

- Reading from external datasets (e.g., text files, CSV files, or databases):

```
JavaRDD<String> textRDD = sc.textFile(
    "path/to/file.txt");
```

- Transforming existing RDDs:

```
JavaRDD<Integer> squaredRDD = rdd.map(x -> x * x);
```

RDDs support two types of operations – transformations and actions.

- **Transformations**: These are operations that create a new RDD from an existing one, such as `map()`, `filter()`, `flatMap()`, and `reduceByKey()`. Transformations are lazily evaluated.

- **Actions**: These are operations that trigger computation and return a result to the driver program, or write data to an external storage system. Examples include `collect()`, `count()`, `first()`, and `saveAsTextFile()`.

By leveraging RDDs and their distributed nature, Spark enables developers to process and analyze large-scale datasets efficiently across a cluster of machines.

Let's look at the following code snippet:

```
// Create an RDD from a text file
JavaRDD<String> lines = spark.sparkContext().textFile(
    "path/to/data.txt", 1);

// Map transformation to parse integers from lines
JavaRDD<Integer> numbers = lines.map(Integer::parseInt);

// Filter transformation to find even numbers
JavaRDD<Integer> evenNumbers = numbers.filter(
    n -> n % 2 == 0);

// Action to count the number of even numbers
long count = evenNumbers.count();

// Print the count
System.out.println("Number of even numbers: " + count);
```

This code demonstrates the usage of RDDs. It performs the following steps:

- It creates an RDD called `lines` by reading a text file located at `"path/to/data.txt"`, using `spark.sparkContext().textFile()`. The second argument, 1, specifies the minimum number of partitions for the RDD.

- It applies a map transformation to the lines RDD using `Integer::parseInt`. This transformation converts each line of text into an integer, resulting in a new RDD called `numbers`.

- It applies a filter transformation to the numbers RDD using `n -> n % 2 == 0`. This transformation keeps only the even numbers in the RDD, creating a new RDD called `evenNumbers`.

- It performs an action on the `evenNumbers` RDD using `count()`, which returns the number of elements in the RDD. The result is stored in the `count` variable.

- Finally, it prints the count of even numbers using `System.out.println()`.

This code showcases the basic usage of RDDs in Spark, demonstrating how to create an RDD from a text file, apply transformations (map and filter) to the RDD, and perform an action (count) to retrieve a result. The transformations are lazily evaluated, meaning they are not executed until an action is triggered.

Spark programming with Java – unleashing the power of DataFrames and RDDs

In this section, we'll explore the commonly used *transformations* and *actions* within Spark's Java API, focusing on both DataFrames and RDDs.

Spark's DataFrame API – a comprehensive guide

DataFrames have become the primary data abstraction in Spark 2.0 and above, offering a more efficient and user-friendly way to work with structured and semi-structured data. Let's explore the *DataFrame API* in detail, including how to create DataFrames, perform transformations and actions, and leverage SQL-like querying.

First up is creating DataFrames.

There are several ways to create DataFrames in Spark; here is an example to create one from an existing RDD:

```
// Create an RDD from a text file
JavaRDD<String> textRDD = spark.sparkContext().textFile(
    "path/to/data.txt", 1);

// Convert the RDD of strings to an RDD of Rows
JavaRDD<Row> rowRDD = textRDD.map(line -> {
    String[] parts = line.split(",");
    return RowFactory.create(parts[0],
        Integer.parseInt(parts[1]));
});
```

```
// Define the schema for the DataFrame
StructType schema = new StructType()
    .add("name", DataTypes.StringType)
    .add("age", DataTypes.IntegerType);

// Create the DataFrame from the RDD and the schema
Dataset<Row> df = spark.createDataFrame(rowRDD, schema);
```

This code creates a DataFrame from an existing RDD. It starts by creating an RDD of strings (textRDD) from a text file. Then, it converts textRDD to an RDD of row objects (rowRDD), using map(). The schema for the DataFrame is defined using StructType and StructField. Finally, the DataFrame is created using spark.createDataFrame(), passing rowRDD and the schema.

Next, we'll encounter DataFrame transformations.

DataFrames provide a wide range of transformations for data manipulation and processing. Some common transformations include the following:

- **Filtering rows**:

```
Dataset<Row> filteredDf = df.filter(col(
    "age").gt(25));
```

- **Selecting columns**:

```
Dataset<Row> selectedDf = df.select("name", "age");
```

- **Adding or modifying columns**:

```
Dataset<Row> newDf = df.withColumn("doubledAge", col(
    "age").multiply(2));
```

- **Aggregating data**:

```
Dataset<Row> aggregatedDf = df.groupBy("age").agg(
    count("*").as("count"));
```

Now, we will move on to DataFrame actions.

Actions trigger the computation on DataFrames and return the results to the driver program. Some common actions include the following:

- **Collecting data in the driver**:

```
List<Row> collectedData = df.collectAsList();
```

- **Counting rows**:

```
long count = df.count();
```

- **Saving data to a file or data source**:

```
df.write().format("parquet").save("path/to/output");
```

- **SQL-like querying with DataFrames**: One of the powerful features of DataFrames is the ability to use SQL-like queries for data analysis and manipulation. Spark SQL provides a SQL interface to query structured data stored as DataFrames:

  - **Registering a DataFrame as a temporary view**:

```
df.createOrReplaceTempView("people");
```

  - **Executing SQL queries**:

```
Dataset<Row> sqlResult = spark.sql(
    "SELECT * FROM people WHERE age > 25");
```

  - **Joining DataFrames**:

```
Dataset<Row> joinedDf = df1.join(df2,
    df1.col("id").equalTo(df2.col("personId")));
```

These examples demonstrate the expressiveness and flexibility of the DataFrame API in Spark. By leveraging DataFrames, developers can perform complex data manipulations, transformations, and aggregations efficiently, while also benefiting from the optimizations provided by the Spark SQL engine.

By mastering these operations and understanding when to use DataFrames versus RDDs, developers can build efficient and powerful data processing pipelines in Spark. The Java API's evolution continues to empower developers to tackle big data challenges effectively with both structured and unstructured data.

Performance optimization in Apache Spark

Optimizing performance in Spark applications involves understanding and mitigating several key issues that can affect scalability and efficiency. This section covers strategies to handle data shuffling, manage data skew, and optimize data collection in the driver, providing a holistic approach to performance tuning:

- **Managing data shuffling**: Data shuffling is a critical operation in Spark that can significantly impact performance. It occurs when operations such as groupBy(), join(), or reduceByKey() require data to be redistributed across partitions. Shuffling involves disk I/O and network I/O and can lead to substantial resource consumption.

- **Strategies to minimize shuffling**:

  - **Optimize transformations**: Avoid unnecessary shuffling by choosing transformations that minimize data movement. For example, using map() before reduceByKey() can reduce the amount of data shuffled.

- **Adjust partitioning**: Use `repartition()` or `coalesce()` to optimize the number of partitions and distribute data more evenly across the cluster.

- **Handling data skew**: Data skew occurs when one or more partitions receive significantly more data than others, leading to uneven workloads and potential bottlenecks.

- **Strategies to handle data skew**:

  - **Salting keys**: Modify the keys that cause skew by adding a random prefix or suffix to distribute the load more evenly

  - **Custom partitioning**: Implement a custom partitioner that distributes data more evenly, based on your application's specific characteristics

  - **Filter and split**: Identify skewed data, process it separately, and then merge the results to prevent overloaded partitions

- **Optimizing data collection in the driver**: Collecting large datasets in the driver with operations such as `collect()` can lead to out-of-memory errors and degrade overall performance.

- **Safe practices for data collection**:

  - **Limit data retrieval**: Use operations such as `take()`, `first()`, or `show()` to retrieve only necessary data samples instead of an entire dataset

  - **Aggregate in a cluster**: Perform aggregations or reductions in a cluster as much as possible before collecting results, minimizing the data volume moved to the driver

  - **Use foreachPartition**: Instead of collecting data in the driver, process it within each partition using `foreachPartition()` to apply operations, such as database writes or API calls, directly within each partition

Here is an example of efficient data handling:

```
// Example of handling skewed data
JavaPairRDD<String, Integer> skewedData = rdd.mapToPair(
    s -> new Tuple2<>(s, 1))
    .reduceByKey((a, b) -> a + b);

// Custom partitioning to manage skew
JavaPairRDD<String, Integer> partitionedData = skewedData
    .partitionBy(new CustomPartitioner());

// Reducing data transfer to the driver
List<Integer> aggregatedData = partitionedData.map(
    tuple -> tuple._2())
    .reduce((a, b) -> a + b)
    .collect();
```

This example showcases two techniques to manage data skew and optimize data collection in Apache Spark:

- **Custom partitioning**: The code uses a custom partitioner (`CustomPartitioner`) to distribute skewed data more evenly across the cluster. By calling `partitionBy()` with the custom partitioner on the skewed data, it creates a new RDD (`partitionedData`) with a more balanced data distribution, mitigating the impact of skew.

- **In-cluster reduction before collection**: To minimize data transfer to the driver, the code performs aggregation operations on the partitioned data within the cluster before collecting the results. It uses `map()` to extract the values and `reduce` to sum them up across partitions. By aggregating the data before `collect()`, it reduces the amount of data sent to the driver, optimizing data collection and minimizing network overhead.

These techniques help improve the performance and scalability of Spark applications when dealing with skewed data distributions and large result sets.

Spark optimization and fault tolerance – advanced concepts

Understanding some advanced concepts such as the execution **Directed Acyclic Graph** (**DAG**), *caching*, and *retry* mechanisms is essential for a deeper understanding of Spark's optimization and fault tolerance capabilities. Integrating these topics can enhance the effectiveness of Spark application development. Let's break down these concepts and how they relate to the DataFrame API.

Execution DAG in Spark

The DAG in Spark is a fundamental concept that underpins how Spark executes workflows across a distributed cluster. When you perform operations on a DataFrame, Spark constructs a DAG of stages, with each stage consisting of tasks based on transformations applied to the data. This DAG outlines the steps that Spark will execute across the cluster.

The following are the key points:

- **DAG scheduling**: Spark's DAGScheduler divides the operators into stages of tasks. A stage contains tasks based on transformations that can be performed without shuffling data. Boundaries of stages are generally determined by operations that require shuffling data, such as `groupBy()`.

- **Lazy evaluation**: Spark operations are lazily evaluated, meaning computations are delayed until an action (such as `show()`, `count()`, or `save()`) is triggered. This allows Spark to optimize the entire data processing pipeline, consolidating tasks and stages efficiently.

- **Optimization**: Through the Catalyst optimizer, Spark converts this logical execution plan (the DAG) into a physical plan that optimizes the execution, by rearranging operations and combining tasks.

Caching and persistence

Caching in Spark is critical for optimizing the performance of iterative algorithms and interactive data analysis, where the same dataset is queried repeatedly. Caching can be used as follows:

- **DataFrame caching**: You can persist a DataFrame in memory using the `cache()` or `persist()` methods. This is particularly useful when data is accessed repeatedly, such as when tuning machine learning models or running multiple queries on the same subset of data.

- **Storage levels**: The `persist()` method can take a storage level parameter (`MEMORY_ONLY`, `MEMORY_AND_DISK`, etc.), allowing you finer control over how your data is stored.

Retry mechanisms and fault tolerance

Spark provides robust fault tolerance through its distributed architecture and by rebuilding lost data, using the lineage of transformations (DAG):

- **Task retries**: If a task fails, Spark automatically retries it. The number of retries and the conditions for a retry can be configured in Spark's settings.

- **Node failure**: In case of node failures, Spark can recompute lost partitions of data from the original source, as long as the source data is still accessible and the lineage is intact.

- **Checkpointing**: For long-running and complex DAGs, checkpointing can be used to truncate the RDD lineage and save the intermediate state to a reliable storage system, such as HDFS. This reduces recovery time if there are failures.

Here's an example demonstrating these concepts in action:

```
public class SparkOptimizationExample {
    public static void main(String[] args) {
        SparkSession spark = SparkSession.builder()
            .appName("Advanced Spark Optimization")
            .master("local")
            .getOrCreate();

        // Load and cache data
        Dataset<Row> df = spark.read().json(
            "path/to/data.json").cache();

        // Example transformation with explicit caching
        Dataset<Row> processedDf = df
            .filter("age > 25")
            .groupBy("occupation")
            .count();
```

```
            // Persist the processed DataFrame with a specific storage
    level
            processedDf.persist(
                StorageLevel.MEMORY_AND_DISK());

            // Action to trigger execution
            processedDf.show();

            // Example of fault tolerance: re-computation from cache after
    failure
            try {
                // Simulate data processing that might fail
                processedDf.filter("count > 5").show();
            } catch (Exception e) {
                System.out.println("Error during processing,
                        retrying...");
                processedDf.filter("count > 5").show();
            }

            spark.stop();
        }
    }
```

This code demonstrates how Spark's advanced features can be used to optimize complex data processing tasks:

- **Execution DAG and lazy evaluation**: When you load data using `spark.read().json(...)`, Spark builds an `execution DAG` that represents the data processing pipeline. This DAG outlines the stages of operations on the data. Spark utilizes *lazy evaluation*, delaying computations until an action such as `show()` is triggered. This allows Spark to analyze the entire DAG and optimize the execution plan.

- **Caching and persistence**: Spark's caching capabilities are leveraged in this example. The initial `data (df)` is *cached* using `cache()`. This stores the data in memory, allowing faster access for subsequent transformations. Additionally, the transformed data (`processedDf`) is *persisted* with `persist(StorageLevel.MEMORY_AND_DISK())`. This ensures the that processed data remains available even after the triggering action, (`show()`), potentially improving performance for future operations that rely on it. Specifying the `MEMORY_AND_DISK` storage level keeps the data in memory for faster access, while also persisting it to disk for fault tolerance.

- **Fault tolerance with retries**: The code demonstrates Spark's robust fault tolerance mechanisms. The simulated error during data processing showcases how Spark can leverage cached data. Even if the initial filtering operation on `processedDf` fails (due to a potentially non-existent column), Spark can still complete the operation by recomputing the required data from the already cached `processedDf`. This highlights Spark's ability to handle failures and ensure successful completion of tasks.

By effectively utilizing execution DAGs, caching, persistence, and retry mechanisms, this code exemplifies how Spark can optimize performance, improve data processing efficiency, and ensure robust execution of complex workflows even in the face of potential failures.

Spark versus Hadoop – choosing the right framework for the job

Spark and Hadoop are two powerful big data processing frameworks that have gained widespread adoption in the industry. While both frameworks are designed to handle large-scale data processing, they have distinct characteristics and excel in different scenarios. In this section, we'll explore the strengths of Spark and Hadoop and discuss situations where each framework is best suited.

Scenarios where Hadoop's MapReduce excels include the following:

- **Batch processing**: MapReduce is highly efficient for large-scale batch processing tasks where data can be processed in a linear, map-then-reduce manner.

- **Data warehousing and archiving**: Hadoop is often used to store and archive large datasets, thanks to its cost-effective storage solution, HDFS. It's suitable for scenarios where data doesn't need to be accessed in real-time.

- **Highly scalable processing**: For tasks that are not time-sensitive and can benefit from linear scalability, MapReduce can efficiently process petabytes of data across thousands of machines.

- **Fault tolerance on commodity hardware**: Hadoop's infrastructure is designed to reliably store and process data across potentially unreliable commodity hardware, making it a cost-effective solution for massive data storage and processing.

Scenarios where Apache Spark excels include the following:

- **Iterative algorithms in machine learning and data mining**: Spark's in-memory data processing capabilities make it significantly faster than MapReduce for iterative algorithms, which are common in machine learning and data mining tasks.

- **Real-time stream processing**: Spark Streaming allows you to process real-time data streams. It's ideal for scenarios where data needs to be processed immediately as it arrives, such as in log file analysis and real-time fraud detection systems.

- **Interactive data analysis and processing**: Spark's ability to cache data in memory across operations makes it an excellent choice for interactive data exploration, analysis, and processing tasks. Tools such as Apache Zeppelin and Jupyter integrate well with Spark for interactive data science work.

- **Graph processing**: GraphX, a component of Spark, enables graph processing and computation directly within the Spark ecosystem, making it suitable for social network analysis, recommendation systems, and other applications that involve complex relationships between data points.

In practice, Spark and Hadoop are not mutually exclusive and often used together. Spark can run on top of HDFS and even integrate with Hadoop's ecosystem, including YARN for resource management. This integration leverages Hadoop's storage capabilities while benefiting from Spark's processing speed and versatility, providing a comprehensive solution for big data challenges.

Hadoop and Spark equivalents in major cloud platforms

While Apache Hadoop and Apache Spark are widely used in on-premises big data processing, major cloud platforms offer managed services that provide similar capabilities without the need to set up and maintain the underlying infrastructure. In this section, we'll explore the equivalent services to Hadoop and Spark in AWS, Azure, and GCP:

- **Amazon Web Services (AWS)**:

  - **Amazon Elastic MapReduce**: Amazon **Elastic MapReduce** (**EMR**) is a managed cluster platform that simplifies running big data frameworks, including Apache Hadoop and Apache Spark. It provides a scalable and cost-effective way to process and analyze large volumes of data. EMR supports various Hadoop ecosystem tools such as Hive, Pig, and HBase. It also integrates with other AWS services such as Amazon S3 for data storage and Amazon Kinesis for real-time data streaming.

  - **Amazon Simple Storage Service**: Amazon **Simple Storage Service** (**S3**) is an object storage service that provides scalable and durable storage for big data workflows. It can be used as a data lake to store and retrieve large datasets, serving as an alternative to HDFS. S3 integrates seamlessly with Amazon EMR and other big data processing services.

- **Microsoft Azure**:

  - **Azure HDInsight**: Azure HDInsight is a managed Apache Hadoop, Spark, and Kafka service in the cloud. It allows you to easily provision and manage Hadoop and Spark clusters on Azure. HDInsight supports a wide range of Hadoop ecosystem components, including Hive, Pig, and Oozie. It integrates with Azure Blob Storage and Azure Data Lake Storage to store and access big data.

- **Azure Databricks**: Azure Databricks is a fully managed Apache Spark platform optimized for the Microsoft Azure cloud. It provides a collaborative and interactive environment to run Spark workloads. Databricks offers seamless integration with other Azure services and supports various programming languages, such as Python, R, and SQL.

- **Google Cloud Platform (GCP)**:

  - **Google Cloud Dataproc**: Google Cloud Dataproc is a fully managed Spark and Hadoop service. It allows you to quickly create and manage Spark and Hadoop clusters on GCP. Dataproc integrates with other GCP services such as Google Cloud Storage and BigQuery. It supports various Hadoop ecosystem tools and provides a familiar environment to run Spark and Hadoop jobs.

  - **Google Cloud Storage**: Google Cloud Storage is a scalable and durable object storage service. It serves as a data lake to store and retrieve large datasets, similar to Amazon S3. Cloud Storage integrates with Google Cloud Dataproc and other GCP big data services.

Major cloud platforms offer managed services that provide equivalent functionality to Apache Hadoop and Apache Spark, simplifying the provisioning and management of big data processing clusters. These services integrate with their respective cloud storage solutions for seamless data storage and access.

By leveraging these managed services, organizations can focus on data processing and analysis without the overhead of managing the underlying infrastructure. Developers and architects can utilize their existing skills and knowledge while benefiting from the scalability, flexibility, and cost-effectiveness of cloud-based big data solutions.

Now that we've covered the fundamentals, let's see how Java and big data technologies work together to solve real-world problems.

Real-world Java and big data in action

Delving beyond the theoretical, we'll delve into three practical use cases that showcase the power of this combination.

Use case 1 – log analysis with Spark

Let's consider a scenario where an e-commerce company wants to analyze its web server logs to extract valuable insights. The logs contain information about user requests, including timestamps, requested URLs, and response status codes. The goal is to process the logs, extract relevant information, and derive meaningful metrics. We will explore log analysis using Spark's DataFrame API, demonstrating efficient data filtering, aggregation, and joining techniques. By leveraging DataFrames, we can easily parse, transform, and summarize log data from CSV files:

```java
public class LogAnalysis {
    public static void main(String[] args) {
```

```java
SparkSession spark = SparkSession.builder()
    .appName("Log Analysis")
    .master("local")
    .getOrCreate();
try {
    // Read log data from a file into a DataFrame
    Dataset<Row> logData = spark.read()
        .option("header", "true")
        .option("inferSchema", "true")
        .csv("path/to/log/data.csv");
    // Filter log entries based on a specific condition
    Dataset<Row> filteredLogs = logData.filter(
        functions.col("status").geq(400));
    // Group log entries by URL and count the occurrences
    Dataset<Row> urlCounts = filteredLogs.groupBy(
        "url").count();
    // Calculate average response time for each URL
    Dataset<Row> avgResponseTimes = logData
        .groupBy("url")
        .agg(functions.avg("responseTime").alias(
            "avgResponseTime"));
    // Join the URL counts with average response times
    Dataset<Row> joinedResults = urlCounts.join(
        avgResponseTimes, "url");
    // Display the results
    joinedResults.show();
} catch (Exception e) {
    System.err.println(
        "An error occurred in the Log Analysis process: " +
        e.getMessage());
    e.printStackTrace();
} finally {
    spark.stop();
}
    }
}
```

This Spark code snippet is designed for log analysis, using Apache Spark's *DataFrame API*, an effective tool for handling structured data processing. The code performs several operations on server log data, which is assumed to be stored in the CSV format:

- **Data loading**: The `spark.read()` function is used to load log data from a CSV file into a DataFrame, with `header` set to `true` to use the first line of the file as column names, and `inferSchema` set to `true` to automatically deduce the data types of each column.

- **Data filtering**: The DataFrame is filtered to include only log entries where the status code is `400` or higher, typically indicating client errors (such as `404 Not Found`) or server errors (such as `500 Internal Server Error`).

- **Aggregation**: The filtered logs are grouped by URL, and the occurrences of each URL are counted. This step helps to identify which URLs are frequently associated with errors.

- **Average calculation**: A separate aggregation calculates the average response time for each URL across all logs, not just those with errors. This provides insights into the performance characteristics of each endpoint.

- **Join operation**: The URL counts from the error logs, and the average response times are joined on the URL field, merging the error frequency with performance metrics into a single dataset.

- **Result display**: Finally, the combined results are displayed, showing each URL along with its count of error occurrences and average response time. This output is useful for diagnosing issues and optimizing server performance.

This example demonstrates how to use Spark to efficiently process and analyze large datasets, leveraging its capabilities for filtering, aggregation, and joining data to extract meaningful insights from web server logs.

Use case 2 – a recommendation engine

This code snippet demonstrates how to build and evaluate a recommendation system using Apache Spark's **Machine Learning Library** (**MLlib**). Specifically, it utilizes the **Alternating Least Squares** (**ALS**) algorithm, which is popular for collaborative filtering tasks such as movie recommendations:

```
// Read rating data from a file into a DataFrame
Dataset<Row> ratings = spark.read()
    .option("header", "true")
    .option("inferSchema", "true")
.csv("path/to/ratings/data.csv");
```

This code reads the rating data from a CSV file into a DataFrame called `ratings`. The `spark.read()` method is used to read the data, and the `option` method is used to specify the following options:

- `"header", "true"`: Indicates that the first line of the CSV file contains the column names

- `"inferSchema", "true"`: Instructs Spark to infer the data types of the columns based on the data

The csv() method specifies the path to the CSV file containing the rating data:

```
// Split the data into training and testing sets
Dataset<Row>[] splits = ratings.randomSplit(new double[]{
    0.8, 0.2});
Dataset<Row> trainingData = splits[0];
Dataset<Row> testingData = splits[1];
```

This code splits the ratings DataFrame into training and testing datasets, using the randomSplit() method. The new double[]{0.8, 0.2} argument specifies the proportions of the split, with 80% of the data going into the training set and 20% into the testing set. The resulting datasets are stored in the trainingData and testingData variables, respectively:

```
// Create an ALS model
ALS als = new ALS()
    .setMaxIter(10)
    .setRegParam(0.01)
    .setUserCol("userId")
    .setItemCol("itemId")
    .setRatingCol("rating");
```

This code creates an instance of the ALS model using the ALS class. The model is configured with the following parameters:

- setMaxIter(10): Sets the maximum number of iterations to 10

- setRegParam(0.01): Sets the regularization parameter to 0.01

- setUserCol("userId"): Specifies the column name for user IDs

- setItemCol("itemId"): Specifies the column name for item IDs

- setRatingCol("rating"): Specifies the column name for ratings

```
// Train the model
ALSModel model = als.fit(trainingData);
```

The preceding code trains the ALS model using the fit() method, passing the trainingData DataFrame as the input. The trained model is stored in the model variable.

```
// Generate predictions on the testing data
Dataset<Row> predictions = model.transform(testingData);
```

The preceding code generates predictions on the `testingData` DataFrame using the trained model. The `transform()` method applies the model to the testing data and returns a new DataFrame, called `predictions`, which contains the predicted ratings.

```
// Evaluate the model
RegressionEvaluator evaluator = new RegressionEvaluator()
    .setMetricName("rmse")
    .setLabelCol("rating")
    .setPredictionCol("prediction");
double rmse = evaluator.evaluate(predictions);
System.out.println("Root-mean-square error = " + rmse);
```

The preceding code evaluates the performance of the trained model using the `RegressionEvaluator` class. `evaluator` is configured to use the **root mean squared error** (**RMSE**) metric, with the actual ratings stored in the `"rating"` column and the predicted ratings stored in the `"prediction"` column. The `evaluate()` method calculates the RMSE on the `predictions` DataFrame, and the result is printed to the console.

```
// Generate top 10 movie recommendations for each user
Dataset<Row> userRecs = model.recommendForAllUsers(10);
userRecs.show();
```

The preceding code generates the top 10 movie recommendations for each user using the trained model. The `recommendForAllUsers()` method is called with an argument of 10, specifying the number of recommendations to generate per user. The resulting recommendations are stored in the `userRecs` DataFrame, and the `show` method is used to display the recommendations.

This example is typical for scenarios where businesses need to recommend products or content to users based on their past interactions. It demonstrates the process of building a movie recommendation engine using Apache Spark's DataFrame API and the ALS algorithm. The ALS algorithm is particularly well-suited for this purpose, due to its scalability and effectiveness in handling sparse datasets that are typical of user-item interactions.

Use case 3 – real-time fraud detection

Fraud detection involves analyzing transactions, user behavior, and other relevant data to identify anomalies that could signify fraud. The complexity and evolving nature of fraudulent activities necessitates the use of advanced analytics and machine learning. Our objective is to monitor transactions in real-time and flag those with a high likelihood of being fraudulent, based on historical data and `patterns.models`, and massive data processing capabilities.

This code demonstrates a real-time fraud detection system using Apache Spark Streaming. It reads transaction data from a `.CSV` file, applies a pre-trained machine learning model to predict the likelihood of fraud for each transaction, and outputs the prediction results to the console. Here is a sample code snippet:

```java
public class FraudDetectionStreaming {
    public static void main(String[] args) throws
StreamingQueryException {
        SparkSession spark = SparkSession.builder()
            .appName("FraudDetectionStreaming")
            .getOrCreate();

        PipelineModel model = PipelineModel.load(
            "path/to/trained/model");

        StructType schema = new StructType()
            .add("transactionId", "string")
            .add("amount", "double")
            .add("accountNumber", "string")
            .add("transactionTime", "timestamp")
            .add("merchantId", "string");

        Dataset<Row> transactionsStream = spark
            .readStream()
            .format("csv")
            .option("header", "true")
            .schema(schema)
            .load("path/to/transaction/data");

        Dataset<Row> predictionStream = model.transform(
            transactionsStream);

        predictionStream = predictionStream
            .select("transactionId", "amount",
                "accountNumber", "transactionTime",
                "merchantId","prediction", "probability");

        StreamingQuery query = predictionStream
            .writeStream()
            .outputMode("append")
            .format("console")
            .start();

        query.awaitTermination();
```

```
    }
}
```

Here is the code explanation:

- The main() method is defined, which is the entry point of the application.

- A SparkSession is created with the application name FraudDetectionStreaming.

- A pre-trained machine learning model is loaded, using PipelineModel.load(). The path to the trained model is specified as "path/to/trained/model".

- The schema for the transaction data is defined using StructType. It includes fields such as transactionId, amount, accountNumber, transaction Time, and merchantId.

- A streaming DataFrame transactionsStream is created, using spark.readStream() to read data from a CSV file. The file path is specified as "path/to/transaction/data". The header option is set to "true" to indicate that the CSV file has a header row, and the schema is provided using the schema() method.

- The pre-trained model is applied to transactionsStream using model.transform(), resulting in a new DataFrame predictionStream that includes the predicted fraud probabilities.

- The relevant columns are selected from predictionStream using select(), including transactionId, amount, accountNumber, transactionTime, merchantId, prediction, and probability.

- A StreamingQuery is created, using predictionStream.writeStream() to write the prediction results to the console. The output mode is set to "append", and the format is set to "console".

- The streaming query starts using query.start(), and the application waits for the query to terminate using query.awaitTermination().

This code demonstrates the basic structure of real-time fraud detection using Spark Streaming. You can further enhance it by incorporating additional data preprocessing, handling more complex schemas, and integrating with other systems to alert or take actions, based on the detected fraudulent transactions.

Having explored the potential of Java and big data technologies in real-world scenarios, such as log analysis, recommendation engines, and fraud detection, this chapter showcased the versatility and power of this combination to tackle a wide range of data-driven challenges.

Summary

In this chapter, we embarked on an exhilarating journey, exploring the realm of big data and how Java's prowess in concurrency and parallel processing empowers us to conquer its challenges. We began by unraveling the essence of big data, characterized by its immense volume, rapid velocity, and diverse variety – a domain where traditional tools often fall short.

As we ventured further, we discovered the power of Apache Hadoop and Apache Spark, two formidable allies in the world of distributed computing. These frameworks seamlessly integrate with Java, enabling us to harness the true potential of big data. We delved into the intricacies of this integration, learning how Java's concurrency features optimize big data workloads, resulting in unparalleled scalability and efficiency.

Throughout our journey, we placed a strong emphasis on the DataFrame API, which has become the de facto standard for data processing in Spark. We explored how DataFrames provide a more efficient, optimized, and user-friendly way to work with structured and semi-structured data compared to RDDs. We covered essential concepts such as transformatieons, actions, and SQL-like querying using DataFrames, enabling us to perform complex data manipulations and aggregations with ease.

To ensure a comprehensive understanding of Spark's capabilities, we delved into advanced topics such as the Catalyst optimizer, execution DAG, caching, and persistence techniques. We also discussed strategies to handle data skew and minimize data shuffling, which are critical for optimizing Spark's performance in real-world scenarios.

Our adventure led us through three captivating real-world scenarios – log analysis, recommendation systems, and fraud detection. In each of these scenarios, we showcased the immense potential of Java and big data technologies, leveraging the DataFrame API to solve complex data processing tasks efficiently.

Armed with the knowledge and tools acquired in this chapter, we stand ready to build robust and scalable big data applications using Java. We have gained a deep understanding of the core characteristics of big data, the limitations of traditional data processing approaches, and how Java's concurrency features and big data frameworks such as Hadoop and Spark enable us to overcome these challenges.

We are now equipped with the skills and confidence to tackle the ever-expanding world of big data. Our journey will continue in the next chapter, as we explore how Java's concurrency features can be harnessed for efficient and powerful machine learning tasks.

Questions

1. What are the core characteristics of big data?

    A. Speed, accuracy, and format

    B. Volume, velocity, and variety

    C. Complexity, consistency, and currency

    D. Density, diversity, and durability

2. Which component of Hadoop is primarily designed for storage?

    A. **Hadoop Distributed File System (HDFS)**

    B. **Yet Another Resource Negotiator (YARN)**

    C. MapReduce

    D. HBase

3. What is the primary advantage of using Spark over Hadoop for certain big data tasks?

    A. Spark is more cost-effective than Hadoop.

    B. Spark provides better data security than Hadoop.

    C. Spark offers faster in-memory data processing capabilities.

    D. Spark supports a wider variety of data formats than Hadoop.

4. Which of the following is NOT a true statement about Apache Spark?

    A. Spark can only process structured data.

    B. Spark allows for in-memory data processing.

    C. Spark supports real-time stream processing.

    D. Spark uses **Resilient Distributed Datasets (RDDs)** for fault-tolerant storage.

5. What is a key benefit of applying concurrency to big data tasks?

    A. It simplifies the code base for big data applications.

    B. It ensures data processing tasks are executed sequentially.

    C. It helps to break down large datasets into smaller, manageable chunks for processing.

    D. It reduces the storage requirements for big data.

7

Concurrency in Java for Machine Learning

The landscape of **machine learning** (**ML**) is rapidly evolving, with the ability to process vast amounts of data efficiently and in real time becoming increasingly crucial. Java, with its robust concurrency framework, emerges as a powerful tool for developers navigating the complexities of ML applications. This chapter delves into the synergistic potential of Java's concurrency mechanisms when applied to the unique challenges of ML, exploring how they can significantly enhance performance and scalability in ML workflows.

Throughout this chapter, we will provide a comprehensive understanding of Java's concurrency tools and how they align with the computational demands of ML. We'll explore practical examples and real-world case studies that illustrate the transformative impact of employing Java's concurrent programming paradigms in ML applications. From leveraging parallel streams for efficient data preprocessing to utilizing thread pools for concurrent model training, we'll showcase strategies to achieve scalable and efficient ML deployments.

Furthermore, we'll discuss best practices for thread management and reducing synchronization overhead, ensuring optimal performance and maintainability of ML systems built with Java. We'll also explore the exciting intersection of Java concurrency and generative AI, inspiring you to push the boundaries of what's possible in this emerging field.

By the end of this chapter, you'll be equipped with the knowledge and skills needed to harness the power of Java's concurrency in your ML projects. Whether you're a seasoned Java developer venturing into the world of ML or an ML practitioner looking to leverage Java's concurrency features, this chapter will provide you with insights and practical guidance to build faster, scalable, and more efficient ML applications.

So, let's dive in and unlock the potential of Java's concurrency in the realm of ML!

Technical requirements

You'll need to have the following software and dependencies set up in your development environment:

- **Java Development Kit (JDK)** 8 or later
- Apache Maven for dependency management
- An IDE of your choice (e.g., IntelliJ IDEA or Eclipse)

For detailed instructions on setting up **Deeplearning4j (DL4J)** dependencies in your Java project, please refer to the official DL4J documentation:

```
https://deeplearning4j.konduit.ai/
```

The code in this chapter can be found on GitHub:

```
https://github.com/PacktPublishing/Java-Concurrency-and-Parallelism
```

An overview of ML computational demands and Java concurrency alignment

ML tasks often involve processing massive datasets and performing complex computations, which can be highly time-consuming. Java's concurrency mechanisms enable the execution of multiple parts of these tasks in parallel, significantly speeding up the process and improving the efficiency of resource utilization.

Imagine working on a cutting-edge ML project that deals with terabytes of data and intricate models. The data preprocessing alone could take days, not to mention the time needed for training and inference. However, by leveraging Java's concurrency tools, such as threads, executors, and futures, you can harness the power of parallelism at various stages of your ML workflow, tackling these challenges head-on and achieving results faster than ever before.

The intersection of Java concurrency and ML demands

The intersection of Java concurrency mechanisms and the computational demands of modern ML applications presents a promising frontier. ML models, especially those involving large datasets and deep learning, require significant resources for data preprocessing, training, and inference. By leveraging Java's multithreading capabilities, parallel processing, and distributed computing frameworks, ML practitioners can tackle the growing complexity and scale of ML tasks. This synergy between Java concurrency and ML enables optimized resource utilization, accelerated model development, and high-performance solutions that keep pace with the increasing sophistication of ML algorithms and the relentless growth of data.

Parallel processing – the key to efficient ML workflows

The secret to efficient ML workflows lies in **parallel processing** – the ability to execute multiple tasks simultaneously. Java's concurrency features allow you to parallelize various stages of your ML pipeline, from data preprocessing to model training and inference.

For instance, by dividing the tasks of data cleaning, feature extraction, and normalization among multiple threads, you can significantly reduce the time spent on data preprocessing. Similarly, model training can be parallelized by distributing the workload across multiple cores or nodes, making the most of your computational resources.

Handling big data with ease

In the era of big data, ML models often require processing massive datasets that can be challenging to handle efficiently. Java's Fork/Join framework provides a powerful solution to this problem by enabling a divide-and-conquer approach. This framework allows you to split large datasets into smaller, more manageable subsets that can be processed in parallel across multiple cores or nodes.

With Java's data parallelism capabilities, handling terabytes of data becomes as manageable as processing kilobytes, unlocking new possibilities for ML applications.

An overview of key ML techniques

To understand how Java's concurrency features can benefit ML workflows, let's explore some prominent ML techniques and their computational demands.

Neural networks

Neural networks are essential components in many ML applications. They consist of layers of interconnected artificial neurons that process information and learn from data. The training process involves adjusting the weights of connections between neurons based on the difference between predicted and actual outputs. This process is typically done using algorithms such as backpropagation and gradient descent.

Java's concurrency features can significantly speed up neural network training by parallelizing data preprocessing and model updates. This is especially beneficial for large datasets. Once trained, neural networks can be used for making predictions on new data, and Java's concurrency features enable parallel inference on multiple data points, enhancing the efficiency of real-time applications.

For further study, you can explore these resources:

- *Wikipedia's Neural Network Overview* (`https://en.wikipedia.org/wiki/Neural_network`) provides a comprehensive introduction to both biological and artificial neural networks, covering their structure, function, and applications

- *Artificial Neural Networks* (https://www.analyticsvidhya.com/blog/2024/04/decoding-neural-networks/) offers detailed explanations of how neural networks work, including concepts such as forward propagation, backpropagation, and the differences between shallow and deep neural networks

These resources will give you a deeper understanding of neural networks and their applications in various fields.

Convolutional neural networks

Convolutional neural networks (**CNNs**) are a specialized type of neural network designed to handle grid-like data, such as images and videos. They are particularly effective for tasks such as image recognition, object detection, and segmentation. CNNs are composed of several types of layers:

- **Convolutional layers**: These layers apply convolution operations to the input data using filters or kernels, which help in detecting various features such as edges, textures, and shapes.

- **Pooling layers**: These layers perform downsampling operations, reducing the dimensionality of the data and thereby reducing computational load. Common types include max pooling and average pooling.

- **Fully connected layers**: After several convolutional and pooling layers, the final few layers are fully connected, similar to traditional neural networks, to produce the output.

Java's concurrency features can be effectively utilized to parallelize the training and inference processes of CNNs. This involves distributing the data preprocessing tasks and model computations across multiple threads or cores, leading to faster execution times and improved performance, especially when handling large datasets.

For further study, you can explore these resources:

- `Wikipedia's Convolutional Neural Network Overview` provides a comprehensive introduction to CNNs, explaining their structure, function, and applications in detail

- Analytics Vidhya's `CNN Tutorial` offers an intuitive guide to understanding how CNNs work, with practical examples and explanations of key concepts

These resources will provide you with a deeper understanding of CNNs and their applications in various fields.

Other relevant ML techniques

Here's a brief overview of other commonly used ML techniques, along with their relevance to Java concurrency:

- **Support vector machines (SVMs)**: These are powerful tools for classification tasks. They can benefit from parallel processing during training data preparation and model fitting. More information can be found at `https://scikit-learn.org/stable/modules/svm.html`.

- **Decision trees**: These are tree-like structures used for classification and regression. Java concurrency can be used for faster data splitting and decision tree construction during training. More information can be found at `https://en.wikipedia.org/wiki/Decision_tree`.

- **Random forests**: These are ensembles of decision trees, improving accuracy and robustness. Java concurrency can be leveraged for parallel training of individual decision trees. More information can be found at `https://scikit-learn.org/stable/modules/generated/sklearn.ensemble.RandomForestClassifier.html`.

These are just a few examples. Many other ML techniques can benefit from Java concurrency in various aspects of their workflows.

The intersection of Java's concurrency mechanisms and the computational demands of ML presents a powerful opportunity for developers to create efficient, scalable, and innovative ML applications. By leveraging parallel processing, handling big data with ease, and understanding the synergy between Java's concurrency features and various ML techniques, you can embark on a journey where the potential of ML is unleashed, and the future of data-driven solutions is shaped.

Case studies – real-world applications of Java concurrency in ML

The power of Java concurrency in enhancing ML workflows is best demonstrated through real-world applications. These case studies not only showcase the practical implementation but also highlight the transformative impact on performance and scalability. Next, we explore notable examples where Java's concurrency mechanisms have been leveraged to address complex ML challenges, complete with code demos to illustrate key concepts.

Case study 1 – Large-scale image processing for facial recognition

A leading security company aimed to improve the efficiency of its facial recognition system, tasked with processing millions of images daily. The challenge was to enhance the throughput of image preprocessing and feature extraction phases, which are critical for accurate recognition.

Solution

By employing Java's Fork/Join framework, the company parallelized the image processing workflow. This allowed for recursive task division, where each subtask processed a portion of the image dataset concurrently, significantly speeding up the feature extraction process.

Here is the code snippet:

```java
public class ImageFeatureExtractionTask extends RecursiveTask<Void> {
    private static final int THRESHOLD = 100;
// Define THRESHOLD here
    private List<Image> imageBatch;

    public ImageFeatureExtractionTask(
        List<Image> imageBatch) {
            this.imageBatch = imageBatch;
        }
    @Override
    protected Void compute() {
        if (imageBatch.size() > THRESHOLD) {
            List<ImageFeatureExtractionTask> subtasks =
            createSubtasks();
            for (ImageFeatureExtractionTask subtask :
                subtasks) {
                    subtask.fork();
                }
            } else {
                processBatch(imageBatch);
            }
            return null;
        }

    private List<ImageFeatureExtractionTask> createSubtasks() {
        List<ImageFeatureExtractionTask> subtasks = new ArrayList<>();

        // Assume we divide the imageBatch into two equal parts
        int mid = imageBatch.size() / 2;

        // Create new tasks for each half of the imageBatch
        ImageFeatureExtractionTask task1 = new
        ImageFeatureExtractionTask(
            imageBatch.subList(0, mid));
        ImageFeatureExtractionTask task2 = new
        ImageFeatureExtractionTask(
            imageBatch.subList(mid, imageBatch.size()));
```

```
        // Add the new tasks to the list of subtasks
        subtasks.add(task1);
        subtasks.add(task2);

        return subtasks;
    }
    private void processBatch(List<Image> batch) {
        // Perform feature extraction on the batch of images
    }
}
```

The provided code demonstrates the implementation of a task-based parallel processing approach using Java's Fork/Join framework for extracting features from a batch of images. Here's a description of the code:

- The `ImageFeatureExtractionTask` class extends `RecursiveTask<Void>`, indicating that it represents a task that can be divided into smaller subtasks and executed in parallel.

- The class has a constructor that takes a list of `Image` objects called `imageBatch`, representing the batch of images to process.

- The `compute()` method is the main entry point for the task. It checks whether the size of the `imageBatch` constructor exceeds a defined THRESHOLD value.

- If the `imageBatch` size is above the THRESHOLD value, the task divides itself into smaller subtasks using the `createSubtasks()` method. It creates two new `ImageFeatureExtractionTask` instances, each responsible for processing half of the `imageBatch`.

- The subtasks are then forked (executed asynchronously) using the `fork()` method, allowing them to run concurrently.

- If the `imageBatch` size is below the THRESHOLD value, the task directly processes the entire batch using the `processBatch()` method, which is assumed to perform the actual feature extraction on the images.

- The `createSubtasks()` method is responsible for dividing `imageBatch` into two equal parts and creating new `ImageFeatureExtractionTask` instances for each half. These subtasks are added to a list and returned.

- The `processBatch()` method is a placeholder for the actual feature extraction logic, which is not implemented in the provided code.

This code showcases a divide-and-conquer approach using the Fork/Join framework, where a large batch of images is recursively divided into smaller subtasks until a threshold is reached. Each subtask processes a portion of the images independently, allowing for parallel execution and potentially improving the overall performance of the feature extraction process.

Case study 2 – Real-time data processing for financial fraud detection

A financial services firm needed to enhance its fraud detection system, which analyzes vast streams of transactional data in real time. The goal was to minimize detection latency while handling peak load efficiently.

Solution

Utilizing Java's executors and futures, the firm implemented an asynchronous processing model. Each transaction was processed in a separate thread, allowing for concurrent analysis of incoming data streams.

Here's a simplified code example highlighting the use of executors and futures for concurrent transaction processing:

```java
public class FraudDetectionSystem {
    private ExecutorService executorService;
    public FraudDetectionSystem(int numThreads) {
        executorService = Executors.newFixedThreadPool(
            numThreads);
    }

    public Future<Boolean> analyzeTransaction(Transaction transaction)
    {
        return executorService.submit(() -> {
            // Here, add the logic to determine if the transaction is
fraudulent
            boolean isFraudulent = false;
// This should be replaced with actual fraud detection logic

            // Assuming a simple condition for demonstration, e.g.,
high amount indicates potential fraud
            if (transaction.getAmount() > 10000) {
                isFraudulent = true;
            }
            return isFraudulent;
        });
    }
    public void shutdown() {
        executorService.shutdown();
    }
}
```

The Transaction class, which is used in the code example, represents a financial transaction. It encapsulates the relevant information about a transaction, such as the transaction ID, amount, timestamp, and other necessary details. Here's a simple definition of the Transaction class:

```
public class Transaction {
    private String transactionId;
    private double amount;
    private long timestamp;

    // Constructor
    public Transaction(String transactionId, double amount, long
    timestamp) {
            this.transactionId = transactionId;
            this.amount = amount;
            this.timestamp = timestamp;
        }

    // Getters and setters
    // ...
}
```

Here's a description of the code:

- The FraudDetectionSystem class represents the fraud detection system. It utilizes an ExecutorService to manage a thread pool for concurrent transaction processing.

- The analyzeTransaction() method submits a task to the ExecutorService to perform fraud detection analysis on a transaction. It returns a Future<Boolean> representing the asynchronous result of the analysis.

- The shutdown() method is used to gracefully shut down the ExecutorService when it is no longer needed.

- The Transaction class represents a financial transaction, containing relevant data fields such as the transaction ID and amount. Additional fields can be added based on the specific requirements of the fraud detection system.

To use FraudDetectionSystem, you can create an instance with the desired number of threads and submit transactions for analysis:

```
FraudDetectionSystem fraudDetectionSystem = new
FraudDetectionSystem(10);
// Create a sample transaction with a specific amount
Transaction transaction = new Transaction(15000);

  // Submit the transaction for analysis
```

```
Future<Boolean> resultFuture = fraudDetectionSystem.
analyzeTransaction(transaction);
try {
    // Perform other tasks while the analysis is being performed
asynchronously
    // Retrieve the analysis result
    boolean isFraudulent = resultFuture.get();

    // Process the result
    System.out.println(
        "Is transaction fraudulent? " + isFraudulent);
        // Shutdown the fraud detection system when no longer needed
        fraudDetectionSystem.shutdown();
            } catch (Exception e) {
                e.printStackTrace();
            }
```

This code creates a FraudDetectionSystem instance with a thread pool of 10 threads, creates a sample Transaction object, and submits it for asynchronous analysis using the analyzeTransaction() method. The method returns a Future<Boolean> representing the future result of the analysis.

These case studies underscore the vital role of Java concurrency in addressing the scalability and performance challenges inherent in ML workflows. By parallelizing tasks and employing asynchronous processing, organizations can achieve remarkable improvements in efficiency and responsiveness, paving the way for innovation and advancement in ML applications.

Java's tools for parallel processing in ML workflows

Parallel processing has become a cornerstone for ML workflows, enabling the handling of complex computations and large datasets with increased efficiency. Java, with its robust ecosystem, offers a variety of libraries and frameworks designed to support and enhance ML development through parallel processing. This section explores the pivotal role of these tools, with a focus on DL4J for neural networks and Java's concurrency utilities for data processing.

DL4J – pioneering neural networks in Java

DL4J is a powerful open source library for building and training neural networks in Java. It provides a high-level API for defining and configuring neural network architectures, making it easier for Java developers to incorporate deep learning into their applications.

One of the key advantages of DL4J is its ability to leverage Java's concurrency features for efficient training of neural networks. DL4J is designed to take advantage of parallel processing and distributed computing, allowing it to handle large-scale datasets and complex network architectures.

DL4J achieves efficient training through several concurrency techniques:

- **Parallel processing**: DL4J can distribute the training workload across multiple threads or cores, enabling parallel processing of data and model updates. This is particularly useful when training on large datasets or when using complex network architectures.

- **Distributed training**: DL4J supports distributed training across multiple machines or nodes in a cluster. By leveraging frameworks such as Apache Spark or Hadoop, DL4J can scale out the training process to handle massive datasets and accelerate training times.

- **GPU acceleration**: DL4J seamlessly integrates with popular GPU libraries such as CUDA and cuDNN, allowing it to utilize the parallel processing power of GPUs for faster training. This can significantly speed up the training process, especially for computationally intensive tasks such as image recognition or **natural language processing (NLP)**.

- **Asynchronous model updates**: DL4J employs asynchronous model updates, where multiple threads can simultaneously update the model parameters without strict synchronization. This approach reduces the overhead of synchronization and allows for more efficient utilization of computational resources.

By leveraging these concurrency techniques, DL4J enables Java developers to build and train neural networks efficiently, even when dealing with large-scale datasets and complex architectures. The library abstracts away many of the low-level details of concurrency and distributed computing, providing a high-level API that focuses on defining and training neural networks.

To get started with DL4J, let's take a look at a code snippet that demonstrates how to create and train a simple feedforward neural network for classification using the Iris dataset:

```
public class IrisClassification {
    public static void main(String[] args) throws IOException {
        // Load the Iris dataset
        DataSetIterator irisIter = new IrisDataSetIterator(
            150, 150);

        // Build the neural network
        MultiLayerConfiguration conf = new NeuralNetConfiguration.
        Builder()
            .updater(new Adam(0.01))
            .list()
            .layer(new DenseLayer.Builder().nIn(4).nOut(
                10).activation(Activation.RELU).build())
            .layer(new OutputLayer.Builder(
                LossFunctions.LossFunction.NEGATIVELOGLIKELIHOOD)
                .activation(Activation.SOFTMAX).nIn(
                    10).nOut(3).build())
```

```
                    .build();

        MultiLayerNetwork model = new MultiLayerNetwork(
            conf);
        model.init();
        model.setListeners(new ScoreIterationListener(10));

        // Train the model
        model.fit(irisIter);

        // Evaluate the model
        Evaluation eval = model.evaluate(irisIter);
        System.out.println(eval.stats());

        // Save the model
        ModelSerializer.writeModel(model, new File(
            "iris-model.zip"), true);
    }
}
```

To compile and run this code, make sure you have the following dependencies in your project's pom.xml file:

```xml
<dependencies>
    <dependency>
        <groupId>org.deeplearning4j</groupId>
        <artifactId>deeplearning4j-core</artifactId>
        <version>1.0.0-beta7</version>
    </dependency>
    <dependency>
        <groupId>org.nd4j</groupId>
        <artifactId>nd4j-native-platform</artifactId>
        <version>1.0.0-beta7</version>
    </dependency>
</dependencies>
```

This code demonstrates a complete workflow for building, training, and evaluating a neural network for classifying the Iris dataset using DL4J. It involves configuring a neural network, training it on the dataset, evaluating its performance, and saving the model for future use.

Here is the code description:

- **Load the Iris dataset**: `IrisDataSetIterator` is a utility class (likely custom-built or provided by DL4J) to load the famous Iris flower dataset and iterate over it in batches. The dataset consists of 150 samples, with each sample having 4 features (sepal length, sepal width, petal length, and petal width) and a label indicating the species.

- **Build the neural network**: `NeuralNetConfiguration.Builder ()` sets up the network's architecture and training parameters:

  - `updater(new Adam(0.01))`: Uses the Adam optimization algorithm for efficient learning, with a learning rate of `0.01`.

  - `list ()`: Indicates we're creating a multilayer (feedforward) neural network.

  - `layer(new DenseLayer...)`: Adds a hidden layer with 10 neurons, using the **rectified linear unit (ReLU)** activation function. ReLU is a common choice for hidden layers due to its computational efficiency and effectiveness in preventing vanishing gradients.

  - `layer(new OutputLayer...)`: Adds the output layer with three neurons (one for each iris species) and the **softmax** activation function. Softmax converts the raw outputs into probabilities, ensuring they sum to 1 and are suitable for classification tasks. The loss function is set to `NEGATIVELOGLIKELIHOOD`, which is a standard choice for multi-class classification.

- **Initialize and train the model**:

  - `MultiLayerNetwork model = new MultiLayerNetwork(conf)`: Creates the network based on the configuration.

  - `model.init ()`: Initializes the network's parameters (weights and biases).

  - `model.setListeners(new ScoreIterationListener(10))`: Attaches a listener to print the score every 10 iterations during training. This helps you monitor progress.

  - `model.fit(irisIter)`: Trains the model on the Iris dataset. The model learns to adjust its internal parameters to minimize the loss function and accurately predict iris species.

- **Evaluate the model**:

  - `Evaluation eval = model.evaluate(irisIter)`: Evaluates the model's performance on the Iris dataset (or a separate test set if you had one).

  - `System.out.println(eval.stats())`: Prints out a comprehensive evaluation report, including accuracy, precision, recall, F1 score, and so on.

- **Save the model:**

  - `ModelSerializer.writeModel(model, new File("iris-model.zip"), true)`: Saves the trained model in a `.zip` file. This allows you to reuse it for predictions later without retraining.

  - The `iris-model.zip` file encapsulates both the learned parameters (weights and biases) of the trained ML model, crucial for accurate predictions, and the model's configuration, including its architecture, layer types, activation functions, and hyperparameters. This comprehensive storage mechanism ensures the model can be seamlessly reloaded and employed for future predictions, eliminating the need for retraining.

This standard Java class can be executed directly from an IDE, packaged as a JAR file using `mvn clean package`, and can be run with Java JAR or deployed to a cloud platform.

Prior to commencing model training, it's advisable to preprocess the input data. Standardizing or normalizing the features can significantly enhance the model's performance. Additionally, experimenting with various hyperparameters such as learning rates, layer sizes, and activation functions is crucial for discovering the optimal configuration. Implementing regularization techniques, such as dropout or L2 regularization, helps prevent overfitting. Finally, utilizing cross-validation provides a more accurate evaluation of the model's effectiveness on new, unseen data.

This example provides a starting point for creating and training a basic neural network using DL4J. For more detailed information, refer to the `DL4J documentation`. This comprehensive resource provides in-depth explanations, tutorials, and guidelines for configuring and working with neural networks using the DL4J framework. You can explore various sections of the documentation to gain a deeper understanding of the available features and best practices.

Java thread pools for concurrent data processing

Java's built-in thread pools provide a convenient and efficient way to handle concurrent data processing in ML workflows. Thread pools allow developers to create a fixed number of worker threads that can execute tasks concurrently, optimizing resource utilization and minimizing the overhead of thread creation and destruction.

In the context of ML, thread pools can be leveraged for various data processing tasks, such as data preprocessing, feature extraction, and model evaluation. By dividing the workload into smaller tasks and submitting them to a thread pool, developers can achieve parallel processing and significantly reduce the overall execution time.

Java's concurrency API, particularly the `ExecutorService` interface and `ForkJoinPool` classes, provide high-level abstractions for managing thread pools. `ExecutorService` allows developers to submit tasks to a thread pool and retrieve the results asynchronously using `Future` objects. `ForkJoinPool`, on the other hand, is specifically designed for divide-and-conquer algorithms, where a large task is recursively divided into smaller subtasks until a certain threshold is reached.

Let's consider a practical example of using Java thread pools for parallel feature extraction in an ML workflow. Suppose we have a large dataset of images, and we want to extract features from each image using a pre-trained CNN model. CNNs are a type of deep learning neural network particularly well-suited for analyzing images and videos. We can leverage a thread pool to process multiple images concurrently, improving the overall performance.

Here is the code snippet:

```java
// Define the CNNModel class
class CNNModel {
    // Placeholder method for feature extraction
    public float[] extractFeatures(Image image) {
    // Implement the actual feature extraction logic here
        // For demonstration purposes, return a dummy feature array
        return new float[]{0.1f, 0.2f, 0.3f};
    }
}

// Define the Image class
class Image {
    // Placeholder class representing an image
}
public class ImageFeatureExtractor {
    private ExecutorService executorService;
    private CNNModel cnnModel;
    public ImageFeatureExtractor(
        int numThreads, CNNModel cnnModel) {
            this.executorService = Executors. newFixedThreadPool(
                numThreads);
            this.cnnModel = cnnModel;
    }
    public List<float[]> extractFeatures(List<Image> images) {
        List<Future<float[]>> futures = new ArrayList<>();
        for (Image image : images) {
            futures.add(executorService.submit(() ->
                cnnModel.extractFeatures(image)));
        }
        List<float[]> features = new ArrayList<>();
        for (Future<float[]> future : futures) {
            try {
                features.add(future.get());
            } catch (Exception e) {
                // Handle exceptions
            }
```

```
        }
        return features;
    }
    public void shutdown() {
        executorService.shutdown();
    }
}
```

In this code snippet, we define three classes:

- The CNNModel class contains an extractFeatures(Image image) method that, in a real scenario, would implement the logic for extracting features from an image. Here, it returns a dummy array of floats representing extracted features for demonstration purposes.

- The Image class serves as a placeholder representing an image. In practice, this class would include properties and methods relevant to handling image data.

- The ImageFeatureExtractor class is designed to manage the concurrent feature extraction process:

    - Constructor: Accepts the number of threads (numThreads) and an instance of CNNModel. It initializes ExecutorService with a fixed thread pool size based on numThreads, which controls the concurrency level of the feature extraction process.

    - extractFeatures(List<Image> images): Takes a list of Image objects and uses the executor service to submit feature extraction tasks concurrently for each image. Each task calls the extractFeatures() method of the CNNModel on a separate thread. The method collects the futures returned by these tasks into a list and waits for all futures to complete. It then retrieves the extracted features from each future and compiles them into a list of float arrays.

    - shutdown(): Shuts down the executor service, stopping any further task submissions and allowing the application to terminate cleanly.

This approach demonstrates the efficient handling of potentially CPU-intensive feature extraction tasks by distributing them across multiple threads, thus leveraging modern multi-core processors to speed up the processing of large sets of images.

Practical examples – utilizing Java's parallel streams for feature extraction and data normalization

Let's dive into some practical examples of utilizing Java's parallel streams for feature extraction and data normalization in the context of ML workflows.

Example 1 – Feature extraction using parallel streams

Suppose we have a dataset of text documents, and we want to extract features from these documents using the **Term Frequency-Inverse Document Frequency (TF-IDF)** technique. We can leverage Java's parallel streams to process the documents concurrently and calculate the TF-IDF scores efficiently.

Here is the Document class, which represents a document with textual content:

```java
class Document {
    private String content;
    // Constructor, getters, and setters
    public Document(String content) {
        this.content = content;
    }
    public String getContent() {
        return content;
    }
}
```

Here is the `FeatureExtractor` class, which processes a list of documents to extract TF-IDF features for each document:

```java
public class FeatureExtractor {
    private List<Document> documents;
    public FeatureExtractor(List<Document> documents) {
        this.documents = documents;
    }
    public List<Double[]> extractTfIdfFeatures() {
        return documents.parallelStream()
            .map(document -> {
                String[] words = document.getContent(
                    ).toLowerCase().split("\\s+");
                return Arrays.stream(words)
                    .distinct()
                    .mapToDouble(word -> calculateTfIdf(
                        word, document))
                    .boxed()
                    .toArray(Double[]::new);
            })
            .collect(Collectors.toList());
    }
    private double calculateTfIdf(String word, Document document) {
        double tf = calculateTermFrequency(word, document);
        double idf = calculateInverseDocumentFrequency(
            word);
```

```
        return tf * idf;
    }
    private double calculateTermFrequency(String word, Document
    document) {
        String[] words = document.getContent().toLowerCase(
            ).split("\\s+");
        long termCount = Arrays.stream(words)
            .filter(w -> w.equals(word))
            .count();
        return (double) termCount / words.length;
    }
    private double calculateInverseDocumentFrequency(String word) {
        long documentCount = documents.stream()
            .filter(document -> document.getContent(
                ).toLowerCase().contains(word))
            .count();
        return Math.log((double) documents.size() / (
            documentCount + 1));
    }
}
}
```

Here's the code breakdown:

- The FeatureExtractor class extracts TF-IDF features from a list of Document objects using parallel streams

- The extractTfIdfFeatures() method does the following:

  - Processes the documents concurrently using parallelStream()

  - Calculates the TF-IDF scores for each word in each document

  - Returns the results as a list of Double[] arrays

- The calculateTermFrequency() and calculateInverseDocumentFrequency() methods are helper methods:

  - calculateTermFrequency() computes the term frequency of a word in a document

  - calculateInverseDocumentFrequency() computes the inverse document frequency of a word

- The Document class represents a document with its content

- Parallel streams are utilized to efficiently parallelize the feature extraction process

- Multi-core processors are leveraged to speed up the computation of TF-IDF scores for large datasets

Integrating this feature extraction code into a larger ML pipeline is straightforward. You can use the `FeatureExtractor` class as a preprocessing step before feeding the data into your ML model.

Here's an example of how you can integrate it into a pipeline:

```
// Assuming you have a list of documents
List<Document> documents = // ... load or generate documents

// Create an instance of FeatureExtractor
FeatureExtractor extractor = new FeatureExtractor(documents);

// Extract the TF-IDF features
List<Double[]> tfidfFeatures = extractor.extractTfIdfFeatures();

// Use the extracted features for further processing or model training
// ...
```

By extracting the TF-IDF features using the `FeatureExtractor` class, you can obtain a numerical representation of the documents, which can be used as input features for various ML tasks such as document classification, clustering, or similarity analysis.

Example 2 – Data normalization using parallel streams

Data normalization is a common preprocessing step in ML to scale the features to a common range. Let's say we have a dataset of numerical features, and we want to normalize each feature using the min-max scaling technique. We can utilize parallel streams to normalize the features concurrently.

Here is the code snippet:

```
import java.util.Arrays;
import java.util.stream.IntStream;
public class DataNormalizer {
    private double[][] data;
    public DataNormalizer(double[][] data) {
        this.data = data;
    }
    public double[][] normalizeData() {
        int numFeatures = data[0].length;
        return IntStream.range(0, numFeatures)
            .parallel()
            .mapToObj(featureIndex -> {
                double[] featureValues = getFeatureValues(
                    featureIndex);
                double minValue = Arrays.stream(
                    featureValues).min().orElse(0.0);
```

```
                double maxValue = Arrays.stream(
                    featureValues).max().orElse(1.0);
                return normalize(featureValues, minValue,
                    maxValue);
            })
            .toArray(double[][]::new);
    }
    private double[] getFeatureValues(int featureIndex) {
        return Arrays.stream(data)
                .mapToDouble(row -> row[featureIndex])
                .toArray();
    }
    private double[] normalize(double[] values, double
        minValue, double maxValue) {
            return Arrays.stream(values)
                .map(value -> (value - minValue) / (
                    maxValue - minValue))
                .toArray();
    }
}
```

The main components of the DataNormalizer class are as follows:

- The normalizeData() method uses IntStream.range(0, numFeatures).parallel() to process each feature concurrently

- For each feature, the mapToObj() operation is applied to perform the following steps:

  - Retrieve the feature values using the getFeatureValues() method

  - Calculate the minimum and maximum values of the feature using Arrays.stream(featureValues).min() and Arrays.stream(featureValues).max(), respectively

  - Normalize the feature values using the normalize() method, which applies the min-max scaling formula

- The normalized feature values are collected into a 2D array using toArray(double[][]::new)

- The getFeatureValues() and normalize() methods are helper methods used to retrieve the values of a specific feature and apply the min-max scaling formula, respectively

Integrating data normalization into an ML pipeline is crucial to ensure that all features are on a similar scale, which can improve the performance and convergence of many ML algorithms. Here's an example of how you can use the `DataNormalizer` class in a pipeline:

```
// Assuming you have a 2D array of raw data
double[][] rawData = // ... load or generate raw data

// Create an instance of DataNormalizer
DataNormalizer normalizer = new DataNormalizer(rawData);

// Normalize the data
double[][] normalizedData = normalizer.normalizeData();

// Use the normalized data for further processing or model training
// ...
```

By normalizing the raw data using the `DataNormalizer` class, you ensure that all features are scaled to a common range, typically between 0 and 1. This preprocessing step can significantly improve the performance and stability of many ML algorithms, especially those based on gradient descent optimization.

These examples demonstrate how you can easily integrate the `FeatureExtractor` and `DataNormalizer` classes into a larger ML pipeline. By using these classes as preprocessing steps, you can efficiently perform feature extraction and data normalization in parallel, leveraging the power of Java's parallel streams. The resulting features and normalized data can then be used as input for subsequent steps in your ML pipeline, such as model training, evaluation, and prediction.

As we conclude this section, we have explored a variety of Java tools that significantly enhance the parallel processing capabilities essential for modern ML workflows. Utilizing Java's robust parallel streams, executors, and the Fork/Join framework, we've seen how to tackle complex, data-intensive tasks more efficiently. These tools not only facilitate faster data processing and model training but also enable scalable ML deployments capable of handling the increasing size and complexity of datasets.

Understanding and implementing these concurrency tools is crucial because they allow ML practitioners to optimize computational resources, thereby reducing execution times and improving application performance. This knowledge ensures that your ML solutions can keep pace with the demands of ever-growing data volumes and complexity.

Next, we will transition from the foundational concepts and practical applications of Java's concurrency tools to a discussion on achieving scalable ML deployments using Java's concurrency APIs. In this upcoming section, we'll delve deeper into strategic implementations that enhance the scalability and efficiency of ML systems using these powerful concurrency tools.

Achieving scalable ML deployments using Java's concurrency APIs

Before delving into the specific strategies for leveraging Java's concurrency APIs in ML deployments, it's essential to understand the critical role these APIs play in the modern ML landscape. ML tasks often require processing vast amounts of data and performing complex computations that can be highly time-consuming. Java's concurrency APIs enable the execution of multiple parts of these tasks in parallel, significantly speeding up the process and improving the efficiency of resource utilization. This capability is indispensable for scaling ML deployments, allowing them to handle larger datasets and more sophisticated models without compromising performance.

To achieve scalable ML deployments using Java's concurrency APIs, we can consider the following strategies and techniques:

- **Data preprocessing**: Leverage parallelism to preprocess large datasets efficiently. Utilize Java's parallel streams or custom thread pools to distribute data preprocessing tasks across multiple threads.

- **Feature extraction**: Employ concurrent techniques to extract features from raw data in parallel. Utilize Java's concurrency APIs to parallelize feature extraction tasks, enabling faster processing of high-dimensional data.

- **Model training**: Implement concurrent model training approaches to accelerate the learning process. Utilize multithreading or distributed computing frameworks to train models in parallel, leveraging the available computational resources.

- **Model evaluation**: Perform model evaluation and validation concurrently to speed up the assessment process. Utilize Java's concurrency primitives to parallelize evaluation tasks, such as cross-validation or hyperparameter tuning.

- **Pipeline parallelism**: Implement a pipeline where different stages of the ML model training (e.g., data loading, preprocessing, and training) can be executed in parallel. Each stage of the pipeline can run concurrently on separate threads, reducing overall processing time.

Best practices for thread management and reducing synchronization overhead

When dealing with Java concurrency, effective thread management and reducing synchronization overhead are crucial for optimizing performance and maintaining robust application behavior.

Here are some best practices that can help achieve these objectives:

- **Use concurrency utilities instead of low-level synchronization**:

  - **Avoid synchronized overhead, where possible**: Utilize high-level concurrency utilities from the `java.util.concurrent` package such as `ConcurrentHashMap`, `Semaphore`, and `ReentrantLock`, which offer extended capabilities and better performance compared to traditional synchronized methods and blocks.

  - **Leverage thread-safe collections:** Replace synchronized wrappers around standard collections with concurrent collections. For example, use `ConcurrentHashMap` instead of `Collections.synchronizedMap(new HashMap<...>())`.

- **Minimize lock contention**:

  - **Reduce lock scope**: Acquire locks for the shortest possible duration and release them as soon as the critical section is executed, to minimize the time other threads wait for the lock.

  - **Use fine-grained locks**: Instead of using a single lock for a shared object, use multiple locks to guard different parts of the object if they are independent of each other.

  - **Opt for ReadWriteLock when applicable**: When read operations greatly outnumber write operations, `ReadWriteLock` can offer better throughput by allowing multiple threads to read the data concurrently while still ensuring mutual exclusion during writes.

- **Optimize task granularity**:

  - **Balance granularity and overhead**: Too fine a granularity can lead to higher overhead in terms of context switching and scheduling. Conversely, too coarse a granularity might lead to underutilization of CPU resources. Strike a balance based on the task and system capabilities.

  - **Use partitioning strategies**: In cases such as batch processing or data-parallel algorithms, partition the data into chunks that can be processed independently and concurrently, but are large enough to ensure that the overhead of thread management is justified by the performance gain.

- **Use asynchronous programming techniques**:

  - **Use CompletableFuture**: Asynchronous operations with `CompletableFuture` can help avoid blocking threads, allowing them to perform other tasks or to be returned to the thread pool, reducing the need for synchronization and the number of threads required.

  - **Employ event-driven architectures**: In scenarios such as I/O operations, use event-driven, non-blocking APIs to free up threads from waiting for operations to complete, thus enhancing scalability and reducing the need for synchronization.

- **Efficient use of thread pools:**

  - **Right-size thread pools:** Customize the number of threads in the pool based on the hardware capabilities and the nature of tasks. Use the `Executors` factory methods to create thread pools that match your application's specific needs.

  - **Avoid thread leakage:** Ensure that threads are properly returned to the pool after task completion. Watch out for tasks that can block indefinitely or hang, which can exhaust the thread pool.

  - **Monitor and tune performance:** Regular monitoring and tuning based on actual system performance and throughput can help in optimally configuring thread pools and concurrency settings.

- **Consider new concurrency features in Java:**

  - **Project Loom:** Stay informed about upcoming features such as Project Loom, which aims to introduce lightweight concurrency constructs such as fibers, offering a potential reduction in overhead compared to traditional threads.

Implementing these best practices allows for more efficient thread management, reduces the risks of deadlock and contention, and improves the overall scalability and responsiveness of Java applications in concurrent execution environments.

As we leverage Java's concurrency features to optimize ML deployments and implement best practices for efficient thread management, we stand at the forefront of a new era in AI development. In the next section, we will explore the exciting possibilities that arise when combining Java's robustness and scalability with the cutting-edge field of generative AI, opening up a world of opportunities for creating intelligent, creative, and interactive applications.

Generative AI and Java – a new frontier

Generative AI encompasses a set of technologies that enable machines to understand and generate content with minimal human intervention. This can include generating text, images, music, and other forms of media. The field is primarily dominated by ML and deep learning models.

Generative AI includes these key areas:

- **Generative models:** These are models that can generate new data instances that resemble the training data. Examples include **generative adversarial networks** (**GANs**), **variational autoencoders** (**VAEs**), and Transformer-based models such as **Generative Pre-trained Transformer** (**GPT**) and DALL-E.

- **Deep learning:** Most generative AI models are based on deep learning techniques that use neural networks with many layers. These models are trained using a large amount of data to generate new content.

- **NLP**: This is a pivotal area within AI that deals with the interaction between computers and humans through natural language. The field has seen a transformative impact through generative AI models, which can write texts, create summaries, translate languages, and more.

For Java developers, understanding and incorporating generative AI concepts can open up new possibilities in software development.

Some of the key areas where generative AI can be applied in Java development include the following:

- **Integration in Java applications**: Java developers can integrate generative AI models into their applications to enhance features such as chatbots, content generation, and customer interactions. Libraries such as *DL4J* or the *TensorFlow* Java API make it easier to implement these AI capabilities in a Java environment.

- **Automation and enhancement**: Generative AI can automate repetitive coding tasks, generate code snippets, and provide documentation, thereby increasing productivity. Tools such as *GitHub Copilot* are paving the way, and Java developers can benefit significantly from these advancements.

- **Custom model training**: While Java is not traditionally known for its AI capabilities, frameworks such as *DL4J* allow developers to train their custom models directly within Java. This can be particularly useful for businesses that operate on Java-heavy infrastructure and want to integrate AI without switching to Python.

- **Big data and AI**: Java continues to be a strong player in big data technologies (such as *Apache Hadoop* and *Apache Spark*). Integrating AI into these ecosystems can enhance data processing capabilities, making predictive analytics and data-driven decision-making more efficient.

As AI continues to evolve, its integration into Java environments is expected to grow, bringing new capabilities and transforming how traditional systems are developed and maintained. For Java developers, this represents a new frontier that holds immense potential for innovation and enhanced application functionalities.

Leveraging Java's concurrency model for efficient generative AI model training and inference

When training and deploying generative AI models, handling massive datasets and computationally intensive tasks efficiently is crucial. Java's concurrency model can be a powerful tool to optimize these processes, especially in environments where Java is already an integral part of the infrastructure.

Let us explore how Java's concurrency features can be utilized for enhancing generative AI model training and inference.

Parallel data processing – using the Stream API

For AI, particularly during data preprocessing, parallel streams can be used to perform operations such as filtering, mapping, and sorting concurrently, reducing the time needed for preparing datasets for training.

Here is an example:

```
List<Data> dataList = dataList.parallelStream()
.map(data -> preprocess(data))
.collect(Collectors.toList());
```

The code snippet uses *parallel stream* processing to preprocess a list of Data objects concurrently. It creates a parallel stream from dataList, applies the preprocess method to each object, and collects the preprocessed objects into a new list, which replaces the original dataList. This approach can potentially improve performance when dealing with large datasets by utilizing multiple threads for concurrent execution.

Concurrent model training – ExecutorService for asynchronous execution

You can use ExecutorService to manage a pool of threads and submit training tasks concurrently. This is particularly useful when training multiple models or performing cross-validation, as these tasks are inherently parallelizable.

Here is a code example:

```
ExecutorService executor = Executors.newFixedThreadPool(
    10); // Pool of 10 threads
for (int i = 0; i < models.size(); i++) {
    final int index = i;
    executor.submit(() -> trainModel(models.get(index)));
}
executor.shutdown();
executor.awaitTermination(1, TimeUnit.HOURS);
```

The code uses ExecutorService with a fixed thread pool of 10 to execute model training tasks concurrently. It iterates over a list of models, submitting each training task to ExecutorService using submit(). The shutdown() method is called to initiate the shutdown of ExecutorService, and awaitTermination() is used to wait for all tasks to be completed or until a specified timeout is reached. This approach allows for Concurrent model training parallel execution of model training tasks, potentially improving performance when dealing with multiple models or computationally intensive training.

Efficient asynchronous inference

`CompletableFuture` provides a non-blocking way to handle operations, which can be used to improve the response time of AI inference tasks. This is crucial in production environments to serve predictions quickly under high load.

Here is a code snippet:

```
CompletableFuture<Prediction> futurePrediction = CompletableFuture.
supplyAsync(() -> model.predict(input),
    executor);
// Continue other tasks
futurePrediction.thenAccept(prediction -> display(prediction));
```

The code uses `CompletableFuture` for asynchronous inference in AI systems. It creates a `CompletableFuture` that represents an asynchronous prediction computation using `supplyAsync`, which takes a (`model.predict(input)`) supplier function and an `Executor`. The code continues executing other tasks while the prediction is computed asynchronously. Once the prediction is complete, a callback registered with `thenAccept()` is invoked to handle the prediction result. This non-blocking approach improves response times in production environments under high load.

Reducing synchronization overhead – lock-free algorithms and data structures

Utilize concurrent data structures such as `ConcurrentHashMap` and atomic classes such as `AtomicInteger` to minimize the need for explicit synchronization. This reduces overhead and can enhance performance when multiple threads interact with shared resources during AI tasks.

Here is an example:

```
ConcurrentMap<String, Model> modelCache = new ConcurrentHashMap<>();
modelCache.putIfAbsent(modelName, loadModel());
```

The code uses `ConcurrentHashMap` to reduce synchronization overhead in AI tasks. `ConcurrentHashMap` is a thread-safe map that allows multiple threads to read and write simultaneously without explicit synchronization. The code attempts to add a new entry to `modelCache` using `putIfAbsent()`, which ensures that only one thread loads the model for a given `modelName`, while subsequent threads retrieve the existing model from the cache. By using thread-safe concurrent data structures, the code minimizes synchronization overhead and improves performance in multithreaded AI systems.

Case study – Java-based generative AI project illustrating concurrent data generation and processing

This case study outlines a hypothetical Java-based project that leverages the Java concurrency model to facilitate generative AI in concurrent data generation and processing. The project involves a generative model that creates synthetic data for training an ML model in a situation where real data is scarce or sensitive.

The objective is to generate synthetic data that mirrors real-world data characteristics and use this data to train a predictive model efficiently.

It includes the following key components.

Data generation module

This uses a GAN implemented in DL4J. The GAN learns from a limited dataset to produce new, synthetic data points.

The code is designed to produce synthetic data points using a GAN. GANs are a type of neural network architecture where two models (a generator and a discriminator) are trained simultaneously. The generator tries to produce data that is indistinguishable from real data, while the discriminator attempts to differentiate between real and generated data. In practical applications, once the generator is sufficiently trained, it can be used to generate new data points that mimic the characteristics of the original dataset.

Here is the code snippet:

```
ForkJoinPool customThreadPool = new ForkJoinPool(4); // 4 parallel
threads

List<DataPoint> syntheticData = customThreadPool.submit(() ->
    IntStream.rangeClosed(1, 1000).parallel().mapToObj(
        i -> g.generate()).collect(Collectors.toList())
).get();
```

Here's a breakdown of what each part of the code does:

- `ForkJoinPool` is instantiated with a parallelism level of 4, indicating that the pool will use four threads. This pool is designed to efficiently handle a large number of tasks by dividing them into smaller parts, processing them in parallel, and combining the results. The purpose here is to utilize multiple cores of the processor to enhance the performance of data-intensive tasks.

- The `customThreadPool.submit(...)` method submits a task to `ForkJoinPool`. The task is specified as a lambda expression that generates a list of synthetic data points. Inside the lambda, we see the following:

  - `IntStream.rangeClosed(1, 1000)`: This generates a sequential stream of integers from 1 to 1,000, where each integer represents an individual task of generating a data point.

  - `.parallel()`: This method converts the sequential stream into a parallel stream. When a stream is parallel, the operations on the stream (such as mapping and collecting) are performed in parallel across multiple threads.

  - `.mapToObj(i -> g.generate())`: For each integer in the stream (from 1 to 1000), the `mapToObj` function calls the `generate()` method on an instance of a generator, g. This method is assumed to be responsible for creating a new synthetic data point. The result is a stream of `DataPoint` objects.

  - `.collect(Collectors.toList())`: This terminal operation collects the results from the parallel stream into `List<DataPoint>`. The collection process is designed to handle the parallel stream correctly, aggregating the results from multiple threads into a single list.

- Since `submit()` returns a future, calling `get()` on this future blocks the current thread until all the synthetic data generation tasks are complete and the list is fully populated. The result is that `syntheticData` will hold all the generated data points after this line executes.

By utilizing `ForkJoinPool`, this code efficiently manages the workload across multiple processor cores, reducing the time required to generate a large dataset of synthetic data. This approach is particularly advantageous in scenarios where quick generation of large volumes of data is crucial, such as in training ML models where data augmentation is required to improve model robustness.

Data processing module

This applies various preprocessing techniques to both real and synthetic data to prepare it for training. Tasks such as normalization, scaling, and augmentation are applied to enhance the synthetic data.

The use of parallel streams is particularly advantageous for processing large datasets where the computational load can be distributed across multiple cores of a machine, thereby reducing the overall processing time. This is essential in ML projects where preprocessing can often become a bottleneck due to the volume and complexity of the data.

Here is a code snippet:

```
List<ProcessedData> processedData = syntheticData.parallelStream()
    .map(data -> preprocess(data))
    .collect(Collectors.toList());
```

This is the code breakdown:

- **Data source**: The code begins with a list named `syntheticData`, which is the source of the data to be processed. The `ProcessedData` type suggests that the list will hold processed versions of the original data.

- **Parallel processing**: The `.parallelStream()` method creates a parallel stream from the `syntheticData` list. This allows the processing to be divided across multiple processor cores if available, potentially speeding up the operation.

- **Mapping and preprocessing**: The `.map(data -> preprocess(data))` section applies a transformation to each element in the stream:

  - Each element (referred to as `data`) is passed into the `preprocess()` function. The `preprocess()` function (not shown in the snippet) is responsible for modifying or transforming the data in some way. The output of the `preprocess()` function becomes the new element in the resulting stream.

  - `.collect(Collectors.toList())` gathers the processed elements from the stream and places them into a new `List<ProcessedData>` called `processedData`.

This code snippet efficiently takes a list of data, applies preprocessing steps in parallel, and collects the results into a new list of processed data.

Model training module

The model training module leverages the power of DL4J to train a predictive model on processed data. To accelerate training, it breaks down the dataset into batches, allowing the model to be trained on multiple batches simultaneously using `ExecutorService`. Further efficiency is gained by employing `CompletableFuture` to update the model asynchronously after processing each batch; this prevents the main training process from being stalled.

Here is a code snippet:

```
public MultiLayerNetwork trainModel(List<DataPoint> batch) {
    // Configure a multi-layer neural network
    MultiLayerConfiguration conf = ...;
    MultiLayerNetwork model = new MultiLayerNetwork(conf);

    // Train the network on the data batch
    model.fit(batch);

    return model;
}
ExecutorService executorService = Executors.newFixedThreadPool(10);
List<Future<Model>> futures = new ArrayList<>();
```

```
for (List<DataPoint> batch : batches) {
    Future<Model> future = executorService.submit((() ->
        trainModel(batch));
    futures.add(future);
}
List<Model> models = futures.stream().map(
    Future::get).collect(Collectors.toList());
executorService.shutdown();
```

This is an explanation of the key components:

- `trainModel(List<DataPoint> batch)`: This function defines the core model training logic within the DL4J framework. It accepts a batch of data and returns a partially trained model.

- `ExecutorService executorService = Executors.newFixedThreadPool(10)`: A thread pool of 10 threads is created, allowing simultaneous training on up to 10 data batches for improved efficiency.

- `List<Future<Model>> futures = new ArrayList<>(); ... futures.add(future);`: This code snippet stores references to the asynchronous model training tasks. Each `Future<Model>` object represents a model being trained on a specific batch.

- `List<Model> models = futures.stream()...`: This line extracts the trained models from the futures list once they are ready.

- `executorService.shutdown();`: This signals the completion of the training process and releases resources associated with the thread pool.

This project demonstrates a well-structured approach to addressing the challenges of data scarcity in ML. By leveraging a GAN for synthetic data generation, coupled with efficient concurrent processing and a robust DL4J-based training module, it provides a scalable solution for training predictive models in real-world scenarios. The use of Java's concurrency features ensures optimal performance and resource utilization throughout the pipeline.

Summary

This chapter offered an in-depth exploration of harnessing Java's concurrency mechanisms to significantly enhance ML processes. By facilitating the simultaneous execution of multiple operations, Java effectively shortens the durations required for data preprocessing and model training, which are critical bottlenecks in ML workflows. The chapter presented practical examples and case studies that demonstrate how Java's concurrency capabilities can be applied to real-world ML applications. These examples vividly showcased the substantial improvements in performance and scalability that could be achieved.

Furthermore, the chapter outlined specific strategies, such as utilizing parallel streams and custom thread pools, to optimize large-scale data processing and perform complex computations efficiently. This discussion is crucial for developers aiming to enhance the scalability and performance of ML systems. Additionally, the text provided a detailed list of necessary tools and dependencies, accompanied by illustrative code examples. These resources are designed to assist developers in effectively integrating Java concurrency strategies into their ML projects.

The narrative also encouraged forward-thinking by suggesting the exploration of innovative applications at the intersection of Java concurrency and generative AI. This guidance opens up new possibilities for advancing technology using Java's robust features.

In the upcoming chapter, (*Chapter 8, Microservices in the Cloud and Java's Concurrency*), the discussion transitions to the application of Java's concurrency tools within microservices architectures. This chapter aims to further unpack how these capabilities can enhance scalability and responsiveness in cloud environments, pushing the boundaries of what can be achieved with Java in modern software development.

Questions

1. What is the primary benefit of integrating Java's concurrency mechanisms into ML workflows?

    A. To increase the programming complexity

    B. To enhance data security

    C. To optimize computational efficiency

    D. To simplify code documentation

2. Which Java tool is highlighted as crucial for processing large datasets in ML projects quickly?

    A. **Java Database Connectivity (JDBC)**

    B. **Java Virtual Machine (JVM)**

    C. Parallel Streams

    D. JavaFX

3. What role do custom thread pools play in Java concurrency for ML?

    A. They decrease the performance of ML models.

    B. They are used to manage database transactions only.

    C. They improve scalability and manage large-scale computations.

    D. They simplify the user interface design.

4. Which of the following is a suggested application of Java's concurrency in ML as discussed in this chapter?

    A. To handle multiple user interfaces simultaneously

    B. To perform data preprocessing and model training more efficiently

    C. To replace Python in scientific computing

    D. To manage client-server architecture only

5. What future direction does this chapter encourage exploring with Java concurrency?

    A. Decreasing the reliance on multithreading

    B. Combining Java concurrency with generative AI

    C. Phasing out older Java libraries

    D. Focusing exclusively on single-threaded applications

8

Microservices in the Cloud and Java's Concurrency

In today's fast-evolving digital landscape, microservices have emerged as a game-changing architectural style, enabling organizations to enhance scalability, flexibility, and deployment speeds. Java, with its robust ecosystem and powerful concurrency tools, stands at the forefront of this transformation, facilitating the seamless integration of microservices into cloud environments. This chapter delves into how Java's advanced features empower developers to build, deploy, and scale microservices more efficiently, making it an ideal choice for modern cloud-based applications.

By embracing Java-powered microservices, businesses can break down complex applications into manageable, independently deployable components that are tailored to specific business functions. This modularity not only accelerates development cycles but also improves system resilience and maintenance. Furthermore, Java's concurrency utilities play a crucial role in optimizing these services to handle vast scales of operations with ease, ensuring high availability and responsiveness across distributed systems.

In this chapter, we will cover the following key topics:

- **Principles of microservices in the cloud**: Understand the architectural shifts that make microservices a preferred model in modern software development, focusing on their dynamic integration with cloud platforms

- **Java's concurrency essentials**: Dive into Java's concurrency **Application Programming Interface (API)** to discover how these tools can dramatically improve the performance and scalability of your microservices

- **Concurrency patterns and techniques**: Learn about advanced patterns such as circuit breakers and event-driven communication, which are essential for maintaining high service availability and robust error handling

- **Best practices for microservices**: Explore strategic guidelines for deploying and scaling your microservices, ensuring they are optimized for the cloud environment

- **Hands-on design and implementation**: Apply what you've learned through practical case studies and real-world applications that demonstrate effective microservice design using Java

By the end of this chapter, you'll be well-equipped to leverage Java's advanced features for designing, deploying, and managing microservices that are not only scalable and efficient but also resilient and easy to maintain. Prepare to transform theoretical knowledge into practical skills that will advance your capabilities in developing cloud-native applications.

Technical requirements

frameworkframeworkFor detailed setup instructions on microservices frameworks, refer to their official documentation. This chapter will focus on using Spring Boot for the microservice example.

Here are the official documentation sites:

- **Spring Boot**: `https://spring.io/guides/gs/spring-boot`
- **Micronaut**: `https://guides.micronaut.io/latest/creating-your-first-micronaut-app-maven-java.html`
- **Quarkus**: `https://quarkus.io/get-started/`

The code in this chapter can be found on GitHub:

`https://github.com/PacktPublishing/Java-Concurrency-and-Parallelism`

Core principles of microservices – architectural benefits in cloud platforms

Microservices architecture offers a modern approach to software development, particularly in cloud environments. This architecture divides complex systems into smaller, independent services, providing flexibility and scalability. In this section, we will explore the foundational concepts of microservices architecture, its advantages over traditional designs, and how it integrates seamlessly into cloud ecosystems.

Foundational concepts – microservices architecture and its benefits in the cloud

Microservices architecture is a modern approach to software development that focuses on dividing complex systems into small, loosely coupled services. This architecture has proven especially beneficial in cloud environments, where its modularity and flexibility offer numerous advantages:

- **Modularity**: Microservices architecture divides a system into independent, loosely coupled services, each with its own functionality. This modularity allows for granular control over each component, making it easier to develop, test, deploy, and maintain individual services. It also enables teams to work on different services simultaneously, promoting parallel development.

- **Communication**: Microservices communicate via lightweight, standardized protocols such as HTTP/REST or gRPC, allowing for efficient interaction between services. These communication patterns, both synchronous and asynchronous, enable the development of scalable systems that can handle increasing workloads. Services expose well-defined APIs, which act as contracts for communication and facilitate loose coupling.

- **Independent deployment**: Microservices can be deployed, scaled, and updated independently without affecting the functionality of other services. This flexibility reduces downtime, enhances scalability, and simplifies the integration of new features. It also allows for the use of different technologies within each microservice.

- **Resilience and fault isolation**: Microservices architecture promotes resilience by isolating failures within individual services. If one service fails, it doesn't necessarily bring down the entire system. This fault isolation is achieved through design patterns such as circuit breakers and bulkheads.

- **Scalability**: Microservices can be scaled individually based on their specific resource requirements and demand patterns. This granular scalability allows for optimized resource allocation and cost-efficiency.

By leveraging the benefits of modularity, communication, independent deployment, resilience, and scalability, microservices architecture enables the development of flexible, maintainable, and scalable systems in cloud environments.

Let's look at the following image:

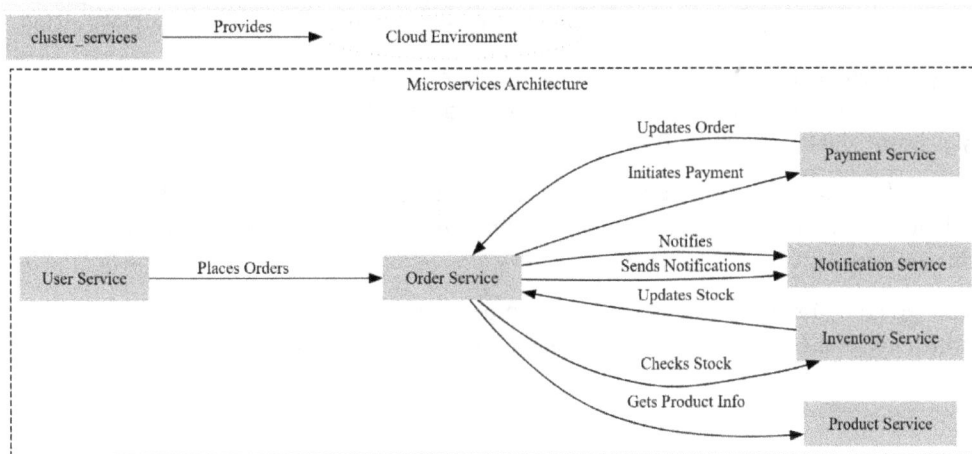

Figure 8.1: A microservices architecture

Figure 8.1 showcases a microservices architecture deployed in a cloud environment, highlighting the interplay between various services that cater to end user functionalities. It includes these key components:

- **User service**: Enables users to place orders

- **Order service**: Processes orders by interacting with the product service for product details and the inventory service for stock checks; if items are available, it proceeds to initiate payment via the payment service

- **Payment service**: Handles payment processing

- **Notification service**: Alerts users about their order status once the payment is confirmed, facilitated by notifications sent from the order service

- **Inventory service**: Updates stock levels post-order to ensure accurate inventory management

The architecture is built to be modular and scalable, with each service dedicated to a specific task and communicating through defined interfaces. This setup allows for flexibility in development and deployment, while the cloud environment supports scalability and robustness, adjusting to workload variations and ensuring system resilience.

Architectural comparison – differences between monolithic and microservices designs

Microservices architecture presents a modern alternative to traditional monolithic designs. However, each has its distinct benefits and drawbacks. Here we focus on contrasting these two architectural styles.

The details surrounding monolithic architecture are as follows:

- **Structure**: Monolithic designs are characterized by a single, unified codebase that houses all functionalities, including the **User Interface (UI)**, business logic, data storage, and processing.

- **Deployment**: In a monolithic architecture, the entire application is deployed as a single entity. Any changes or updates necessitate redeploying the whole application, potentially causing downtime and limiting flexibility.

- **Scalability**: Scaling monolithic applications can be cumbersome as it often involves scaling the whole system even if only specific functionalities require it. This can lead to inefficiency and the unnecessary use of resources.

- **Advantages**: Monolithic architectures are simpler in terms of development and deployment, which makes them ideal for smaller projects or for organizations that lack extensive resources.

Let's look at *Figure 8.2*:

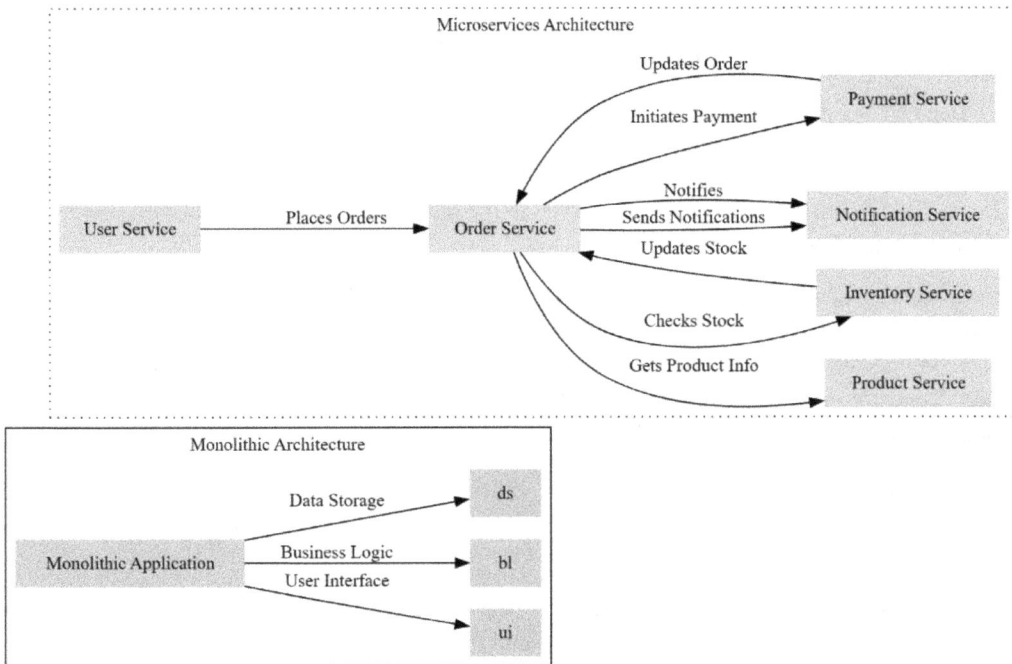

Microservices Architecture

Updates Order

Payment Service

Initiates Payment

Notifies

User Service ——— Places Orders → Order Service   Sends Notifications   Notification Service

Updates Stock

Inventory Service

Checks Stock

Gets Product Info

Product Service

Monolithic Architecture

Data Storage → ds

Monolithic Application   Business Logic → bl
User Interface

ui

Figure 8.2: A comparative overview of monolithic and microservices architectures

In the monolithic architecture, the application is constructed as a single unit, integrating UI, business logic, and data storage within one codebase. This approach is initially simpler for development and deployment but can become cumbersome as the application scales. Changes in one area can affect the entire system, and scaling necessitates deploying the whole application.

Conversely, microservices architecture decomposes the application into small, autonomous services, each handling a distinct function. These services interact via defined interfaces, supporting loose coupling and independent development, deployment, and scaling. While this structure enhances agility and scalability, it also brings complexity in coordinating services and increases operational demands.

Let's look at some comparison and transition considerations:

- **Complexity**: Monolithic architectures are simpler, but microservices increase in complexity with the need for detailed service orchestration. This can be mitigated with specific tools and frameworks.

- **Adoption**: Transitioning from monolithic to microservices architecture involves segmenting the existing codebase into separate services and establishing new patterns for inter-service communication.

The choice between these architectures should be guided by the application's requirements for scale and complexity, as well as the organization's strategic objectives. Often, businesses start with a monolithic design and evolve toward microservices as their operational needs grow.

Real-world examples – Netflix's evolution and Amazon's flexibility

Netflix's journey to microservices architecture began in 2009 when it faced rapid growth and scaling challenges. This transition was driven by the need to handle massive workloads and deliver content seamlessly to a global audience.

Netflix introduced adaptive streaming, allowing it to deliver video content with varying resolutions based on users' internet speeds and devices. Microservices architecture allowed services such as video encoding, content delivery, and user profiling to work independently yet cohesively. Netflix's microservices design also enabled the incorporation of a recommendation engine that suggests content based on users' viewing history, enhancing user engagement.

Over time, Netflix's architecture has allowed for the integration of additional services, such as multilingual support, regional content libraries, and offline viewing, demonstrating how microservices architecture accommodates evolving functionalities.

Amazon's transition to a microservices architecture has enabled it to scale its e-commerce platform efficiently, accommodating diverse functionalities and third-party services. Amazon's microservices architecture enables integration with various third-party services, including payment gateways, analytics platforms, and customer review systems, allowing it to accommodate diverse user needs and incorporate new features.

Amazon's microservices design allows for the integration of different technologies and tools, enabling it to evolve its technology stack smoothly. The independent deployment capabilities of Amazon's microservices architecture allow it to iterate rapidly, ensuring its e-commerce platform stays competitive and adapts to changing market demands.

Netflix and Amazon serve as powerful examples of how microservices architectures can be leveraged to address real-world challenges and drive business success. However, it's important to note that these benefits are not limited to these tech giants, and that businesses across various industries are embracing microservices to build scalable, flexible, and resilient applications.

In this section, we delved into microservices architecture in the cloud, discussing its core principles, such as modularity and scalability, and contrasting these with monolithic designs. We highlighted how companies such as Netflix and Amazon have leveraged microservices to enhance business outcomes through real-world case studies. Moving forward, we will examine Java's concurrency tools, which are essential for developing scalable and resilient microservices, and how they address the unique demands of cloud-based microservices architecture.

Essential Java concurrency tools for microservice management

Java's concurrency tools are vital for managing microservices in cloud environments, enabling efficient task management and **parallel processing**. In this section, we'll explore how these tools facilitate the development of responsive and scalable microservices architectures, integrating seamlessly into modern cloud ecosystems.

Concurrency tools – an exploration of Java's concurrency tools that are tailored for microservices

In Java, concurrency tools such as **ExecutorService**, **parallel streams**, **CompletableFuture**, and the **Fork/Join frameworkframework** play crucial roles in microservices architectures. ExecutorService manages pools of worker threads for efficient task execution, while parallel streams expedite data processing tasks by operating concurrently, thereby enhancing performance. CompletableFuture supports asynchronous programming, facilitating non-blocking tasks and inter-service communication. The Fork/Join frameworkframework helps divide and conquer large tasks by breaking them down into smaller, parallelizable units, thus optimizing execution times. These tools are foundational for developing scalable and efficient microservices, and we will further explore their practical applications in enhancing cloud-based microservices management in upcoming sections.

Task parallelism – using Java's concurrency mechanisms to manage microservices efficiently

Task parallelism is an essential aspect of managing microservices efficiently. Java's concurrency mechanisms offer practical solutions to distribute workloads, handle multiple tasks concurrently, and ensure responsive microservices.

Let's look at a code snippet:

```
// Concurrent processing tasks
List<Future<?>> tasks = new ArrayList<>();

tasks.add(executorService.submit(() ->
    inventoryService.deductProductQuantity(order)));
tasks.add(executorService.submit(() ->
    invoiceService.generateInvoice(order)));
tasks.add(executorService.submit(() ->
    emailService.sendOrderConfirmation(order)));

// Wait for all tasks to complete
for (Future<?> task : tasks) {
    try {
```

```
                    task.get(); // Wait for each task to finish
                } catch (Exception e) {
                    System.err.println("Error processing order: " + order.
                    getId());
                    throw e; // Rethrow exception after logging
            }
        }
    }
```

The provided code snippet demonstrates task parallelism using Java's concurrency mechanisms within a microservices architecture. Here's a concise analysis:

- **Asynchronous task management**:

  - A List<Future<?>> collects tasks submitted to an executorService for asynchronous execution. Tasks include inventory adjustments, invoice processing, and email confirmations related to an order.

  - Using Future<?> allows for tracking task outcomes and synchronizing them upon completion.

- **Executing and synchronizing tasks**: Tasks are submitted concurrently:

  - **Inventory management**: Adjusts stock levels

  - **Invoice generation**: Processes financial transactions

  - **Email confirmation**: Sends transaction feedback to the customer

  The code waits for each task to complete using get(), ensuring that all operations finish before proceeding. This synchronization is critical for maintaining consistency and reliability in service response.

- **Handling failures**: Exceptions from tasks are caught, logged, and rethrown, demonstrating robust error handling that allows the system to maintain high fault tolerance

In a microservices architecture, task parallelism enables different microservices to work concurrently, each focusing on its specific responsibility. This approach allows for efficient processing of requests and optimizes the overall performance of the system. By leveraging task parallelism, microservices can handle multiple tasks simultaneously, leading to faster response times and improved throughput.

However, task parallelism is just one aspect of achieving high performance in microservices. Another important concept is parallel processing, which involves breaking down a large task into smaller, independent subtasks that can be processed concurrently. In the next section, we will explore how parallel processing can be applied in microservices using Java's parallel streams and the Fork/Join framework.

Parallel processing for responsive microservices

Parallel processing is a powerful technique for improving the performance and responsiveness of microservices. By breaking down large tasks into smaller, independent subtasks and processing them concurrently, microservices can handle data-intensive operations and computationally expensive tasks more efficiently.

Java provides several tools for parallel processing, including parallel streams and the Fork/Join framework. Let's explore how these tools can be used in a microservices context.

Let's look at a parallel stream example:

```java
@Service
public class DataProcessingService {

    public List<Result> processData(List<Data> dataList) {
        return dataList.parallelStream()
            .map(this::processDataItem)
            .collect(Collectors.toList());
    }

    private Result processDataItem(Data dat{
        // Perform complex data processing logic
        // ...
    }
}
```

In this example, the `processData()` method receives a list of Data objects. Instead of processing the data sequentially, it uses the `parallelStream()` method to create a parallel stream. The `map()` operation is applied to each data item, invoking the `processDataItem()` method concurrently. Finally, the processed results are collected into a list.

By using parallel streams, the data processing can be distributed across multiple threads, allowing for faster execution and improved microservice responsiveness.

The Fork/Join framework is another powerful tool for parallel processing in Java. It is designed to efficiently handle recursive algorithms and divide-and-conquer scenarios.

Here's an example of using the Fork/Join framework in a microservice to perform a complex computation:

```java
@Service
public class ComplexComputationService {

    @Autowired
    private ForkJoinPool forkJoinPool;
```

```java
// Dependency injection of ForkJoinPool

    public Result computeResult(Problem problem) {
        return forkJoinPool.invoke(
            new ComplexComputationTask(problem));
    }


    private static class ComplexComputationTask extends
    RecursiveTask<Result> {
        private final Problem problem;
        public ComplexComputationTask(
        Problem problem) {
            this.problem = problem;
        }


        @Override
        protected Result compute() {
            if (problem.isSimple()) {
                return solveSimpleProblem(problem);
            } else {
            List<ComplexComputationTask> subtasks = problem.
            decompose()
                .map(ComplexComputationTask::new)
                .collect(Collectors.toList());

            subtasks.forEach(ForkJoinTask::fork);

            return subtasks.stream()
                .map(ForkJoinTask::join)
                .reduce(Result::combine)
                .orElse(Result.EMPTY);
            }
        }

        private Result solveSimpleProblem(Problem problem){
            // Logic to solve a simple problem directly
            // Placeholder implementation:
            return new Result();
            // Replace with actual logic
        }
    }
}
```

In this example, the `ComplexComputationService` uses the Fork/Join framework to perform a complex computation. The `computeResult()` method receives a `Problem` object and submits a `ComplexComputationTask` to the `ForkJoinPool`.

The `ComplexComputationTask` extends `RecursiveTask` and implements the `compute()` method. If the problem is simple, it solves it directly. Otherwise, it decomposes the problem into smaller subtasks, forks them for parallel execution, and then joins the results using the `join()` method. The results are combined using the `reduce()` operation.

By utilizing the Fork/Join framework, the microservice can efficiently solve complex problems by recursively dividing them into smaller subproblems and processing them in parallel.

These examples demonstrate how parallel processing techniques, such as parallel streams and the Fork/Join framework, can be applied in microservices to achieve better performance and responsiveness. By leveraging the power of parallel processing, microservices can handle large-scale data processing and complex computations more efficiently, resulting in improved user experience and faster response times.

In this section, we examined Java's concurrency tools and their role in microservices. We discussed how thread pools, parallel streams, and the Fork/Join framework enhance microservice performance through task parallelism, improving throughput and responsiveness. While beneficial, Java's concurrency mechanisms also present challenges. Next, in the *Challenges and solutions in microservices concurrency* section, we will address common issues with concurrency in microservices and outline effective strategies and practices.

Challenges and solutions in microservices concurrency

Microservices architectures offer unparalleled flexibility and scalability for modern applications, yet their concurrent nature presents unique challenges. This section delves into critical aspects of microservices concurrency, exploring potential bottlenecks, strategies for ensuring data consistency, approaches to achieving resilience, and practical solutions to these challenges through Java's concurrency mechanisms.

Bottlenecks – diagnosing potential challenges in concurrent microservices architectures

The introduction of concurrency in microservices architectures often leads to challenges and potential bottlenecks. Efficiently identifying and resolving these bottlenecks is crucial for maintaining the performance and smooth operation of concurrent microservices. This section outlines tools and strategies for effectively diagnosing and mitigating these issues, with a focus on cloud-based utilities.

First, let us look at **API Gateway**.

API Gateway acts as the central hub for incoming requests. It manages the flow of traffic efficiently, ensuring smooth operation and preventing bottlenecks:

- **Request throttling**: Imposes rate limits on requests to prevent service overload and ensure consistent performance
- **Traffic routing**: Directs traffic efficiently to the appropriate services, distributing loads evenly and reducing coordination and communication bottlenecks
- **Caching**: By caching responses to frequently accessed endpoints, the gateway lessens the load on backend services and enhances response times
- **Metrics collection**: Collects critical metrics such as response times, error rates, and request volumes, which are crucial for identifying and addressing bottlenecks

Next, we will explore monitoring and logging tools.

These tools are vital for diagnosing and resolving bottlenecks in microservices architectures:

- **AWS CloudWatch**: This offers real-time monitoring and logging, enabling the tracking of metrics such as resource utilization and response times. Alarms can be configured to alert threshold breaches, helping promptly identify and address emerging bottlenecks.
- **Azure Monitor**: This provides comprehensive monitoring, alerting, and log analytics features, offering insights into potential contention points and communication delays.
- **Google Cloud Logging**: This captures logs from various microservices, offering insights into service interactions and identifying areas of latency or overhead. Log-based metrics help track specific bottleneck-inducing events.

These solutions enable ongoing tracking and analysis of performance metrics, revealing trends that can pinpoint bottlenecks. They also guide necessary architectural adjustments, such as implementing caching strategies, sharding databases, or modifying communication patterns to boost efficiency.

By integrating API Gateways with robust monitoring tools, microservices architectures can proactively diagnose and resolve bottlenecks, thus ensuring enhanced performance, scalability, and resilience. This integrated approach ensures that concurrency challenges are managed effectively, fostering a robust environment for microservices operation.

Consistency – ensuring data consistency and smooth inter-service communication

Ensuring consistency in microservices architecture, particularly given its distributed nature, is critical. This section delves into how distributed databases and message brokers are fundamental in achieving consistency across services.

We'll start with distributed databases. Selecting the right distributed databases such as Amazon RDS, Google Cloud SQL, and Azure Database for PostgreSQL is key. These services ensure transactional consistency and **Atomicity, Consistency, Isolation, Durability (ACID)** compliance, which is crucial for operations that require reliable data handling. They manage data integrity across microservices by ensuring complete transactions before committing, and if a transaction fails, it is fully rolled back to maintain consistency.

These databases enhance scalability with features such as read replicas and sharding. They support robust data replication across zones or regions for improved availability and disaster recovery. Fully managed solutions reduce operational overhead, allowing teams to focus on core functionalities. Alternatives such as Apache Cassandra and Google Cloud Spanner, while offering less stringent consistency, excel in scenarios needing high scalability and low-latency access across geographic regions.

Next, let's consider message brokers. Tools such as AWS SQS, Google Pub/Sub, Apache Kafka, and Azure Service Bus streamline inter-service communication by managing asynchronous message queues. They enhance consistency in the following ways:

- **Decoupling services**: These brokers allow services to operate independently, improving system uptime by maintaining functionality even when parts fail.

- **Reliable delivery**: They ensure that messages accurately reach intended services, supporting high-volume conditions. Kafka, for instance, is known for its durability, while Azure Service Bus offers reliability within its ecosystem.

- **Event-driven architecture support**: They aid services in dynamically responding to changes, essential for maintaining consistency across services reacting to the same events.

From a design perspective, the choice between using a **Relational Database Service (RDS)** or a message broker depends on the specific requirements of your application:

- **Use RDS** for transactional data needs requiring ACID properties, complex data relationships needing strong integrity, or centralized data management, as well as when complex queries are necessary for analytics

- **Use message brokers** for asynchronous communication needs, event-driven architectures, scalability under varying loads, efficient high-volume traffic handling, or complex workflow orchestration across multiple microservices

Often, the strengths of RDSs and message brokers complement each other in a microservices architecture, and they are not mutually exclusive. For example, you might use an RDS to manage transactional data integrity while using a message broker to handle events that result from changes in the data, thus combining reliable data management with reactive service orchestration. This approach leverages the strengths of both technologies to create a robust, scalable, and resilient architecture.

Let's look at *Figure 8.3*:

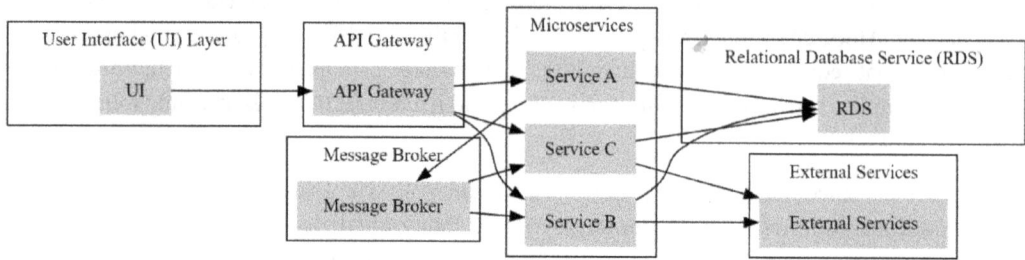

Figure 8.3: A microservice architecture with an API Gateway, message broker, and RDS

This figure depicts a microservice architecture design leveraging an RDS and a message broker to facilitate communication and data persistence.

Key components of this design include the following:

1. **UI layer**: Users interact here
2. **API Gateway**: Routes requests to microservices
3. **Microservices**: Handle specific functionalities
4. **RDS**: Stores data persistently (relational tables)
5. **Message broker**: Enables asynchronous communication between microservices

Here's how it works:

- A user initiates a request through UI.
- The API Gateway routes requests to the relevant microservice(s).
- The microservice interacts with the RDS or publishes a message to the message broker.
- Other microservices subscribed to the message broker receive and process the message.
- Data persistence might occur in an RDS.
- The microservice generates a response and sends it back to the user through the API gateway.

Its benefits are as follows:

- **Decoupling**: Microservices are loosely coupled and can scale independently
- **Data consistency**: Using RDS maintains data integrity across services

In essence, the message broker fosters asynchronous communication, while the RDS offers persistent storage.

Resilience – achieving system resilience and fault tolerance

Achieving robustness in microservices involves implementing the following strategies that enhance system resilience and fault tolerance:

- **Circuit breakers**: Utilizing tools such as Netflix Hystrix or Resilience4j, circuit breakers help manage service failures gracefully. They prevent cascading failures by halting the propagation of faults across services, thus maintaining system functionality during partial outages.

- **Load balancers**: Employing cloud-native load balancers assists in evenly distributing incoming traffic among available services. This not only enhances fault tolerance by avoiding overloading any single service but also helps in preventing bottlenecks, thus ensuring smoother operation and better response times across the system.

Circuit breakers and load balancers can work together to build resilient microservices. Load balancers distribute traffic, preventing bottlenecks and single points of failure. Circuit breakers provide additional protection by isolating failing services and preventing cascading failures.

This section has outlined the pivotal role of concurrency management in microservices, delving into the challenges and solutions related to potential bottlenecks and ensuring data consistency. We examined tools and strategies for mitigating issues such as traffic congestion and maintaining data integrity across distributed services, utilizing API gateways for traffic management, and utilizing message brokers for seamless inter-service communication. By integrating distributed databases and robust messaging systems, microservices can achieve enhanced performance, scalability, and resilience.

Moving forward, we will transition from theoretical concepts to practical applications. The upcoming section, *Hands-on – designing concurrent microservices in Java*, will provide a detailed guide on implementing these concurrency principles in Java.

Practical design and implementation – building effective Java microservices

This section dives into practical Java code examples, showcasing how to tackle concurrency challenges in a microservices architecture using cloud utilities and mechanisms such as message brokers, distributed databases, and circuit breakers.

Use case 1 – e-commerce application – processing orders

In an e-commerce application with a microservice for processing orders, concurrency challenges can arise due to multiple order requests trying to deduct from the same balance simultaneously, leading to inconsistencies and data integrity issues. To address these challenges, we can leverage the optimistic locking that is offered by most distributed databases.

Optimistic locking uses a version number associated with the user's account balance. When an update query is executed, it includes the expected version number. If the version in the database doesn't match the expected version, it indicates that another transaction might have modified the balance first, causing the update to fail. This prevents race conditions and ensures data consistency. Here are steps involved in the code snippet:

1. Open the pom.xml file in the project's root directory and add the following dependencies:

```xml
<dependency>
    <groupId>org.springframework.boot</groupId>
    <artifactId>spring-boot-starter-data-jpa</artifactId>
    <version>2.5.0</version>
</dependency>
<dependency>
    <groupId>org.springframework.boot</groupId>
    <artifactId>spring-boot-starter-web</artifactId>
    <version>2.5.0</version>
</dependency>
<dependency>
    <groupId>mysql</groupId>
    <artifactId>mysql-connector-java</artifactId>
    <version>8.0.23</version>
</dependency>
</dependencies>
```

2. Create the UserAccount class:

```java
Entity
public class UserAccount {

    @Id
    private Long id;
    private Long balance;

    @Version
    private Long version;
  // Getters and setters omitted for brevity
}
```

This code defines a UserAccount JPA entity. It has a @Version number (version) for optimistic locking, ensuring data consistency during updates.

3. Create the `AccountRepository` interface in the same package. This interface should extend `JpaRepository` and define the `deductBalance()` method:

```
public interface AccountRepository extends
JpaRepository<UserAccount, Long> @Modifying
    @Query("UPDATE UserAccount ua
        SET ua.balance = ua.balance - :amount,
            ua.version = ua.version + 1
        WHERE ua.id = :userId AND ua.version =
            :expectedVersion")
    int deductBalance(@Param("userId") Long userId,
    @Param("amount") Long amount,
    @Param("expectedVersion") Long expectedVersion);
}
```

4. Create the `AccountService` class in the same package and inject an `AccountRepository` instance into it:

```
@Repository
interface AccountRepository {
    UserAccount findById(
        Long userId) throws IllegalArgumentException;
        int deductBalance(Long userId, Long amount,
            Long version);
}

public class AccountService {
    private AccountRepository accountRepository;
    public AccountService(AccountRepository accountRepository) {
        this.accountRepository = accountRepository;
    }

    public void deductBalance(Long userId,
    Long amount) throws InsufficientBalanceException {
        UserAccount account = accountRepository. findById(
            userId);
        if (account == null) {
            throw new IllegalArgumentException(
                "User not found");
        }
        if (account.getBalance() < amount) {
            throw new InsufficientBalanceException(
                "Insufficient balance");
        }
        Long expectedVersion = account.getVersion();
```

```
        int rowsUpdated = accountRepository.
        deductBalance(userId, amount,
            expectedVersion);
        if (rowsUpdated != 1) {
            throw new OptimisticLockingException(
                "Balance update failed, retry");
        }
    }
}
```

This code snippet demonstrates the `deductBalance()` method within the `AccountService` class. The method first attempts to retrieve a user account by ID through the `accountRepository`. If the user account is not found, or if the account's balance is insufficient for the deduction, relevant exceptions are thrown to indicate these errors.

For optimistic locking, the method retrieves the current version number of the account being updated. It then invokes `accountRepository.deductBalance()` using the user ID, the amount to be deducted, and the expected version number. After this operation, the method checks the number of rows that were updated (`rowsUpdated`). A successful update — which is indicated by exactly one row being updated — allows the process to proceed. If the update affects either no rows or more than one row, it suggests that the account may have been concurrently modified by another process. In this case, an `OptimisticLockingException` is thrown, indicating that the update failed due to outdated data, prompting a retry to maintain data consistency.

5. Next, we can use a message broker for asynchronous communication:

```
@Component
public class MessageProducer {
    private final AmazonSQS sqsClient;
    private final String queueUrl;
    private final ObjectMapper objectMapper;
// ObjectMapper to serialize messages

    public MessageProducer(@Value("${
        aws.sqs.queueUrl}") String queueUrl) {
            this.sqsClient = AmazonSQSClientBuilder.standard().
            build();
            this.queueUrl = queueUrl;
            this.objectMapper = new ObjectMapper();
// Initialize ObjectMapper
    }
    //Sends a serialized message to the SQS queue.
    public String sendMessage(String string) {
        try {
            String messageBody = objectMapper.
```

```
                writeValueAsString(string);
    // Serialize message to JSON
                SendMessageRequest sendMsgRequest = new
                SendMessageRequest()
                        .withQueueUrl(queueUrl)
                        .withMessageBody(messageBody);

                SendMessageResult result = sqsClient.
                sendMessage(sendMsgRequest);
                return result.getMessageId();
    // Return the message ID on successful send
            } catch (Exception e) {
                System.err.println("Error sending message to SQS: "
                + e.getMessage());
                throw new RuntimeException("Failed to send message
                to SQS", e);
            }
        }
    }
}
```

6. Finally, we can create the OrderService class and inject a MessageProducer instance into it:

```
@Service
public class OrderService {

    @Autowired
    private MessageProducer messageProducer;

    public void processOrder(Order order) throws
    InsufficientBalanceException {
        // Validate order and deduct balance
        deductBalance(order.getId(),
            order.getAmount());

        // Publish order confirmation message
        OrderConfirmationMessage confirmation = new
        OrderConfirmationMessage(order.getId());
        messageProducer.sendMessage(
            confirmation.getMessage());

        // Publish order fulfillment message
        publishFulfillmentMessage(order);
    }
```

The order processing microservice publishes a message to the message broker after successful validation and balance deduction. Separate services subscribed to the broker can then handle order confirmation and fulfillment asynchronously. This ensures that the order processing microservice isn't blocked by these downstream tasks.

These examples showcase how Java code can leverage cloud functionalities to address concurrency challenges in microservices. By combining optimistic locking and message brokers, you can build a more robust and scalable e-commerce application. These are simplified examples. Real-world implementations might involve additional error handling, logging, and configuration.

Use case 2 – building a data processing pipeline with microservices

This case study delves into designing and implementing a data processing pipeline using a microservices architecture:

1. The first step is to design the microservices. We'll construct the pipeline with three distinct microservices:

    *   **Data ingestion service**: This service acts as the entry point, which is responsible for receiving and validating incoming data from external sources. Once validated, it publishes the data to an Amazon SQS queue for further processing. The service depends on the Amazon SQS client library.

    *   **Data processing service**: This service subscribes to the Amazon SQS queue used by the data ingestion service. It consumes the data, applies business logic for transformation, and publishes the processed data to another SQS queue for persistence. This service relies on both the Amazon SQS client library and the AWS Glue SDK.

    *   **Data persistence service**: The final service consumes the processed data from the second SQS queue. Its primary function is to store the data persistently in Amazon RDS for long-term accessibility. This service utilizes both the Amazon SQS client library and the Amazon RDS client library.

    By leveraging AWS services, we can build a scalable and efficient data processing solution that benefits from the modularity and flexibility inherent in a microservices architecture.

2. The next step is to set up the AWSs:

    *   **Two AWS Simple Queue Service (SQS) queues** will be set up:

        *   **Initial data queue**: Create a queue intended for receiving initial unprocessed data

        *   **Processed data queue**: Set up another queue for holding processed data ready for further actions or storage

- **AWS RDS instance**: Set up an RDS instance to provide persistent storage for your application. You can choose MySQL, PostgreSQL, or any other available RDS database engine depending on your application requirements. This database will be used to store and manage the data processed by your application.

- **AWS Simple Notification Service (SNS)**: Create an SNS topic to facilitate the notification process. This topic will be used to publish messages notifying subscribers of successful data processing events and other important notifications. Determine the subscribers to this topic, which could include email addresses, SMS, HTTP endpoints, or even other AWS services such as Lambda or SQS, depending on your notification requirements.

3. The third step is to set up a Maven project. Create a new Maven project for each microservice (DataIngestionService, DataProcessingLambda, and DataPersistenceService) in your preferred **Integrated Development Environment (IDE)** or using the command line. Open the pom.xml file in each project's root directory and add the related dependencies.

4. The fourth step is to implement the data ingestion service:

```
@Service
public class DataIngestionService {
    private final AmazonSQS sqsClient;

    public DataIngestionService(AmazonSQS sqsClient) {
        this.sqsClient = sqsClient;
    }

    public void ingestData(Data dat{
        // Validate the incoming data
        if (isValid(data)) {
            // Publish the data to Amazon SQS
            SendMessageRequest sendMessageRequest = new
            SendMessageRequest()
                    .withQueueUrl("data-ingestion-queue-url")
                    .withMessageBody(data.toString());
            sqsClient.sendMessage(sendMessageRequest);
        }
    }

    private boolean isValid(Data dat{
        boolean isValid = true;
        // Implement data validation logic
        // ...
        return isValid;
    }
```

The code represents the implementation of the data ingestion service, which is responsible for receiving incoming data, validating it, and publishing it to Amazon SQS for further processing.

The `DataIngestionService` class is annotated with `@Service`, indicating that it is a Spring service component. It has a dependency on the `AmazonSQS client`, which is injected through the constructor.

The `ingestData()` method takes a `data object` as input and performs data validation by calling the `isValid()` method. If the data is valid, it creates a `SendMessageRequest` object with the specified SQS queue URL and the data payload as the message body. The message is then sent to the SQS queue using the `sqsClient.sendMessage()` method.

5. The fifth step is to implement the data processing service using AWS Lambda:

```
public class DataProcessingLambda implements
RequestHandler<SQSEvent, Void> {
    private final AmazonSQS sqsClient;
    public DataProcessingLambda() {
        this.sqsClient = AmazonSQSClientBuilder.defaultClient();
    }


    @Override
    public Void handleRequest(SQSEvent event,
        Context context) {
            for (SQSEvent.SQSMessage message :
                event.getRecords()) {
                    String data = message.getBody();

    // Transform the data within the Lambda function
                String transformedData= transformData(
                    data);

                // Publish the transformed data to another Amazon
    SQS for persistence or further
                // processing
                sqsClient.sendMessage(
                    new SendMessageRequest()
                        .withQueueUrl(
                            "processed-data-queue-url")
                        .withMessageBody(transformedData));
        }
        return null;
    }

    /**
     * Simulate data transformation.
```

```
     * In a real scenario, this method would contain logic to
transform data based
     * on specific rules or operations.
     *
     * @param data the original data from the SQS message
     * @return transformed data as a String
     */
    private String transformData(String dat{
        // Example transformation: append a timestamp or modify
the string in some way
        return "Transformed: " + data + " at " + System.
        currentTimeMillis();
    }
}
```

This Lambda function, `DataProcessingLambda`, processes data from an Amazon SQS queue by implementing the `RequestHandler` interface to handle `SQSEvent` events. It initializes an Amazon SQS client in the constructor and uses it to send transformed data to another SQS queue for further processing or storage.

The `handleRequest()` method, serving as the function's entry point, processes each `SQSMessage` from the `SQSEvent`, extracting the data and transforming it directly within the function through the `transformData()` method. Here, the transformation appends a timestamp to the data as a simple example, but typically this would involve more complex operations tailored to specific data processing requirements.

Following the data transformation, the function sends the processed data to a specified SQS queue by invoking the `sendMessage()` method on the SQS client.

6. The next step is to create a Spring-managed service that handles storing processed data in a database and notifies subscribers via AWS SNS upon successful persistence:

```
@Service
public class DataPersistenceService {
    private final AmazonSNS snsClient;
    private final DataRepository dataRepository;

    public DataPersistenceService(DataRepository dataRepository)
    {
        // Initialize the AmazonSNS client
        this.snsClient = AmazonSNSClientBuilder.standard().
        build();
        this.dataRepository = dataRepository;
    }

    public void persistData(String data{
        // Assume 'data' is the processed data received
```

```
        // Store the processed data in a database
        Data dataEntity = new Data();
        dataEntity.setProcessedData(data);
        dataRepository.save(dataEntity);

        // Send notification via SNS after successful
persistence
        sendNotification("Data has been successfully persisted
        with the following content: " + data);
    }


    private void sendNotification(String message) {
        // Define the ARN of the SNS topic to send notification
        to
        String topicArn = "arn:aws:sns:region:account-id:your-
        topic-name";

        // Create the publish request
        PublishRequest publishRequest = new PublishRequest()
                .withTopicArn(topicArn)
                .withMessage(message);

        // Publish the message to the SNS topic
        snsClient.publish(publishRequest);
    }
}
```

DataPersistenceService is a Spring-managed bean responsible for handling data persistence and notifying other components or services via Amazon SNS. Here's a step-by-step description of its functionality:

- **Service initialization**: Upon instantiation, it initializes an AmazonSNS client used for sending notifications.

- **Data persistence**: The persistData() method takes a String data parameter, which is the processed data. It creates a Data entity, sets the processed data, and saves it to the database using the DataRepository.

- **Sending notifications**: After successfully saving the data, it calls sendNotification() to notify other parts of the application. It constructs a PublishRequest with a topic ARN (Amazon Resource Name) and the message detailing the successful persistence. The message is then published to the specified SNS topic.

This service is particularly useful in microservice architectures where decoupled components must communicate state changes or updates. Using SNS for notifications enhances the reliability of the system by ensuring not only that data is persisted but also that relevant services or components are informed of the update through a robust, scalable messaging system.

This section details the practical application of Java to manage concurrency in a microservices architecture, particularly for an e-commerce application processing order. It explains how using optimistic locking with version numbers in a distributed database can prevent data inconsistencies during concurrent order processing. Additionally, the use of message brokers is discussed as a method for asynchronous communication, which aids in keeping microservices from being blocked by downstream tasks, thereby improving efficiency and scalability.

Moving forward, the next section will cover strategic best practices for deploying and scaling microservices. This includes leveraging cloud-native services and architectures to optimize performance, scalability, and reliability, as well as providing a comprehensive guide for developers and architects on how to effectively manage microservices in a cloud environment.

Strategic best practices – deploying and scaling microservices

When designing, deploying, and scaling microservices in a cloud environment, it's essential to utilize cloud-native services and architectures to maximize performance, scalability, and reliability.

Here's a straightforward guide on best practices, tailored for developers and architects:

- **Load balancing**:

  - **Purpose**: Distribute incoming traffic evenly across multiple microservice instances to enhance reliability and availability

  - **How to implement**:

    - Use cloud-managed load balancers such as AWS **Elastic Load Balancing** (**ELB**), Azure Load Balancer, or Google Cloud Load Balancing, which can automatically adjust to traffic demands

    - Integrate service discovery tools (e.g., AWS Cloud Map, Azure Service Discovery, or Google Cloud Service Directory) to dynamically manage service instances

- **Caching solutions**

  - **Purpose**: Reduce database load and speed up response times by caching frequently accessed data

- **How to implement**:

  - Opt for managed caching services such as Amazon ElastiCache, Azure Redis Cache, or Google Cloud Memorystore, which offer distributed caching capabilities

  - Choose an appropriate caching strategy (local, distributed, or hybrid) and ensure proper management of cache coherence and expiration

- **Managed databases**

  - **Purpose**: Simplify database management tasks (scaling, backups, patching), allowing developers to concentrate on building functionalities

  - **How to implement**:

    - Implement a database-per-service model using **Database as a Service (DBaaS)** solutions such as Amazon RDS, Azure SQL Database, or Google Cloud SQL to ensure resource isolation and optimized performance

    - Leverage automated features within DBaaS for scaling, backups, and ensuring high availability

- **Microservices architecture considerations**

  - Maintain loose coupling among services to enable independent development, deployment, and scaling

  - Apply **Domain-Driven Design (DDD)** principles by organizing microservices around business capabilities and defining clear bounded contexts

- **Deployment and scaling**

  - **Containers and orchestration**: Deploy microservices using containerization with Kubernetes, which is supported by AWS EKS, Azure AKS, and Google GKE, to manage container lifecycles and automate scaling

  - **Scalability**: Implement auto-scaling based on CPU, memory usage, or custom metrics aligned with your application's needs

- **Monitoring and logging**

  - **Observability**: Implement comprehensive monitoring and logging to keep track of microservice performance and operational health; also, utilize tools such as AWS CloudWatch, Azure Monitor, or Google's Operations Suite for real-time monitoring, performance tracking, and alert management

Adhering to these best practices leverages the strengths of cloud computing, enhancing the resilience, performance, and scalability of your microservices architecture. This strategic approach not only ensures robust service delivery but also maintains the agility needed for continuous innovation and growth.

Advanced concurrency patterns – enhancing microservice resilience and performance

When developing microservices in Java, it is essential to employ concurrency patterns and techniques that enhance the application's responsiveness, fault tolerance, and scalability. These patterns help manage the complexities inherent in distributed systems.

Here's a discussion on key concurrency and data management patterns applicable to microservices.

Data management patterns

Understanding and implementing effective data management patterns is crucial for designing robust microservices. Let's look at them one by one.

Command Query Responsibility Segregation

Command Query Responsibility Segregation or **CQRS** separates the read and write operations of a data store to optimize performance, scalability, and security. This pattern allows reads and writes to be scaled independently. Let's look at the details:

- **Use case**: Useful in complex domains where the read operations significantly outnumber the write operations, or when they can be clearly differentiated

- **Implementation details**: *Figure 8.4* shows a system using CQRS, which separates data updates (commands) from data retrieval (queries)

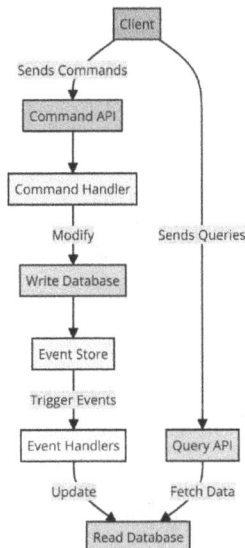

Figure 8.4: CQRS architecture flow

- **Command side**: Handles updates with a Command API, CommandHandler, and Write Database
- **Query side**: Handles reads with a Query API and a separate Read Database optimized for fast reads

This separation improves performance and scalability. Each side can be optimized for its task and scaled independently. Additionally, the Query Side can stay available during updates on the Write Side.

Event sourcing

Event sourcing is a design pattern in which changes to the state of an application are stored as a sequence of events. Instead of storing just the current state of the data in a domain, event sourcing stores a sequence of state-changing events. Whenever the state of a business entity changes, a new event is appended to the list of events associated with that entity. This sequence of events serves as the principal source of truth and can be used to reconstruct past states of an entity. Let's take a closer look:

- **Use case**: Imagine a banking application that requires a robust mechanism for tracking the movement of funds between accounts, ensuring compliance with auditing standards, and the ability to revert or reconstruct account states during disputes or investigations.

- **Implementation details**: Let's look at this diagram:

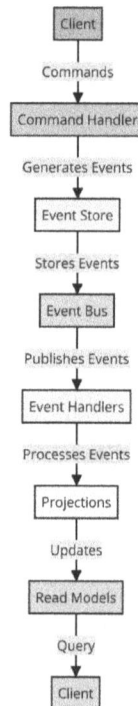

Figure 8.5: The Event Sourcing pattern

Figure 8.5 illustrates the Event Sourcing pattern in software architecture using a horizontal multi-level layout. Here's a description of its components and flow:

- **Client**: Initiates the process by sending commands to the system

- **Command handler**: Receives commands from the client and processes them; generates events based on the commands received

- **Event store**: Captures and stores these events; this storage acts as the authoritative source of truth about the state of the system

- **Event bus**: Distributes the stored events to appropriate handlers

- **Event handlers**: React to the events by processing them and potentially generating new events or commands

- **Projections**: Update the read models based on the events processed by the event handlers

- **Read models**: Provide the updated state of the system back to the client based on the projections

- **Client**: May query the read models to retrieve the current state or results of operations

The Event sourcing pattern allows the system to maintain a full historical record of state changes, which is crucial for auditing and compliance. It also supports scalability by decoupling command processing from state storage and enabling asynchronous event processing.

API versioning

API versioning is a strategy employed to manage changes to an API. It allows for new features and changes without disrupting the existing user experience or requiring clients to make immediate upgrades. This approach is particularly crucial when introducing breaking changes that would otherwise compromise backward compatibility. Here's a more detailed look:

- **Use case**: Imagine a scenario where a financial services API needs to add new fields to a response object that could disrupt existing client applications. By introducing a new version of the API, the service can provide these enhancements while still supporting the older version, ensuring that existing applications continue to function without modification until they opt to upgrade.

- **Implementation details** are as follows:

  - **URI versioning**: Include the version number in the API endpoint path, such as `/api/v1/users` for the first version and `/api/v2/users` for the second version. This method is transparent and easy to understand.

  - **Parameter versioning**: Send the version as a query string or header in the API request, such as `GET /api/users?version=1`. This keeps the URI clean and allows for more flexibility but can be less intuitive.

- **Header versioning**: Include the desired version in the headers of the API requests, such as `Accept:application/vnd.myapi.v1+json`. This approach is less obtrusive and separates versioning from the business logic of the API.

Here's a basic example of how to implement API versioning in a web application using Spring Boot:

```
@RestController
@RequestMapping("/api")
public class UserController {

    // Version 1 of the API
    @GetMapping(value = "/users",
        headers = "X-API-Version=1")
    public List<User> getUsersV1() {
        return userService.findAllUsers();
    }

    // Version 2 of the API
    @GetMapping(value = "/users",
        headers = "X-API-Version=2")
    public List<UserDto> getUsersV2() {
        return userService.findAllUsersV2();
    }
}
```

In this example, different versions of the same endpoint are triggered based on the custom `X-API-Version` request header. This allows clients to specify which version of the API they wish to interact with, enabling backward compatibility while new features are rolled out.

Saga pattern

The **Saga pattern** is a valuable approach for managing data consistency across multiple microservices in distributed transactions. It provides a way to handle long-running business processes that span multiple services, ensuring that each step is executed successfully or compensated if an error occurs. Let's find out more:

- **Use case**: The Saga pattern excels in coordinating long-running microservice workflows where each step requires confirmation. This is ideal for scenarios such as order processing (inventory, payment, and shipping) or hotel booking (reservation, payment, and confirmation). It ensures that the entire process succeeds or is rolled back if a step fails.

- **Implementation details**: Let's look at *Figure 8.6*.

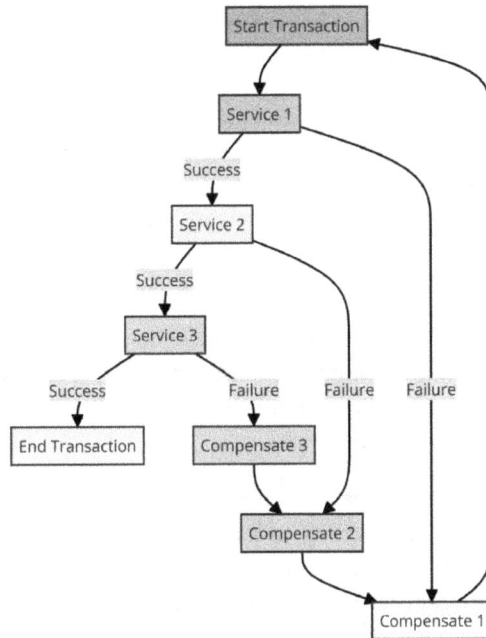

Figure 8.6: The Saga pattern

This activity diagram demonstrates the Saga pattern, showing the flow of a transaction with potential compensating actions:

- **Start transaction**: The process begins
- **Service 1**: The first service is called:
  - If Service 1 succeeds, it proceeds to Service 2
  - If Service 1 fails, it triggers Compensate 1 and returns to the start
- **Service 2**: The second service is called:
  - If Service 2 succeeds, it proceeds to Service 3
  - If Service 2 fails, it triggers Compensate 2 and returns to Compensate 1
- **Service 3**: The third service is called:
  - If Service 3 succeeds, the transaction ends
  - If Service 3 fails, it triggers Compensate 3 and returns to Compensate 2

- **End transaction**: The process completes successfully

  - Compensation steps are taken to revert to previous steps when a failure occurs, ensuring that the system maintains consistency

The Saga pattern allows for the coordination of complex transactions across multiple microservices while maintaining loose coupling and independence between the services. Each service performs its own local transaction and publishes events to trigger the next step in the Saga pattern. If any step fails, compensating actions are executed to roll back the previous steps, guaranteeing eventual consistency.

Database per service

The **database per service** pattern is an architectural approach in which each microservice has its own dedicated database. Instead of sharing a single database across multiple services, each service owns and manages its own data store. This pattern promotes loose coupling, autonomy, and scalability in microservices architectures. Let's take a closer look:

- **Use case**: The Database per service pattern thrives when microservices have diverse data requirements. It empowers each service to leverage the most suitable database technology, optimize data models and queries, and scale independently based on its specific load. This approach fosters polyglot persistence and ensures strict data isolation, making it ideal for multi-tenant architectures and compliance-driven scenarios.

- **Implementation strategies**: Let's look at *Figure 8.7*:

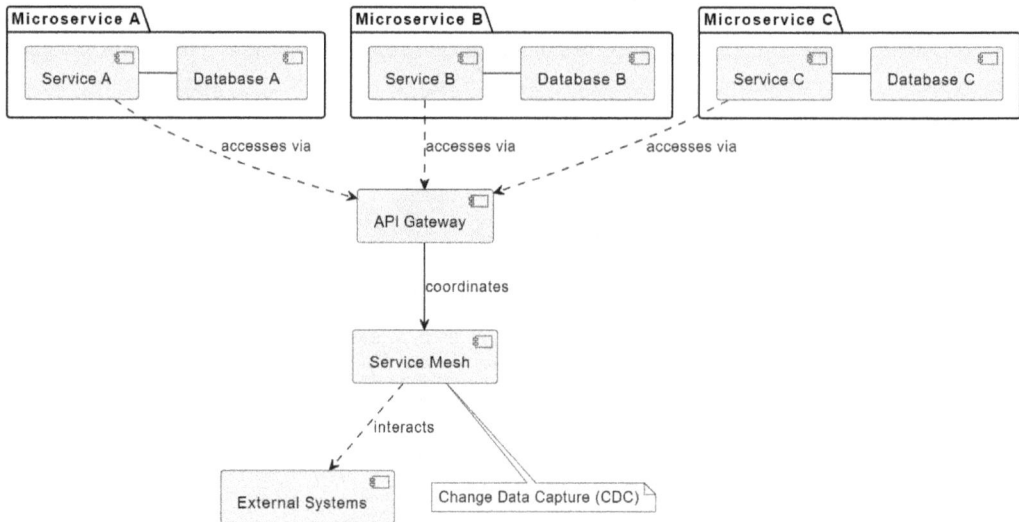

Figure 8.7: The Database per service pattern

This component diagram illustrates the *Database per service* architectural pattern, where each microservice operates with its own dedicated database. This design emphasizes loose coupling, autonomy, and scalability within a microservices architecture. Here's a breakdown of the key components shown in the diagram:

- **Microservices (A, B, and C)**: Each microservice is shown with its respective database. For example, Microservice A uses Database A, Microservice B uses Database B, and Microservice C uses Database C. This design ensures that each microservice can operate independently, manage its own data store, and use a database technology that best suits its needs.

- **API Gateway**: The API Gateway acts as an intermediary that external clients interact with. It abstracts the underlying microservices and provides a single point of entry into the system. Each microservice accesses the API Gateway, allowing for a simplified client interaction model and centralizing some cross-cutting concerns such as authentication and rate limiting.

- **Service mesh**: The service mesh is represented as facilitating communication between the microservices, API Gateway, and external systems. It helps manage service-to-service communications, ensuring reliable data transfer, and implements resilience patterns such as retries and circuit breakers.

- **Change Data Capture (CDC)**: A note on CDC is included to indicate its role in the architecture. CDC tools such as Debezium can be used to capture changes in each microservice's database and propagate these changes to other services or external systems. This setup supports maintaining eventual consistency across the distributed data stores.

- **Interactions**: The diagram shows the flow of data and interactions. Each microservice interacts with the API Gateway, which in turn communicates with the service mesh. The service mesh coordinates further interactions, potentially with external systems, and facilitates the implementation of CDC.

This diagram effectively communicates the separation of concerns and the independence of each microservice within the system, highlighting the advantages of the Database per service pattern in supporting diverse data requirements and scalability.

Shared database pattern

The **shared database pattern** is an architectural approach in which each microservice has its own dedicated database. Instead of sharing a single database across multiple services, each service owns and manages its data store. This pattern promotes loose coupling, autonomy, and scalability in microservices architectures:

- **Use case**: Shared databases excel in enterprises, real-time systems, and compliance-driven environments. They provide a single source of truth for organizations (e.g., customer data across sales, HR, and finance) and ensure consistent data for real-time applications. In regulated industries (such as finance and healthcare), they enforce compliance standards across services, simplifying audits. This pattern thrives where data consistency and integrity are paramount.

- **Implementation strategies**: Let's look at *Figure 8.8*:

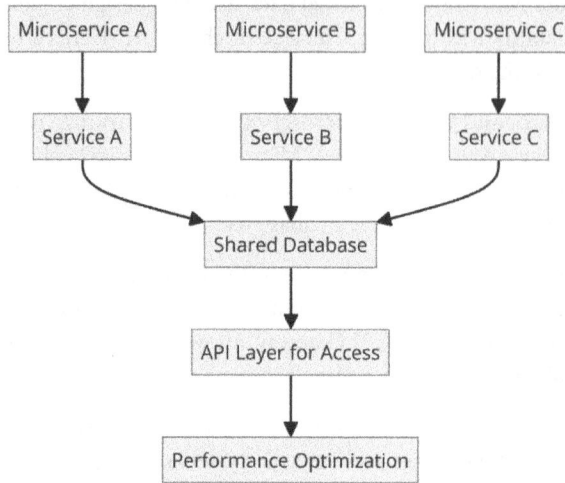

Figure 8.8: The shared database pattern

The diagram represents the shared database pattern where multiple services (microservices A, B, and C) utilize a single, central database. This architecture is commonly adopted to maintain a unified source of truth across different parts of an organization, facilitating consistency and compliance in data management. Its key components include the following:

- **A shared database**: All microservices access this single database for both reading and writing operations. This setup ensures data consistency and simplifies transaction management across services.

- **Microservices (A, B, and C)**: These services are independent in functionality but share the same database for data operations. They represent different business capabilities within the organization.

- **An API layer for access**: An API layer abstracts the database interactions from the services. This layer helps enforce security, manage access patterns, and ensure that changes to the database schema do not directly impact service operations.

- **Performance optimization**: This note suggests the use of strategies such as connection pooling, creating read replicas, and caching to optimize database performance and handle high loads efficiently.

In Java, the implementation of these patterns can be supported by various frameworks and libraries that facilitate asynchronous programming and provide tools for building resilient microservices. By selecting the appropriate patterns for specific challenges, developers can craft robust, scalable, and efficient microservice architectures.

Summary

In this chapter, we explored microservices in the cloud, as well as how Java's concurrency tools can be leveraged to build and manage these services effectively. We discussed the principles of microservices, their advantages over monolithic architectures, and their integration into cloud environments. The key characteristics of microservices, such as modularity and loose coupling, were examined, highlighting their contribution to building resilient and scalable systems.

To develop high-performance microservices, we delved into Java's concurrency essentials, including thread pools, parallel streams, and the Fork/Join framework. These tools enable developers to incorporate parallel processing and efficient task management techniques, thus optimizing microservice performance.

We also addressed potential bottlenecks and concurrency-related issues in microservices architectures, providing practical solutions using Java's concurrent mechanisms. Strategies for ensuring smooth communication, data consistency, and resilience across services were also discussed.

Best practices for designing, deploying, and scaling concurrent microservices in the cloud were covered, including load balancing, caching, and database management. Essential patterns for building resilient and fault-tolerant microservices, such as circuit breakers, bulkheads, and event-driven communication patterns, were explored.

Hands-on exercises and case studies demonstrated how to apply Java's concurrency tools, best practices, and design patterns in real-world scenarios, allowing readers to gain practical experience in designing, deploying, and scaling concurrent microservices in the cloud.

As we conclude this chapter, readers should have a comprehensive understanding of creating scalable, resilient microservices using Java's concurrency tools and design patterns in the cloud. They should be equipped with the knowledge and skills necessary to tackle the challenges of building high-performance, fault-tolerant microservices architectures.

Looking ahead, the next chapter, *Serverless Computing and Java's Concurrent Capabilities*, will explore how Java's concurrency features can be utilized in serverless environments, providing insights and practical guidance on harnessing Java's concurrency tools to build high-performance, event-driven serverless applications.

Questions

1. What is a key advantage of microservices architecture over monolithic architecture?

    A. Increased complexity and coupling

    B. Reduced flexibility and scalability

    C. Independent deployment and scalability

    D. Single point of failure

2. Which Java feature is essential for managing asynchronous tasks within microservices?

    A. **Java Virtual Machine (JVM)**

    B. **Java Database Connectivity (JDBC)**

    C. CompletableFuture

    D. JavaBeans

3. What is the primary role of a load balancer in a microservices architecture?

    A. Encrypting data transfers

    B. Distributing incoming network traffic across multiple instances

    C. Data storage and management

    D. Error logging and handling

4. Which pattern helps prevent a network or service failure from cascading to other parts of the system in microservices?

    A. Singleton pattern

    B. Factory pattern

    C. Circuit breaker pattern

    D. Builder pattern

5. Which best practice is recommended when implementing microservices in the cloud for handling data consistency?

    A. Using a single shared database for all services

    B. Employing different caching strategies for each microservice

    C. Assigning a separate managed database instance for each microservice

    D. Centralizing all data management in one microservice

Serverless Computing and Java's Concurrent Capabilities

Serverless computing has revolutionized application deployment and management, allowing developers to focus on writing code while cloud providers handle the underlying infrastructure. This chapter explores the essentials of serverless computing and how Java's concurrent capabilities can be effectively utilized in this environment. We'll delve into the fundamental concepts of serverless computing, its advantages, specific scenarios where it's particularly beneficial, and the trade-offs involved.

Serverless architectures offer significant benefits in terms of scalability, cost efficiency, and reduced operational overhead, but they also come with challenges such as cold start latency, resource limits, and vendor lock-in. Understanding these trade-offs is crucial for making informed decisions about when and how to use serverless computing.

The chapter will cover the following key topics:

- Fundamentals of serverless computing in Java

- Adapting Java's concurrency model to serverless environments

- Introducing serverless frameworks and services: AWS SAM, Azure Functions Core Tools, Google Cloud Functions, and Oracle Functions

- Industry examples of Java serverless functions with a focus on concurrency

- A practical approach to building with serverless frameworks

We'll explore how Java's concurrency features can be adapted to serverless environments, enabling the development of scalable and efficient applications. Through practical examples and code snippets, you'll learn how to implement concurrency in Java serverless applications, leveraging tools such as ExecutorService, CompletableFuture, and parallel streams.

We'll also discuss best practices for optimizing Java serverless applications, including minimizing cold starts, efficient resource management, and leveraging frameworks such as Spring Cloud Function, Micronaut, and Quarkus.

By the end of this chapter, you'll be equipped with the knowledge to build and optimize Java-based serverless applications, ensuring high performance and responsiveness across various cloud platforms.

Technical requirements

You'll need the AWS **Command Line Interface (CLI)** installed and configured: `https://docs.aws.amazon.com/cli/latest/userguide/getting-started-install.html`.

Here are the instructions for doing so:

1. Visit the official AWS CLI installation page.
2. Choose your operating system (Windows, Mac, or Linux) and download the installer.
3. Run the installer and follow the on-screen instructions.
4. After installation, configure your AWS CLI with credentials using the aws configure command. You'll need your access key ID and secret access key, which can be found in your AWS IAM console.

You'll also need the AWS Serverless Application Model CLI for deploying serverless applications: `https://docs.aws.amazon.com/serverless-application-model/latest/developerguide/install-sam-cli.html`.

Here are the instructions for that:

1. Visit the official AWS SAM CLI installation page.
2. Follow the instructions for your operating system (Windows, Mac, or Linux). These typically involve downloading and running a script or installer.
3. After installation, verify the installation by running sam --version in your terminal. This should display the installed version of the AWS SAM CLI.

The code in this chapter can be found on GitHub:

`https://github.com/PacktPublishing/Java-Concurrency-and-Parallelism`

Fundamentals of serverless computing in java

Serverless computing is a cloud-computing execution model that has revolutionized the way applications are developed and deployed. In this model, the cloud provider dynamically manages the allocation and provisioning of servers, allowing developers to focus on writing code without worrying about the underlying infrastructure. Although applications still run on servers, the management of these servers, including scaling and maintenance, is entirely handled by the cloud provider. This approach marks a significant departure from traditional server-based architectures, where developers are responsible for managing and maintaining the servers that host their applications.

Core concepts of serverless computing

Serverless architectures are built around several core concepts. One of the key concepts is the event-driven approach, where functions are executed in response to various triggers such as HTTP requests, database events, and device activity. This model is particularly well suited for Java developers because Java's concurrency features, such as multithreading and asynchronous processing, align seamlessly with the event-driven nature of serverless computing. Additionally, Java's extensive ecosystem of libraries and tools enhances its integration with cloud functions from providers such as Amazon AWS Lambda, Google Cloud, and Azure Functions. These attributes make Java a powerful choice for developing scalable and efficient serverless applications.

Another crucial concept is statelessness, which means that functions typically execute without retaining any state between invocations. This enables high levels of scalability, as multiple instances of a function can run concurrently without interfering with each other. However, functions can also be designed to be stateful by using external data sources or services, such as AWS Lambda with an external database or Kalix for stateful serverless applications. Serverless platforms also provide automatic scaling based on demand, eliminating the need for manual scaling and ensuring that applications can handle variable workloads efficiently.

Additionally, serverless computing simplifies backend development by offloading routine tasks such as server and database management to the cloud provider, allowing developers to focus on writing business logic. Finally, serverless architectures are highly compatible with microservice architectures, enabling the independent deployment of discrete pieces of functionality.

Traditionally, Java has not been the first choice for serverless computing due to its verbose syntax and slower startup times compared to languages such as Python or JavaScript. However, recent developments in frameworks such as Quarkus and Micronaut have significantly reduced Java's startup time and memory usage, boosting its suitability for serverless environments. While Spring Native initially showed promise in this area, it has since been deprecated in favor of the official native support integrated into Spring Boot 3+. This new native support within Spring Boot 3+ offers enhanced features and capabilities, enabling Java developers to create efficient, scalable serverless applications that fully leverage the advantages of cloud-native architectures.

Advantages of and scenarios for using serverless computing

Serverless computing offers several compelling advantages, particularly in terms of scalability, cost-effectiveness, and reduced operational overhead.

One of the most significant benefits is enhanced scalability. Serverless architectures can instantly and automatically scale up or down based on the demand for the application. This means that during peak traffic periods, the application can seamlessly handle increased loads without any manual intervention. Additionally, the stateless nature of serverless functions allows them to run concurrently and in parallel, enabling high throughput and responsiveness.

Another key advantage of serverless computing is cost-effectiveness. With a pay-per-use pricing model, you only pay for the resources your functions consume during their execution time. This eliminates the costs associated with idle compute resources, making it an attractive option for applications with variable or sporadic workloads. Moreover, serverless computing can reduce the total cost of ownership by minimizing the need for ongoing server maintenance and management.

Serverless computing offers the potential for reduced operational overhead. By offloading server management tasks such as maintenance, patching, and scaling to the cloud provider, developers can focus more on code and functionality. Deployment processes are often simplified, allowing for quicker updates and feature releases without directly managing infrastructure.

However, it's important to note that serverless architectures can introduce their own complexities. Managing multiple runtimes for different functions or services may require additional configuration and monitoring. This is particularly relevant for larger applications where maintaining consistency across diverse environments can become challenging. While serverless platforms often provide built-in high availability and fault tolerance features, ensuring application resilience may still require careful design and optimization.

Furthermore, serverless architectures can easily integrate with other cloud services and be automatically triggered by events from these services, enabling the creation of highly responsive, event-driven applications and automated workflows.

Overall, serverless computing can be a powerful tool, but it's important to evaluate its suitability for your specific use case and understand the potential trade-offs involved.

Drawbacks and trade-offs of serverless computing

While serverless computing offers numerous benefits, it also comes with several trade-offs and potential drawbacks that developers must consider:

- **Cold starts**: One of the most commonly cited drawbacks is the cold start latency. When a serverless function is invoked after a period of inactivity, it can take some time to initialize, leading to delays in response times. This can be particularly problematic for applications requiring low-latency responses.

- **Resource limits**: Serverless platforms impose limits on the execution time, memory, and computational resources available to each function. These constraints can make it challenging to run long-running processes or compute-intensive tasks.

- **Vendor lock-in**: Using serverless architectures often ties developers to a specific cloud provider's ecosystem, making it difficult to migrate applications to another provider without significant rework.

- **Complexity in debugging and monitoring**: Debugging serverless functions can be more complex compared to traditional server-based applications. The ephemeral nature of serverless functions and their distributed execution environments can complicate the debugging process. Additionally, monitoring and maintaining observability across multiple serverless functions requires robust tooling and practices.

- **State management**: Serverless functions are inherently stateless, which can complicate the management of application state across multiple invocations. Developers need to use external storage solutions such as databases or caching services to manage state, which can introduce additional complexity and latency.

- **Cost efficiency**: While serverless computing can be cost-effective for many use cases, it may not always be the most economical option. High-frequency invocations or applications with constant traffic may incur higher costs compared to reserved instances or traditional server-based architectures.

- **Security concerns**: The abstraction of infrastructure management in serverless computing means that developers have less control over the underlying environment. This can introduce security concerns, as vulnerabilities or misconfigurations in the cloud provider's infrastructure can potentially affect the application.

When to use serverless?

Combining the advantages outlined previously, here are additional scenarios where serverless computing is particularly beneficial:

- **Microservice architectures**: Applications designed with microservice architecture, where each service is a small, independently deployable unit.

- **Event-driven applications**: Systems that respond to various events such as data streams, user actions, or IoT signals.

- **Stateless processing**: Applications that perform stateless operations, such as image processing, data transformation, or **Extract, Transform, Load** (ETL) tasks.

However, it's important to note that serverless may not be the best fit for every scenario. Some situations where serverless might not be ideal include the following:

- Applications with long-running processes or high computational requirements

- Workloads that require low-latency responses or have strict performance requirements

- Applications with complex or stateful workflows that require maintaining server state

- Scenarios where you need complete control over the underlying infrastructure and operating system

When deciding whether to use serverless, it's crucial to evaluate your specific application requirements, scalability needs, and cost considerations. Serverless can offer significant benefits in terms of scalability, cost efficiency, and development agility, but it's important to carefully assess whether it aligns with your application's characteristics and goals.

Adapting Java's concurrency model to serverless environments

Serverless computing presents unique challenges for Java's traditional concurrency model. The ephemeral and stateless nature of serverless functions necessitates a shift from long-lived thread pools and shared mutable state to more dynamic and isolated concurrency patterns. In this context, developers must focus on designing concurrency strategies that align with the short-lived, event-driven nature of serverless architectures.

Effective serverless concurrency in Java revolves around maximizing function efficiency within strict time and resource constraints. This involves leveraging asynchronous operations, particularly through CompletableFuture, to handle non-blocking I/O tasks and optimize throughput. Developers should structure their code to process events efficiently, a central tenet of serverless design.

When using CompletableFuture for asynchronous processing in serverless functions, it's crucial to consider the function's execution time limits.

Here's an example:

```
public class ServerlessAsyncFunction implements
RequestHandler<APIGatewayProxyRequestEvent,
APIGatewayProxyResponseEvent> {
    @Override
    public APIGatewayProxyResponseEvent
handleRequest(APIGatewayProxyRequestEvent input, Context context) {
        CompletableFuture<String> dbFuture = CompletableFuture.
supplyAsync(this::queryDatabase);
        CompletableFuture<String> apiFuture = CompletableFuture.
supplyAsync(this::callExternalAPI);
        try {
```

```
            String result = CompletableFuture.allOf(
                dbFuture, apiFuture)
                .thenApply(v -> processResults(
                    dbFuture.join(), apiFuture.join()))
                .get(context.getRemainingTimeInMillis(),
                    TimeUnit.MILLISECONDS);
            return new APIGatewayProxyResponseEvent(
                ).withStatusCode(200).withBody(result);
        } catch (TimeoutException e) {
            return new APIGatewayProxyResponseEvent(
                ).withStatusCode(408).withBody(
                    "Request timed out");
        } catch (Exception e) {
            return new APIGatewayProxyResponseEvent(
                ).withStatusCode(500).withBody(
                    "Internal error");
        }
    }
    private String queryDatabase() {
        // Database query logic
        return "Database result";
    }
    private String callExternalAPI() {
        // API call logic
        return "API result";
    }
    private String processResults(String dbResult, String
        apiResult) {
        // Processing logic
        return "Processed result: " + dbResult + ",
            " + apiResult;
    }
}
```

This demo showcases how to use CompletableFuture for asynchronous processing in a serverless function. By performing database queries and API calls concurrently, the function minimizes the overall execution time and improves responsiveness. This is particularly beneficial in serverless environments where reducing execution time can lead to cost reduction and better scalability.

For data processing, while parallel streams can be beneficial, it's important to consider the trade-offs in a serverless context:

```
public class DataProcessingLambda implements
RequestHandler<List<Data>, List<ProcessedData>> {
    @Override
    public List<ProcessedData> handleRequest(
        List<Data> dataList, Context context) {
            LambdaLogger logger = context.getLogger();
            logger.log("Starting data processing");

  // Use parallel stream only if the data size justifies it
            boolean useParallel = dataList.size() > 100;
// Adjust threshold based on your specific use case
            Stream<Data> dataStream = useParallel ? dataList.
            parallelStream() : dataList.stream();

        List<ProcessedData> processedDataList = dataStream
                .map(this::processDataItem)
                .collect(Collectors.toList());

        logger.log("Data processing completed");
        return processedDataList;
    }

    private ProcessedData processDataItem(Data data) {
        // Ensure this method is thread-safe and efficient
        return new ProcessedData(data);
    }
}
```

This demo illustrates the use of Java's parallel streams to process large datasets in a serverless function. By conditionally using parallel streams based on the data size, the function can efficiently utilize multiple CPU cores to process data concurrently. This approach significantly enhances performance for large datasets, making the function more scalable and responsive in serverless environments.

As serverless computing continues to gain popularity, Java's concurrency features will play a crucial role in enabling developers to build scalable, responsive, and high-performance serverless applications. To further optimize Java serverless applications, let's explore best practices in design and the use of frameworks and libraries.

Designing efficient Java serverless applications

To ensure optimal performance and cost-efficiency in Java serverless applications, it's crucial to follow best practices in design and leverage appropriate frameworks and libraries. Here are the key guidelines and recommendations:

- **Minimizing cold starts**: Cold starts occur when a new instance of a serverless function is provisioned. To reduce cold start times, developers can employ several techniques:

  - **AWS Lambda SnapStart**: SnapStart optimizes the initialization process by taking a snapshot of the initialized execution environment and restoring it when needed, significantly reducing cold start latency. To use SnapStart, enable it in your Lambda function configuration and ensure your code is compatible with serialization.

  - **Enable provisioned concurrency**: This feature keeps functions warm by pre-initializing instances, ensuring quick response times. Set up provisioned concurrency in your Lambda function settings based on expected traffic patterns.

  - **Optimize Java Virtual Machine runtime with custom images**: Tools such as GraalVM Native Image can compile Java applications into native executables, reducing startup time and memory consumption. To use custom images, do the following:

    - Build your application using GraalVM Native Image.

    - Create a custom Lambda runtime using AWS Lambda Runtime Interface Client.

    - Package your native executable with the custom runtime.

    - Deploy the package as a Lambda function.

- **Additional optimizations**:

  - Minimize dependencies in your function to reduce package size

  - Use lazy loading for non-essential resources

  - Implement caching strategies for frequently accessed data

  - Optimize your code for quick startup, moving initialization logic out of the handler method where possible

- **Efficient memory and resource management**:

  - **Right-size function memory**: Allocate sufficient memory to avoid performance bottlenecks while being mindful of costs.

  - **Optimize code execution**: Write efficient code to reduce execution time, avoiding heavy initialization logic within the function handler.

- **Connection pooling**: Use Amazon RDS Proxy for managing database connections effectively, as traditional connection pooling libraries such as HikariCP are not recommended for serverless use.

- **Stateless design**: Design functions to be stateless to ensure scalability and avoid state management issues. Use external storage services such as Amazon **Simple Storage Service** (**S3**), DynamoDB, or Redis for state persistence.

- **Efficient data handling**:

  - **Use streams for large data**: Stream processing helps to handle large datasets without loading all data into memory. Java's Stream API is useful for this purpose.

  - **Optimize serializations**: Use efficient serialization libraries such as Jackson for JSON processing and optimize serialization/deserialization processes.

- **Monitoring and logging**:

  - **Integrated logging**: Use centralized logging services such as AWS CloudWatch Logs and structure logs for easy tracing and debugging.

  - **Performance monitoring**: Tools such as AWS X-Ray are helpful for tracing and monitoring the performance of serverless functions.

Java-specific optimization techniques for AWS Lambda

For Java developers working with AWS Lambda, there are various techniques available to optimize the runtime performance of your functions. These techniques can help you minimize cold start times, reduce memory usage, and improve overall execution speed:

- **Application class data sharing (AppCDS)**: Improves startup time and memory footprint by saving the metadata of loaded classes into an archive file, which can be memory-mapped during subsequent JVM startups

- **Optimizing JVM settings**: Adjust JVM parameters, such as using `-XX:+TieredCompilation -XX:TieredStopAtLevel=1`, to balance startup time and long-term performance

- **Leveraging GraalVM native image**: Compiling your Java application into a native executable can significantly reduce cold start times and memory usage

Frameworks and libraries

Spring Cloud Function simplifies serverless development by enabling developers to write cloud-agnostic functions using standard Java interfaces and Spring annotations. This allows for local creation, testing, and deployment, followed by seamless execution on various cloud infrastructures without code modifications. It supports an event-driven architecture, handling triggers such as HTTP requests, message queues, and timers, while its auto-discovery and registration of functions streamline the development process.

To illustrate, consider a simple Spring Boot application:

```
@SpringBootApplication
public class SpringFunctionApp {

    public static void main(String[] args) {
        SpringApplication.run(
            SpringFunctionApp.class, args);
    }    @Bean
    public Function<String, String> uppercase() {
        return value -> value.toUpperCase();
}}
```

Within this Spring Boot application, Spring Cloud Function automatically registers the `uppercase()` method as a function. When triggered, Spring Cloud Function maps incoming requests to the corresponding function, executes it with the provided input, and returns the result. This abstraction layer allows for seamless deployment across diverse serverless environments, freeing developers to focus solely on business logic.

Micronaut, renowned for its fast startup times and minimal memory footprint, presents an ideal choice for building serverless functions. Designed for creating lightweight and modular JVM-based applications, Micronaut seamlessly integrates with popular serverless platforms.

Let's take a look at a simple Micronaut function:

```
@FunctionBean("helloFunction")
public class HelloFunction extends FunctionInitializer implements
Function<APIGatewayV2HTTPEvent, APIGatewayV2HTTPResponse> {
    @Override
    public APIGatewayV2HTTPResponse apply(
        APIGatewayV2HTTPEvent request) {
            return APIGatewayV2HTTPResponse.builder()
                    .withStatusCode(200)
                    .withBody("Hello World")
                    .build();
    }}
```

In this code, the @FunctionBean("helloFunction") annotation designates the HelloFunction class as a Micronaut function bean. The HelloFunction class extends FunctionInitializer and implements the Function interface, specifically for processing API Gateway HTTP events (APIGatewayV2HTTPEvent).

The overridden apply() method efficiently handles incoming requests, returning an APIGatewayV2HTTPResponse with a status code of 200 and the message "Hello World" in the body. This setup allows for easy deployment of the function to various serverless platforms, including AWS Lambda.

Quarkus, known for its developer-friendly experience and exceptional performance, is a Kubernetes-native Java framework optimized for both GraalVM and OpenJDK HotSpot. Its design prioritizes fast boot times and low memory consumption, making it a compelling choice for serverless applications.

Let's look at a basic Quarkus function:

```
public class GreetingLambda {
    public APIGatewayProxyResponseEvent handleRequest(
        APIGatewayProxyRequestEvent input) {
            return new APIGatewayProxyResponseEvent()
                .withStatusCode(200)
                .withBody("Hello, " + input.getBody());
```

This code snippet showcases a serverless function built with Quarkus. The GreetingLambda class features a handleRequest() method specifically crafted to manage incoming AWS API Gateway requests (APIGatewayProxyRequestEvent). This method processes the request and constructs an APIGatewayProxyResponseEvent, returning a 200 status code along with a personalized greeting that incorporates the request body.

While this function is inherently tailored for deployment on AWS Lambda, adapting it for Azure Functions or Google Cloud Functions necessitates modifications to accommodate the distinct mechanisms these platforms employ for handling HTTP requests.

AWS Lambda Java libraries

Amazon provides a set of libraries specifically designed for building serverless applications on AWS Lambda (https://docs.aws.amazon.com/lambda/latest/dg/lambda-java.html), which greatly simplifies the process of integrating with other AWS services. These libraries are tailored to streamline the development of Lambda functions, ensuring they can efficiently interact with various AWS resources and services.

Let's examine a simple AWS Lambda function:

```
public class S3ObjectProcessor implements RequestHandler<S3Event,
String> {

    @Override
    public String handleRequest(S3Event event,Context context){
        // Get the first S3 record from the event
        S3EventNotificationRecord record = event.getRecords().get(0);
        // Extract the S3 object key from the record
        String objectKey = record.getS3().getObject().getKey();
        // Log the object key
        context.getLogger().log(
            "S3 Object uploaded: " + objectKey);
        return "Object processed successfully: " + objectKey;
    }}
```

This code demonstrates how to utilize AWS Lambda Java libraries to build a serverless function that responds to S3 events. The `RequestHandler` interface and `S3Event` class are provided by the `aws-lambda-java-core` and `aws-lambda-java-events` libraries respectively. The `context` object offers runtime information and a logger for the Lambda function.

By adhering to best practices and leveraging the right frameworks and libraries, developers can build efficient and scalable Java serverless applications. These practices ensure reduced latency, optimal resource usage, and easier maintenance, while the frameworks and libraries provide powerful tools to streamline development and deployment processes. Moving forward, applying these principles will help in achieving high-performance serverless applications that meet modern cloud computing demands.

Introducing serverless frameworks and services – AWS SAM, Azure Functions Core Tools, Google Cloud Functions, and Oracle Functions

To effectively manage and deploy serverless applications across different cloud platforms, it is crucial to understand the frameworks provided by AWS, Azure, and Google Cloud. These frameworks simplify the process of defining, deploying, and managing serverless resources, making it easier for developers to build and maintain scalable applications.

AWS Serverless Application Model

AWS **Serverless Application Model (SAM)** is a framework for building serverless applications on AWS. It extends AWS CloudFormation to provide a simplified way of defining serverless resources such as AWS Lambda functions, API Gateway APIs, DynamoDB tables, and more. This is illustrated in *Figure 9.1*:

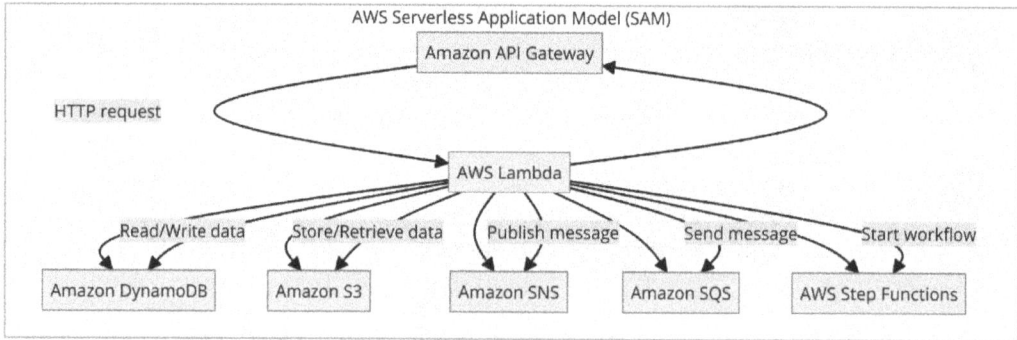

Figure 9.1: AWS Serverless Application Model

The AWS SAM framework diagram illustrates the interactions and components involved in a typical serverless architecture using AWS services. Here's a breakdown of the key components:

- **AWS Lambda**: A serverless computing service. It lets you run code without provisioning or managing servers.

- **Amazon API Gateway**: A fully managed service that allows you to create HTTP APIs that act as the front door for your serverless applications. Clients can invoke your Lambda functions through these APIs.

- **Events**: Events trigger the execution of your Lambda functions. These events can originate from different sources:

  - HTTP requests via API Gateway

  - Changes in data sources such as S3 buckets or DynamoDB tables

  - Scheduled triggers based on time

- **Amazon DynamoDB**: A NoSQL database service for storing and retrieving data at any scale. Your Lambda functions can interact with Amazon DynamoDB to store or retrieve data.

- **Amazon S3**: A scalable object storage service. Your Lambda functions can interact with Amazon S3 to store or retrieve files.

- **Amazon SNS/SQSimple**: **Simple Notification Service (SNS)** is a pub/sub messaging service, while **Simple Queue Service (SQS)** is a message queuing service. Your Lambda functions can use SNS to publish messages to SQS queues or subscribe to receive messages from them.

- **AWS Step Functions**: A service for orchestrating workflows made up of multiple Lambda functions. It allows you to define the order of execution and handle errors.

Figure *9.1* depicts a serverless application on AWS built with the AWS SAM framework. Users initiate interactions through an API Gateway endpoint, which then directs those requests to AWS Lambda functions. These Lambda functions can access and process data from various sources such as DynamoDB (database), S3 (storage), SNS (messaging), and SQS (queuing). Optionally, AWS Step Functions can be used to orchestrate complex workflows involving multiple Lambda functions. By leveraging AWS SAM and its templates, developers can create scalable and cost-effective serverless applications on AWS.

SAM also allows for local testing of Lambda functions. This is a valuable feature that can help developers to debug and troubleshoot their code before deploying it to production. There are two ways to test Lambda functions locally: using AWS Toolkits or running AWS SAM in debug mode. AWS Toolkits are IDE plugins that allow you to set breakpoints, inspect variables, and execute code one line at a time. SAM also allows you to run AWS SAM in debug mode to attach to third-party debuggers such as ptvsd or Delve.

For more information on local testing with SAM, please refer to the AWS documentation: `https://docs.aws.amazon.com/serverless-application-model/latest/developerguide/serverless-sam-cli-using-debugging.html`.

Azure Functions Core Tools

This **command-line interface (CLI)** provides a local development environment that mimics the Azure Functions runtime. It allows developers to build, test, and debug their functions on their machines before deploying them to Azure. Additionally, **Azure Functions Core Tools** integrates with continuous deployment pipelines using tools such as Azure DevOps or GitHub Actions. Let's look at *Figure 9.2*:

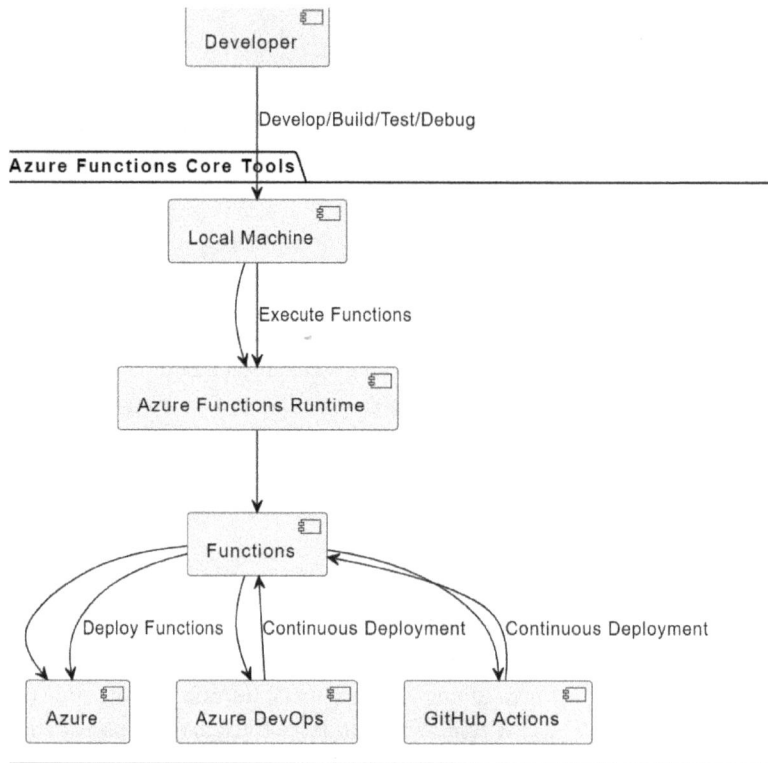

Figure 9.2: Azure Functions Core Tools diagram

Figure *9.2* illustrates the components and workflow involved in developing, testing, and deploying Azure Functions using Azure Functions Core Tools. Here is a detailed description:

- **Developer**: The developer interacts with the local machine to develop, build, test, and debug Azure Functions.

- **Local Machine**: This is the developer's environment where Azure Functions Core Tools are installed. It provides a local development environment that mimics Azure Functions runtime.

- **Azure Functions Runtime**: This component simulates the Azure Functions runtime on the local machine, allowing the developer to execute and test functions locally before deploying them to Azure.

- **Functions**: These are the individual functions created and managed by the developer. They are executed by the Azure Functions runtime and can be deployed to Azure.

- **Azure**: This represents the Azure cloud environment where the functions are deployed. Once the functions are tested locally, they are deployed to Azure for production use.

- **Azure DevOps**: This is a set of development tools and services for managing the entire application lifecycle. It integrates with Azure Functions Core Tools to enable continuous deployment of functions to Azure.

- **GitHub Actions**: This is a **Continuous Integration (CI)** and **Continuous Deployment (CD)** platform provided by GitHub. It integrates with Azure Functions Core Tools to automate the deployment of functions to Azure.

Figure 9.2 represents the complete workflow of using Azure Functions Core Tools for local development, testing, and continuous deployment of Azure Functions.

Google Cloud Functions

Google Cloud Functions is a lightweight, event-driven compute service that allows you to run your code in response to events. It is designed for building and connecting cloud services with simple, single-purpose functions, as shown in *Figure 9.3*:

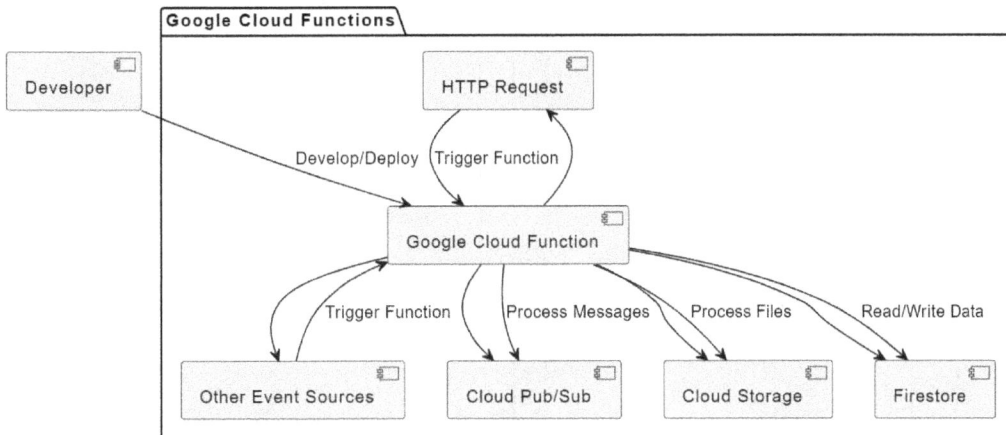

Figure 9.3: Google Cloud Functions

Figure 9.3 illustrates the architecture of Google Cloud Functions, showcasing its components and their interactions. Google Cloud Functions is the core component where the serverless functions reside. These functions can be triggered by various **event sources**, such as the following:

- **HTTP request**: Functions can be triggered by HTTP requests, enabling web-based interactions

- **Cloud Pub/Sub**: Functions can process messages from Cloud Pub/Sub, a messaging service that allows you to send and receive messages between independent applications

- **Cloud storage**: Functions can be triggered by events in Cloud Storage, such as file creation, modification, or deletion

- **Firestore**: Functions can interact with Firestore, a NoSQL document database, to read and write data

- **Other event sources**: Functions can also be triggered by other supported event sources, providing flexibility in handling various types of events

This architecture allows developers to build event-driven applications using serverless functions, eliminating the need to manage server infrastructure. It integrates seamlessly with various Google Cloud services, providing a scalable and flexible environment for running event-driven code.

Oracle Functions

Oracle Functions is a fully managed serverless platform that allows you to run your code without provisioning or managing servers.

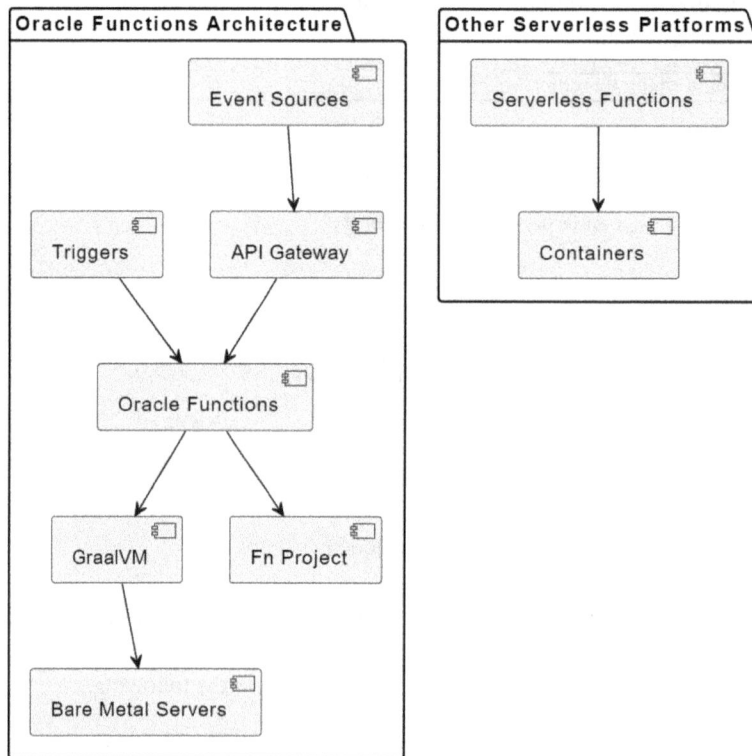

Figure 9.4: Oracle Functions

Figure 9.4 illustrates Oracle Functions' architecture, highlighting key components:

- **Oracle Functions**: Core serverless compute service built on the open source Fn Project
- **GraalVM**: High-performance runtime environment is known for excellent cold start performance
- **Bare Metal Servers**: Underlying infrastructure for enhanced performance and predictability
- **API Gateway**: Manages and routes incoming requests to appropriate functions
- **Event sources and triggers**: Various origins and activators of function execution

The key advantages include the following:

- **Performance**: GraalVM on bare-metal servers reduces cold start times and enhances overall function performance
- **Flexibility**: Versatile function triggering and management through API Gateway and diverse event sources
- **Efficiency**: Better resource utilization by avoiding container overhead

The main differences from traditional serverless architectures are as follows:

- **Runtime**: Uses GraalVM instead of traditional JVM runtimes
- **Deployment**: Functions run on bare-metal servers, not virtualized environments or containers
- **Architecture**: Is container-native but containerless, avoiding common container overhead
- **Foundation**: Based on the open source Fn Project, unlike many proprietary offerings

These features contribute to Oracle Functions' focus on high performance and efficiency, particularly in cold start times and resource utilization.

Oracle Functions' bare-metal deployment excels in performance-critical scenarios where speed and predictable latency are top priorities. However, for cost-conscious applications, containerized options such as AWS Lambda may be more appealing. If you prioritize cold starts in Java workloads, Oracle Functions, powered by GraalVM, offer a potential advantage. For applications processing highly sensitive data, bare-metal's isolation might be preferable.

Industry examples – Java serverless functions with a focus on concurrency

Let's dive into some real-world examples of how companies are using Java serverless functions and how they handle concurrency in their applications. We'll explore industry examples, extract valuable lessons learned, and examine a code example to understand the practical aspects of implementing Java serverless applications with concurrency in mind.

Airbnb – optimizing property listings with serverless solutions

Airbnb, a leading online marketplace for lodging and experiences, employs serverless architectures to enhance the management of property listings and user interactions. Airbnb uses AWS Lambda to achieve this:

- **Image processing**: When a host uploads images of their property, an AWS Lambda function is triggered to process and optimize the images for various device formats and resolutions. This task is performed concurrently to handle *multiple uploads efficiently*.

- **Search indexing**: AWS Lambda functions update search indexes in real time as new properties are listed or existing ones are updated. This ensures that users receive the most accurate and current search results.

By leveraging AWS Lambda, Airbnb ensures high performance, flexibility, and reliability in their serverless architecture. Image processing and search indexing are performed quickly and efficiently, enhancing user experience. The ability to deploy and update serverless functions independently allows for rapid iteration and deployment of new features. Additionally, serverless functions provide high availability and fault tolerance, maintaining a seamless experience for users.

LinkedIn – enhancing data processing with serverless architectures

LinkedIn, the world's largest professional network, utilizes serverless architectures to manage and process vast amounts of data generated by user interactions, job postings, and content sharing. LinkedIn leverages Azure Functions to handle these tasks efficiently:

- **Real-time notifications**: LinkedIn uses Azure Functions to process real-time notifications. When a user receives a connection request or a message, an event triggers Azure Functions to process and deliver the notification promptly.

- **Data analytics**: Azure Functions process data streams in real-time, aggregating metrics and generating insights. This allows LinkedIn to provide users with up-to-date analytics on their profiles, such as profile views and search appearances.

Azure Functions enables LinkedIn to achieve scalability, efficiency, and cost-effectiveness in their serverless architecture. The automatic scaling capabilities of Azure Functions ensure that LinkedIn can handle millions of concurrent user interactions. Serverless functions reduce infrastructure management overhead, allowing LinkedIn's engineering team to focus on developing new features. Moreover, the pay-per-use pricing model helps optimize costs during varying traffic periods.

Expedia – streamlining travel booking with serverless solutions

Expedia, a global travel booking platform, leverages Java-based AWS Lambda functions to handle various aspects of its service, ensuring efficient and reliable operations across its platform:

- **Booking confirmation**: AWS Lambda functions manage booking confirmations in real time. When a user completes a booking, an event triggers a Lambda function to confirm the reservation, update the inventory, and notify the user.

- **Price aggregation**: Expedia uses Lambda to aggregate prices from multiple airlines and hotels concurrently. This ensures that users receive the most competitive rates in real time, enhancing the booking experience.

- **User notifications**: Lambda functions send personalized notifications to users about their bookings, including updates, reminders, and special offers.

AWS Lambda enables Expedia to achieve scalability, efficiency, and improved user experience in its serverless architecture. The automatic scaling capabilities of AWS Lambda allow Expedia to handle peaks in booking volumes seamlessly. Serverless functions streamline complex processes, such as price aggregation, by handling multiple data sources concurrently. Real-time notifications and confirmations enhance the overall user experience, providing timely and relevant information to travelers.

These case studies demonstrate how industry leaders leverage serverless architectures and concurrency management to optimize their applications. By employing serverless solutions, companies can achieve scalability, efficiency, cost-effectiveness, and enhanced user experiences in their respective domains.

Building with serverless frameworks – a practical approach

Serverless frameworks are the developer's toolbox for crafting efficient and robust serverless applications. These frameworks go beyond the core compute services offered by cloud providers, providing a comprehensive set of tools and functionalities. In this section, we'll delve into the importance of serverless frameworks and how they streamline the development process. To solidify this understanding, we'll explore a real-world example using code demonstrations. Specifically, we'll see how AWS SAM simplifies defining and deploying serverless applications on AWS. By the end of this section, you'll be equipped to leverage the power of serverless frameworks in your own projects!

Using AWS SAM to define and deploy a serverless application

We'll design a simulation of a global travel booking platform that leverages Java-based AWS Lambda functions to handle various aspects of their service. This includes booking validation, payment processing, security checks, inventory updates, data processing, and user notifications. We'll use AWS Step Functions to orchestrate these tasks, DynamoDB for data storage, AWS Cognito for security checks, and API Gateway to expose endpoints.

Please look at *Figure 9.5*:

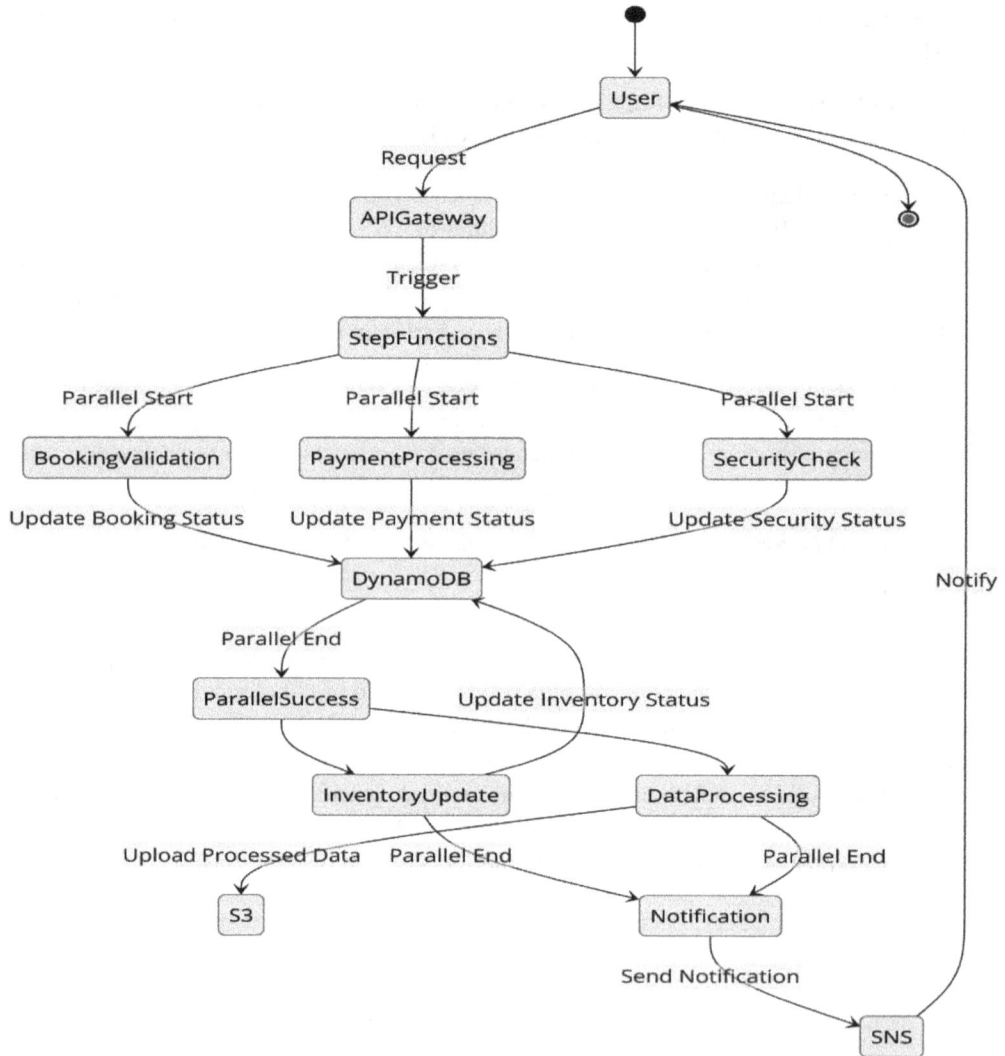

Figure 9.5: A global travel booking system

To accomplish this task, we'll utilize AWS cloud services within a serverless architecture. Our approach involves coordinating multiple Lambda functions using AWS Step Functions, a service designed to orchestrate complex workflows. This allows us to define the order of execution for each function and gracefully handle potential errors.

In our travel booking system, several specialized Lambda functions collaborate within a *Step Functions* workflow. These functions include `BookingValidationFunction`, `PaymentProcessingFunction`, `SecurityCheckFunction`, `InventoryUpdateFunction`, `DataProcessingFunction`, and `SendNotificationFunction`. Each function handles a specific step in the booking process. In this section, we'll focus on `BookingValidationFunction` as an illustrative example. The Java code for this function is presented below, while the code for the remaining functions can be found in the GitHub repository.

By using Step Functions, we gain the ability to create a more robust and manageable system. Step Functions simplify error handling, provide visibility into workflow progress, and enable us to retry failed steps automatically. This results in a more reliable and scalable solution for managing the complexities of the travel booking process.

`BookingValidationFunction` is responsible for validating the booking request data, ensuring that all required fields and data formats are correct. It also verifies the availability of requested items or dates by querying the DynamoDB table.

Here is the Java code for `BookingValidationFunction`:

```java
public class BookingValidationFunctionApplication {

    public static void main(String[] args) {
        SpringApplication.run(
            BookingValidationFunctionApplication.class,
                args);
    }    @Bean
    public Function<Map<String, Object>, Map<String, Object>>
    bookingValidation() {
        return input -> {
            Map<String, Object> response =new HashMap<>();
            // Validate booking details
            if (validateBooking(input)) {
                // Update DynamoDB with booking status
                if (verifyAvailability(input)) {
                    response.put(
                        "status", "Booking Validated");
                } else {
                    response.put("status",
```

```
                            "Booking Not Available");
                }              } else {
                response.put("status",
                      "Invalid Booking Data");
            }return response;
        };}
    private boolean validateBooking(Map<String,Object> input) {
            // Implement validation logic: check for required fields and
            data format
            // Example validation
            return input.containsKey(
                "bookingDate") && input.containsKey("itemId");
        }private boolean verifyAvailability(Map<String, Object> input) {
            // Implement availability check logic by querying the
            BookingTable in DynamoDB
            // This is a placeholder for actual DynamoDB query logic
            // Example query
            String bookingDate = (String) input.get(
                "bookingDate");
            String itemId = (String) input.get("itemId");


            // Assume a service class exists to handle DynamoDB operations
            // return bookingService.isAvailable(bookingDate,
                itemId);


            // For the sake of example, let's assume all bookings are
            available
            return true;
    }}
```

BookingValidationFunctionApplication is a Spring Boot application that serves as a serverless function for validating booking details.

The bookingValidation() method, annotated with @Bean, is the main function that validates the booking details based on the input data. It checks for the presence of required fields, verifies the availability of the booking by querying BookingTable in DynamoDB (placeholder logic), and returns a response map with the validation status.

The class also includes placeholder methods for implementing the validation logic validateBooking() and availability check logic verifyAvailability().

Next, we create the necessary resources in the CloudFormation template. The DynamoDB tables `BookingTable` and `InventoryTable` store and manage the booking and inventory data respectively, enabling efficient and scalable persistence of information related to travel bookings and available inventory items:

```
      # DynamoDB Tables
BookingTable:
    Type: AWS::DynamoDB::Table
    Properties:
        TableName: BookingTable
        AttributeDefinitions:
            - AttributeName: BookingId
            AttributeType: S
        KeySchema:
            - AttributeName: BookingId
            KeyType: HASH
        ProvisionedThroughput:
            ReadCapacityUnits: 5
            WriteCapacityUnits: 5

InventoryTable:
    Type: AWS::DynamoDB::Table
    Properties:
        TableName: InventoryTable
        AttributeDefinitions:
            - AttributeName: ItemId
            AttributeType: S
        KeySchema:
            - AttributeName: ItemId
            KeyType: HASH
        ProvisionedThroughput:
            ReadCapacityUnits: 5
            WriteCapacityUnits: 5
```

In our serverless architecture, we employ Amazon Cognito to handle user authentication and authorization. Cognito is a fully managed service that allows you to easily add user sign-up, sign-in, and access control to your web and mobile apps.

In our case, we'll create a Cognito user pool (`CognitoUserPool`). This user directory serves as the central repository for our application's user identities. When users register or sign in, Cognito securely stores their information and handles the authentication process. It provides features such as the following:

- **User management**: Create, read, update, and delete user profiles
- **Authentication**: Verify user credentials and issue secure tokens (e.g., JWTs) for access
- **Security**: Enforce password policies, multi-factor authentication (MFA), and other security measures
- **Authorization**: Define fine-grained access controls based on user attributes or groups
- **Federation**: Integrate with external identity providers such as Facebook, Google, or enterprise identity systems

By leveraging Cognito, we can offload the complexities of user management and focus on building the core functionality of our travel booking application:

```
# Cognito User Pool
CognitoUserPool:
    Type: AWS::Cognito::UserPool
    Properties:
        UserPoolName: TravelBookingUserPool
```

Once the data processing is complete, the processed data is uploaded to `ProcessedDataBucket`, which is an S3 bucket:

```
# S3 Bucket for processed data
ProcessedDataBucket:
    Type: AWS::S3::Bucket
    Properties:
        BucketName: processed-data-bucket
```

Create an Amazon SNS topic named `BookingNotificationTopic` to facilitate sending notifications related to booking events:

```
# SNS Topic for notifications
BookingNotificationTopic:
    Type: AWS::SNS::Topic
    Properties:
        TopicName: BookingNotificationTopic
```

Define an AWS **Identity and Access Management** (**IAM**) role named `LambdaExecutionRole` with the necessary permissions for the Lambda functions to access and interact with various AWS services, such as DynamoDB, S3, SNS, and Cognito:

```
# IAM Role for Lambda
LambdaExecutionRole:
    Type: AWS::IAM::Role
    Properties:
        AssumeRolePolicyDocument:
            Version: '2012-10-17'
            Statement:
                - Effect: Allow
            Principal:
                Service: lambda.amazonaws.com
                Action: sts:AssumeRole
        Policies:
            - PolicyName: LambdaPolicy
            PolicyDocument:
                Version: '2012-10-17'
            Statement:
                - Effect: Allow
                Action:
                    - dynamodb:PutItem
                    - dynamodb:GetItem
                    - dynamodb:UpdateItem
                    - s3:PutObject
                    - sns:Publish
                    - cognito-idp:AdminGetUser
                Resource: '*'
```

Specify the function names, handlers, runtime, and code location (S3 bucket and key) for each Lambda function. Specify the function names, handlers, runtime, and code location (S3 bucket and key) for each Lambda function:

```
# Lambda Functions
BookingValidationFunction:
    Type: AWS::Lambda::Function
    Properties:
        FunctionName: BookingValidationFunction
        Handler: com.example.
BookingValidationFunctionApplication::apply
        Role: !GetAtt LambdaExecutionRole.Arn
        Runtime: java17
```

```
      Code:
              S3Bucket: your-s3-bucket-name
  S3Key: booking-validation-1.0-SNAPSHOT.jar
```

For each of the other Lambda functions (`PaymentProcessingFunction`, `SecurityCheckFunction`, `InventoryUpdateFunction`, `DataProcessingFunction`, and `SendNotificationFunction`), you would write similar definitions within the `Resources` section of your CloudFormation template, adjusting the `FunctionName`, `Handler`, `CodeUri`, and `Policies` properties to match their respective implementations.

Next, we create the Step Functions state machine in the CloudFormation template. Use the `AWS::StepFunctions::StateMachine` resource type and define the state machine definition in the `DefinitionString` property using Amazon States Language:

```
# Step Function
TravelBookingStateMachine:
    Type: AWS::StepFunctions::StateMachine
    Properties:
        StateMachineName: TravelBookingStateMachine
        RoleArn: !GetAtt LambdaExecutionRole.Arn
        DefinitionString: !Sub |
            {"Comment": "Travel Booking Workflow",
            "StartAt": "ParallelTasks",
            "States": {
                "ParallelTasks": {
                "Type": "Parallel",
                "Branches": [
                    {"StartAt": "BookingValidation",
                        "States": {
                        "BookingValidation": {
                        "Type": "Task",
                    "Resource": "${
                        BookingValidationFunction.Arn}",
                        "End": true
                    }
                }
            }
        },
        {"StartAt": "PaymentProcessing",
            "States": {
                "PaymentProcessing": {
                    "Type": "Task",
                    "Resource": "${
                    PaymentProcessingFunction.Arn}",
                    "End": true
```

```
                    }
                }
            },
            {"StartAt": "SecurityCheck",
                "States": {
                    "SecurityCheck": {
                        "Type": "Task",
                        "Resource": "${
                            SecurityCheckFunction.Arn}",
                        "End": true
                    }
                }
            }
        ],
        "Next": "InventoryUpdate"
        },
        "InventoryUpdate": {
            "Type": "Task",
            "Resource": "${
                InventoryUpdateFunction.Arn}",
            "Next": "DataProcessing"
        },
        "DataProcessing": {
            "Type": "Task",
            "Resource": "${
                DataProcessingFunction.Arn}",
            "Next": "SendNotification"
        },
        "SendNotification": {
            "Type": "Task",
            "Resource": "${
                SendNotificationFunction.Arn}",
            "End": true
        }
    }
}
```

This AWS Step Functions state machine definition describes the workflow for a travel booking system. The workflow starts with a parallel execution of three tasks: `"BookingValidation"`, `"PaymentProcessing"`, and `"SecurityCheck"`. These tasks are performed concurrently using separate Lambda functions.

After the parallel tasks are complete, the workflow proceeds to the `"InventoryUpdate"` task, which updates the inventory using the `"InventoryUpdateFunction"` Lambda function.

Next, the `"DataProcessing"` task is executed using the `"DataProcessingFunction"` Lambda function to perform any necessary data processing.

Finally, the `"SendNotification"` task is triggered, which uses the `"SendNotification Function"` Lambda function to send notifications to the user.

The state machine definition utilizes the previously created Lambda functions and specifies the flow and order in which these tasks are executed. This enables a structured and coordinated workflow for handling travel bookings.

The following code sets up an API gateway for the travel booking application:

```
API Gateway
TravelBookingApi:
    Type: AWS::ApiGateway::RestApi
    Properties:
        Name: TravelBookingApi


TravelBookingResource:
    Type: AWS::ApiGateway::Resource
    Properties:
        ParentId: !GetAtt TravelBookingApi.RootResourceId
        PathPart: booking
        RestApiId: !Ref TravelBookingApi


TravelBookingMethod:
    Type: AWS::ApiGateway::Method
    Properties:
        AuthorizationType: NONE
        HttpMethod: POST
        ResourceId: !Ref TravelBookingResource
        RestApiId: !Ref TravelBookingApi
        Integration:
            IntegrationHttpMethod: POST
            Type: AWS_PROXY
            Uri: !Sub "arn:aws:apigateway:us-west-2:states:action/
StartExecution"
```

The `TravelBookingApi` resource creates a new REST API named `TravelBookingApi`. Now, `TravelBookingResource` defines a new resource path `/booking` under this API. `TravelBookingMethod` sets up a POST method for the `/booking` resource, which is integrated with AWS Step Functions to start the travel booking workflow execution.

Deploy the Java code and the CloudFormation stack (prepare your environment). Ensure you have an AWS account and the necessary permissions to create resources such as DynamoDB tables, API Gateway, Lambda functions, IAM roles, and so on.

Install and configure the AWS CLI if you haven't already. Instructions can be found at `https://docs.aws.amazon.com/cli/latest/userguide/cli-chap-configure.html`.

Package your Lambda function code into JAR files and upload them to an S3 bucket. Note that the S3 bucket name and the keys (paths) where the JAR files are stored. You'll need these when running the CloudFormation script.

After that, modify the CloudFormation script. Ensure that the `S3Bucket: !Ref Bucket Name` and `S3Key: booking-validation-1.0-SNAPSHOT.jar` parameter values match the S3 bucket name where your Lambda code is stored and the JAR file name.

Run the CloudFormation script. Save the CloudFormation script to a file, for example, `travel-booking-template.yaml`.

Open a terminal or command prompt. Run the following command to create a new CloudFormation stack:

```
aws cloudformation create-stack --stack-name TravelBookingStack
--template-body file://travel-booking-template.yaml --parameters
ParameterKey=BucketName,ParameterValue=YOUR_S3_BUCKET_NAME
```

Replace YOUR_S3_BUCKET_NAME with the actual name of your S3 bucket, and the `stack-name` with your stack name.

The `--parameters` flag is used to pass parameters to the CloudFormation template. Parameters are defined in the template and allow you to provide dynamic values at stack creation time. In this case, the `--parameters` flag is used to specify the S3 bucket name where the Lambda code is stored.

You can monitor the progress of your stack creation in the AWS Management Console under CloudFormation. Alternatively, you can use the AWS CLI to check the status:

```
aws cloudformation describe-stacks --stack-name TravelBookingStack
```

Replace the stack name `TravelBookingStack` with your stack name.

Once the stack creation is complete, verify that all the resources (DynamoDB tables, API Gateway, Lambda functions, IAM roles, etc.) have been created successfully. You can navigate to each service in the AWS Management Console to check the resources.

After the resources are deployed, you can test the API Gateway endpoint. The API Gateway URL can be found in either the outputs section of the CloudFormation stack or directly in the API Gateway service in the AWS Management Console.

We can test API Gateway using a tool such as curl or Postman.

First, we use curl:

```
curl -X POST https://YOUR_API_GATEWAY_URL/booking -d '{"bookingId":
"12345", "itemId": "item123", "quantity": 1, "paymentToken":
"token123", "amount": "100"}'
```

Replace `YOUR_API_GATEWAY_URL` with the actual URL of your API Gateway instance.

Next, we use Postman:

- Set the HTTP method to `POST`
- Enter the URL of your API gateway followed by `/booking`
- In the body of the request, use raw JSON to provide the required booking details. Here is an example of a JSON payload:

```
{   "bookingId": "12345",
  "itemId": "item123",
  "quantity": 1,
  "paymentToken": "token123",
  "amount": 100,
  "customerName": "John Doe",
  "customerEmail": "john.doe@example.com"
}
```

Deleting the CloudFormation stack will remove all the resources created by the stack, helping to avoid unnecessary costs and resource usage. To delete the resources created by the CloudFormation stack, run the following command:

```
aws cloudformation delete-stack --stack-name TravelBookingStack
```

Replace the stack name `TravelBookingStack` with your stack name.

This section explored serverless frameworks for building travel booking applications. We highlighted their advantages and how they streamline development. The key takeaways are as follows:

- Serverless frameworks simplify development with comprehensive tools
- We built a Java-based booking validation function using Spring Cloud Function
- AWS Step Functions orchestrated tasks within the booking workflow
- We explored integrating DynamoDB, Cognito, API Gateway, and other AWS services

By understanding these concepts, you're equipped to leverage serverless frameworks for building efficient travel booking applications and explore further functionalities offered by these frameworks and AWS services.

Summary

In this chapter, we explored serverless computing and its integration with Java's concurrency capabilities. We examined the core concepts of serverless architectures, including event-driven processing, statelessness, and automatic scaling, along with their advantages and potential drawbacks. Understanding these concepts is crucial for effectively leveraging serverless computing in your applications.

We discussed adapting Java's concurrency model to serverless environments, focusing on tools such as CompletableFuture and parallel streams. We highlighted the best practices for Java serverless applications, including strategies for minimizing cold starts, efficient resource management, and leveraging frameworks such as Spring Cloud Function, Micronaut, and Quarkus.

The chapter introduced major serverless platforms such as AWS SAM, Azure Functions Core Tools, Google Cloud Functions, and Oracle Functions, highlighting their unique features and how they simplify the development and deployment of serverless applications. We explored Java-specific optimization techniques to enhance performance and reduce latency in serverless environments.

Through real-world examples from companies such as Airbnb, LinkedIn, and Expedia, we saw practical applications of serverless architectures and concurrency management. These case studies illustrated how industry leaders leverage serverless solutions to achieve scalability, efficiency, and enhanced user experiences.

Finally, we provided a hands-on example of building a travel booking application using AWS SAM, demonstrating how to integrate various AWS services and orchestrate workflows using AWS Step Functions. This practical approach equipped you with the knowledge to deploy and manage serverless applications effectively.

By the end of this chapter, you should be well-equipped to leverage Java's concurrency features in serverless environments, apply optimization best practices, and make informed decisions about using serverless architectures in your projects.

As we transition from serverless architectures to broader cloud computing paradigms, our next chapter will explore another crucial aspect of modern application development: auto-scaling. *Chapter 10* will delve into synchronizing Java's concurrency models with cloud auto-scaling dynamics, building upon the concurrent programming concepts we've discussed and applying them to the elastic nature of cloud environments. This knowledge will be essential for developers looking to create robust, scalable Java applications that can efficiently adapt to varying workloads in cloud platforms.

Questions

1. What is the primary advantage of serverless computing over traditional server-based architectures?

   A. Higher server management overhead

   B. Manual scaling of resources

   C. Automatic scaling and reduced operational overhead

   D. Limited integration with cloud services

2. Which Java concurrency feature is particularly useful for performing asynchronous tasks in serverless functions?

   A. **Java Virtual Machine (JVM)**

   B. CompletableFuture

   C. **Java Database Connectivity (JDBC)**

   D. JavaBeans

3. What is the primary purpose of the Fork/Join framework in Java serverless applications?

   A. Encrypting data transfers

   B. Handling single-threaded operations

   C. Managing recursive tasks by dividing them into smaller subtasks

   D. Logging and error handling

4. Which of the following best practices helps minimize cold starts in Java serverless applications?

   A. Use the heaviest deployment package possible

   B. Optimize function size and use provisioned concurrency

   C. Avoid using any form of concurrency

   D. Enable all possible cloud services

5. What is a key benefit of using parallel streams in serverless Java functions?

   A. Blocking the main thread for all tasks

   B. Improved performance through concurrent data processing

   C. Simplifying the deployment process

   D. Reducing the need for error handling

Part 3:
Mastering Concurrency in the Cloud – The Final Frontier

As we reach the culmination of our journey through Java's concurrency landscape, *Part 3* explores the most advanced and forward-looking aspects of concurrent programming in cloud environments. This final section synthesizes the knowledge gained from previous chapters, applying it to the cutting-edge realm of cloud computing and beyond.

Each chapter offers practical, real-world examples and use cases, allowing readers to apply concepts from earlier parts of the book in innovative ways. As we conclude, *Part 3* equips readers with the vision and tools to be at the forefront of concurrent programming in the age of cloud computing and beyond, transforming them from proficient developers into masters of Java concurrency.

This part includes the following chapters:

- *Chapter 10, Synchronizing Java's Concurrency with Cloud Auto-Scaling Dynamics*
- *Chapter 11, Advanced Java Concurrency Practices in Cloud Computing*
- *Chapter 12, The Horizon Ahead*

10

Synchronizing Java's Concurrency with Cloud Auto-Scaling Dynamics

In the era of cloud computing, **auto-scaling** has become a crucial strategy for managing resource utilization and ensuring optimal application performance. As Java remains a prominent language for developing enterprise applications, understanding how to effectively synchronize Java's concurrency models with cloud auto-scaling dynamics is essential.

This chapter delves into the intricacies of leveraging Java's concurrency tools and best practices to build scalable and efficient applications in cloud environments. Through practical examples and real-world case studies, you will learn how to design and optimize Java applications for auto-scaling, implement monitoring and alerting mechanisms, and integrate with popular cloud-native tools and services.

The first practical application explored in this chapter is the development of a Kubernetes-based, auto-scaling Java application that simulates an e-commerce order processing service. Building upon this foundation, the chapter then introduces a second practical example focused on creating a serverless real-time analytics pipeline using Java and AWS services.

By the end of this chapter, readers will have a comprehensive understanding of how to harness the power of Java's concurrency models to build robust, scalable, and cost-effective applications that can seamlessly adapt to the dynamic nature of cloud environments. They will be equipped with the knowledge and skills to tackle the challenges of auto-scaling and ensure optimal performance and resource utilization in their Java-based systems.

In this chapter, we're going to cover the following main topics:

- Fundamentals of cloud auto-scaling – mechanisms and motivations
- Java's concurrency models – alignment with scaling strategies
- Optimizing Java applications for cloud scalability
- Monitoring and managing Java processes during auto-scaling events
- Real-world case studies
- Real-world deployments of Java-based systems in auto-scaling environments
- Advanced topics

Technical requirements

You'll need the following installed to follow along with this chapter:

- **Docker**: `https://docs.docker.com/get-docker/`
- **AWS Command Line Interface (CLI)**: `https://docs.aws.amazon.com/cli/latest/userguide/getting-started-install.html`
- **AWS Serverless Application Model (SAM) CLI**: Installing the AWS SAM CLI - AWS Serverless Application Model (`amazon.com`)
- **Kubernetes CLI (kubectl)**: `https://kubernetes.io/docs/tasks/tools/install-kubectl/`

The code in this chapter can be found on GitHub:

`https://github.com/PacktPublishing/Java-Concurrency-and-Parallelism`

Fundamentals of cloud auto-scaling – mechanisms and motivations

In the ever-evolving landscape of cloud computing, auto-scaling has emerged as a pivotal feature, enabling applications to dynamically adjust their resources to meet varying demands. This section delves into the core concepts and advantages of cloud auto-scaling, providing a comprehensive understanding of how it enhances scalability, cost-effectiveness, and resource utilization.

Definition and core concepts

Cloud auto-scaling automatically adjusts the amount of computational resources in a server farm based on CPU, memory, and network usage, ensuring optimal performance and cost efficiency. **Dynamic resource allocation** is a key concept where resources are added or removed based on real-time demand. Scaling can be done vertically (scaling up/down) by adjusting the capacity of existing instances, or horizontally (scaling out/in) by adding or removing instances to handle changes in workload.

Auto-scaling relies on predefined metrics and threshold-based triggers to initiate scaling actions. Load balancing distributes traffic evenly across instances for improved performance and reliability. Auto-scaling policies define rules for when and how scaling actions occur, either reactively or proactively. Continuous monitoring using tools such as AWS CloudWatch, Google Cloud Monitoring, and Azure Monitor is crucial for triggering scaling actions.

For example, an e-commerce website experiencing a surge in traffic during a holiday sale can leverage auto-scaling to automatically launch additional server instances to handle the increased load and prevent slowdowns or crashes. When the sale ends and traffic decreases, the extra instances are terminated to save costs.

Advantages of cloud auto-scaling

Cloud auto-scaling offers several benefits that enhance the performance, efficiency, and cost effectiveness of applications. Scalability is a key advantage, providing the ability to dynamically adjust resource allocation in response to changing demand through elasticity. Elasticity enables applications to adapt by automatically adjusting allocated resources (scaling up/down) and the number of instances (scaling out/in) based on predefined metrics and thresholds, ensuring optimal performance, cost efficiency, and resource utilization.

Auto-scaling promotes cost effectiveness through a pay-as-you-go model, where resources are only allocated as needed, avoiding costs associated with over-provisioning. Automation reduces the need for manual monitoring and scaling, lowering operational overhead and labor costs. Enhanced resource utilization ensures resources are used efficiently, reducing waste, while integrated load balancing distributes traffic evenly across instances, preventing bottlenecks.

Auto-scaling improves reliability and availability by ensuring there are always enough instances to handle the load, reducing the risk of downtime. It can also improve an application's resilience to localized failures or outages by automatically scaling in different regions or availability zones. Flexibility and agility enable applications to quickly adapt to changes in workload or user demand, crucial for applications with unpredictable traffic patterns, while developers and IT teams can focus on core business activities thanks to auto-scaling's automated nature.

For example, a start-up launching a suddenly popular mobile app can leverage cloud auto-scaling to handle the influx of users without performance degradation, while only incurring costs proportional to actual resource usage. The ability to scale up (vertically) and scale out (horizontally) ensures optimal performance and cost efficiency. By leveraging cloud auto-scaling, businesses can ensure their applications perform optimally, are cost efficient, and can quickly adapt to changing demands, which is essential in today's fast-paced digital landscape.

Triggers and conditions for auto-scaling

Auto-scaling in a cloud environment is driven by various triggers and conditions that ensure applications maintain optimal performance and resource utilization. Understanding these triggers helps in setting up effective auto-scaling policies that respond appropriately to changes in demand.

Common triggers for auto-scaling are as follows:

- **CPU utilization**:

  - **High CPU usage**: When CPU usage exceeds a certain threshold (e.g., 70-80%) over a specified period, additional instances are launched to handle the increased load

  - **Low CPU usage**: When CPU usage drops below a lower threshold (e.g., 20-30%), instances are terminated to save costs

- **Memory utilization**:

  - **High memory usage**: Similar to CPU usage, high memory utilization triggers the addition of more instances to ensure the application remains responsive

  - **Low memory usage**: If memory usage is consistently low, reducing the number of instances helps optimize costs

- **Network traffic**:

  - **Inbound/outbound traffic**: High levels of incoming or outgoing network traffic can trigger scaling actions to ensure sufficient bandwidth and processing power

  - **Latency**: Increased network latency can also be a trigger, prompting the system to scale out to maintain low response times

- **Disk input/output (I/O)**:

  - **High disk I/O operations**: Intensive read/write operations to disk can necessitate scaling out to distribute the load across more instances

  - **Disk space utilization**: Scaling actions can be triggered if available disk space is running low, ensuring that the application does not run into storage issues

- **Custom metrics:**

  - **Application-specific metrics**: Metrics such as the number of active users, requests per second, or transaction rates can be used to trigger scaling actions. These metrics are tailored to the specific needs of the application.

  - **Error rates**: An increase in error rates or failed requests can prompt scaling to handle the load more effectively or to isolate faulty instances.

Let's now look at the conditions for effective auto-scaling:

- **Threshold levels:**

  - **Setting appropriate thresholds**: Define upper and lower thresholds for key metrics to trigger scaling actions. These thresholds should be based on historical data and performance benchmarks.

  - **Hysteresis**: Implementing hysteresis (a delay between scaling actions) helps prevent rapid fluctuations in scaling (thrashing) by adding a buffer time before scaling up or down.

- **Cooldown periods**: After a scaling action is performed, a cooldown period allows the system to stabilize before another scaling action is triggered. This prevents over-scaling and ensures that the metrics accurately reflect the system's needs.

- **Predictive scaling:**

  - **Trend analysis**: Using historical data and **machine learning** (**ML**) algorithms, predictive scaling anticipates future demand and scales resources proactively rather than reactively

- **Scheduled scaling**: Scaling actions can be scheduled based on known patterns, such as increased traffic during business hours or specific events.

- **Resource limits:**

  - **Maximum and minimum limits**: Define the maximum and minimum number of instances to prevent excessive scaling that could lead to resource wastage or insufficient capacity

  - **Resource constraints**: Consider budgetary constraints and ensure that scaling actions do not exceed cost limits

- **Health checks:**

  - **Instance health monitoring**: Regular health checks ensure that only healthy instances are kept in the pool. Unhealthy instances are replaced to maintain application reliability.

  - **Graceful degradation**: Implementing mechanisms for graceful degradation ensures that the application can still function, albeit with reduced performance, when scaling thresholds are reached.

An example scenario would be an online gaming platform experiencing varying levels of user activity throughout the day. During peak hours, CPU and memory utilization increase significantly. By setting auto-scaling policies based on these metrics, the platform can automatically add more instances to handle the load. Conversely, during off-peak hours, the platform scales down to save costs, ensuring optimal resource utilization at all times.

Understanding the triggers and conditions for auto-scaling allows businesses to configure their cloud environments effectively, ensuring applications remain responsive, reliable, and cost efficient. This proactive approach to resource management is essential for maintaining high performance in dynamic and unpredictable usage scenarios.

A guide to setting memory utilization triggers for auto-scaling

Auto-scaling is a critical component for maintaining optimal application performance and resource utilization in cloud environments. This section provides a detailed guide on setting memory utilization triggers for auto-scaling, focusing on two popular auto-scaling solutions: **AWS auto-scaling** services and **Kubernetes-Based Event Driven Autoscaling** (**KEDA**) for Kubernetes. The first part covers the implementation using AWS services, and the second part introduces KEDA, a Kubernetes-based project supported by the Cloud Native Computing Foundation, for event-driven auto-scaling.

AWS auto-scaling services

In this section, we will explore how to set up memory utilization triggers for auto-scaling using AWS auto-scaling services. AWS provides robust tools and services to automatically adjust the number of running instances based on the current demand, ensuring that your application performs optimally while maintaining cost efficiency.

High memory usage

Let's dive in to see how to set high memory usage:

1. **Determine the threshold**:

    I.   **Analyze historical data**: Review past performance metrics to identify typical memory usage patterns. Determine the average and peak memory usage levels.

    II.   **Set a threshold**: A common practice is to set a high memory usage threshold between 70% and 85%. This range helps ensure there is enough buffer to add new instances before memory constraints impact performance.

2. **Configure the auto-scaling policy**:

    I.   **Choose a metric**: Use cloud provider-specific metrics such as Amazon CloudWatch (AWS), Azure Monitor, or Google Cloud Monitoring to track memory usage.

II.   **Set the alarm**: Create an alarm that triggers when memory usage exceeds the defined threshold using the AWS CLI:

```
aws cloudwatch put-metric-alarm --alarm-name HighMemoryUsage \
  --metric-name MemoryUtilization --namespace AWS/EC2
--statistic Average \
  --period 300 --evaluation-periods 2 --threshold 75 \
  --comparison-operator GreaterThanThreshold \
  --dimensions Name=AutoScalingGroupName,Value=your-auto-
scaling-group-name
```

This command creates a CloudWatch alarm with the specified parameters to monitor the memory utilization of instances within a specified auto-scaling group. The alarm is triggered when the average memory utilization exceeds 75% over two consecutive 5-minute periods (300 seconds each).

3.   **Define scaling actions**:

I.   **Scaling up**: Specify the action to take when the threshold is breached, such as adding a specified number of instances.

II.   **Cooldown period**: Set a cooldown period (e.g., 300 seconds) to allow the system to stabilize before evaluating further scaling actions.

Directly run this command using AWS CLI:

```
aws autoscaling put-scaling-policy --auto-scaling-group-name
your-auto-scaling-group-name \
  --policy-name ScaleOutPolicy \
  --scaling-adjustment 2 \
  --adjustment-type ChangeInCapacity \
  --cooldown 300
```

This script defines a scaling policy for an AWS auto-scaling group to address high memory usage. When the average memory utilization across the group's instances exceeds a predefined threshold, the policy triggers a scale-out event, adding two new instances to the group. To maintain the system stability and prevent overly aggressive scaling, a cooldown period of 300 seconds (5 minutes) is enforced after each scale-out event.

Low memory usage

Let's dive in to see how to set low memory usage:

1.   **Determine the threshold**:

I.   **Analyze historical data**: Identify periods of low memory usage and the minimum memory requirements for your application to function correctly.

II.   **Set a threshold**: Set the low memory usage threshold between 20% and 40%. This helps ensure instances are not terminated prematurely, which could affect performance.

2. **Configure the auto-scaling policy**:

I.   **Choose a metric**: Use the same cloud provider-specific metrics to monitor low memory usage.

II.   **Set the alarm**: Create an alarm that triggers when memory usage falls below the defined threshold using the AWS CLI.

Directly run this command using AWS CLI:

```
aws cloudwatch put-metric-alarm --alarm-name LowMemoryUsage \
  --metric-name MemoryUtilization --namespace AWS/EC2
--statistic Average \
  --period 300 --evaluation-periods 2 --threshold 30 \
  --comparison-operator LessThanThreshold \
  --dimensions Name=AutoScalingGroupName,Value=your-auto-
scaling-group-name
```

This script defines a CloudWatch alarm named LowMemoryUsage that monitors the memory utilization of instances within a specified auto-scaling group. The alarm triggers when the average memory utilization falls below 30% over two consecutive 5-minute periods (300 seconds each).

3. **Define scaling actions**:

I.   **Scaling down**: Specify the action to take when the low memory threshold is reached, such as removing a specified number of instances.

II.   **Cooldown period**: Set a cooldown period (e.g., 300 seconds) to allow the system to stabilize before further scaling actions.

Directly run this command using AWS CLI:

```
aws autoscaling put-scaling-policy \
  --auto-scaling-group-name your-auto-scaling-group-name \
  --scaling-adjustment -1 \
  --adjustment-type ChangeInCapacity \
  --cooldown 300 \
  --policy-name scale-in-policy
```

This script configures an auto-scaling policy for an auto-scaling group in AWS. The policy decreases the number of instances in the auto-scaling group by one when triggered. It uses the ChangeInCapacity adjustment type and has a cooldown period of 300 seconds to prevent rapid scaling actions.

KEDA

In addition to the cloud provider-specific auto-scaling services we just discussed, the open-source project KEDA provides a generic and extensible auto-scaling solution for Kubernetes environments. KEDA allows developers to define scalable targets based on various event sources, including cloud services, messaging queues, and custom metrics.

KEDA operates as a Kubernetes operator, running as a deployment on the Kubernetes cluster. It provides a **custom resource definition** (**CRD**) called `ScaledObject`, which defines the scaling behavior for a Kubernetes deployment or service. The `ScaledObject` resource specifies the event source, scaling metrics, and scaling parameters, allowing KEDA to automatically scale the target workload based on the defined criteria.

KEDA supports a wide range of event sources, including the following:

- Cloud services (AWS **Simple Queue Service** or **SQS**, Azure Queue Storage, Google PubSub, etc.)
- Databases (PostgreSQL, MongoDB, etc.)
- Messaging systems (RabbitMQ, Apache Kafka, etc.)
- Custom metrics (Prometheus, Stackdriver, etc.)

By integrating KEDA into your Java-based Kubernetes applications, you can benefit from a generic and extensible auto-scaling solution that seamlessly adapts to the demands of your cloud-native infrastructure. KEDA's event-driven approach and support for a variety of data sources make it a powerful tool for building scalable and responsive Java applications in Kubernetes environments.

Setting up KEDA and auto-scaling in an AWS environment

To demonstrate how KEDA can be used for auto-scaling in an AWS environment, let's walk through a practical example of setting it up in a Kubernetes cluster and integrating it with AWS services:

1. **Install KEDA using Helm**:

```
helm repo add kedacore https://kedacore.github.io/charts
helm repo update
helm install keda kedacore/keda
```

2. **Deploy an example application**: Deploy a simple application that can be scaled based on event metrics. For this example, we'll use a consumer application that processes messages from an AWS SQS queue. You will need to create two YAML files for this purpose.

   This is the first YAML file: `AWS SQS Credentials` (`aws-sqs-credentials.yaml`):

```
apiVersion: v1
kind: Secret
metadata:
```

```yaml
  name: aws-sqs-credentials
type: Opaque
stringData:
  AWS_ACCESS_KEY_ID: "<your-access-key-id>"
  AWS_SECRET_ACCESS_KEY: "<your-secret-access-key>"
```

This is the second YAML file: `Application Deployment` (`sqs-queue-consumer-deployment.yaml`):

```yaml
apiVersion: apps/v1
kind: Deployment
metadata:
  name: sqs-queue-consumer
spec:
  replicas: 1
  selector:
    matchLabels:
      app: sqs-queue-consumer
  template:
    metadata:
      labels:
        app: sqs-queue-consumer
    spec:
      containers:
      - name: sqs-queue-consumer
        image: <your-docker-image>
        env:
        - name: AWS_ACCESS_KEY_ID
          valueFrom:
            secretKeyRef:
              name: aws-sqs-credentials
              key: AWS_ACCESS_KEY_ID
        - name: AWS_SECRET_ACCESS_KEY
          valueFrom:
            secretKeyRef:
              name: aws-sqs-credentials
              key: AWS_SECRET_ACCESS_KEY
        - name: QUEUE_URL
          value: "<your-sqs-queue-url>"
```

3. Define a `ScaledObject` to link the deployment with the KEDA scaler. Create another YAML file for the `ScaledObject` configuration.

This is the `ScaledObject` YAML file (`sqs-queue-scaledobject.yaml`):

```
apiVersion: keda.sh/v1alpha1
kind: ScaledObject
metadata:
  name: sqs-queue-scaledobject
spec:
  scaleTargetRef:
    name: sqs-queue-consumer
  minReplicaCount: 1
  maxReplicaCount: 10
  triggers:
  - type: aws-sqs-queue
    metadata:
      queueURL: "<your-sqs-queue-url>"
      awsAccessKeyID: "<your-access-key-id>"
      awsSecretAccessKey: "<your-secret-access-key>"
      queueLength: "5"
```

4. Deploy the configuration files:

```
Apply the configurations to the Kubernetes cluster.
kubectl apply -f aws-sqs-credentials.yaml
kubectl apply -f sqs-queue-consumer-deployment.yaml
kubectl apply -f sqs-queue-scaledobject.yaml
```

5. **Verify the setup**:

   - **Check KEDA metrics server**: Ensure the KEDA metrics server is running:

   ```
   kubectl get deployment keda-operator -n keda
   ```

   - **Monitor the scaling behavior**: Watch the deployment to see how it scales based on the SQS queue length.

   ```
   kubectl get deployment sqs-queue-consumer -w
   ```

Integrating KEDA into your Kubernetes-based applications provides a powerful and flexible way to manage auto-scaling based on events. This enhances the efficiency and responsiveness of your applications, ensuring they can handle varying workloads effectively.

We've explored the core concepts of cloud auto-scaling, including its mechanisms, advantages, and the triggers and conditions that drive scaling decisions. We've also provided a guide on setting memory utilization triggers. This knowledge is crucial for creating cloud-based applications that can dynamically adapt to changing workloads, ensuring optimal performance while managing costs effectively.

Understanding these principles is essential in today's digital landscape, where applications must handle unpredictable traffic patterns and resource demands. Mastering these concepts equips you to design resilient, efficient, and cost-effective cloud applications.

Next, we'll explore how Java's concurrency models align with these scaling strategies. We'll examine how Java's rich set of concurrency tools can be used to create applications that seamlessly integrate with cloud auto-scaling, enabling efficient resource utilization and improved performance in dynamic environments.

Java's concurrency models – alignment with scaling strategies

Java's concurrency models offer powerful tools that align with auto-scaling strategies, enabling applications to dynamically adjust resource allocation based on real-time demand. Let's explore how Java's concurrency utilities support auto-scaling.

ExecutorService efficiently manages thread pools, allowing dynamic adjustment of active threads to match the workload. CompletableFuture enables asynchronous programming, facilitating non-blocking operations that scale with demand. **Parallel streams** harness the power of multiple CPU cores to process data streams in parallel, enhancing performance.

To demonstrate the practical application of these concurrency tools in auto-scaling, let's walk through a simple example. We will implement an auto-scaling solution that dynamically adjusts the number of worker threads based on the load:

1. Set up ExecutorService:

    ```
    ExecutorService executorService = Executors.
    newFixedThreadPool(initialPoolSize);
    ```

    It initializes ExecutorService with a fixed thread pool size specified by initialPoolSize.

2. Load monitoring:

    ```
    ScheduledExecutorService scheduler = Executors.
    newScheduledThreadPool(1);
    scheduler.scheduleAtFixedRate(() -> {
        int currentLoad = getCurrentLoad();
        adjustThreadPoolSize(executorService,
            currentLoad);
    }, 0, monitoringInterval, TimeUnit.SECONDS);
    ```

This code creates `ScheduledExecutorService`, which periodically checks the current load and adjusts the thread pool size accordingly. The task runs at a fixed interval defined by `monitoringInterval`, starting immediately. It measures the load using `getCurrentLoad()` and adjusts the thread pool size using `adjustThreadPoolSize(executorService, currentLoad)`. This mechanism dynamically scales resources based on real-time workload, ensuring efficient handling of varying demands.

3.   Adjust the thread pool size:

```
public void adjustThreadPoolSize(ExecutorService
executorService, int load) {
    // Logic to adjust the number of threads based on load
    int newPoolSize = calculateOptimalPoolSize(load);
    ((ThreadPoolExecutor) executorService).
    setCorePoolSize(newPoolSize);
    ((ThreadPoolExecutor) executorService).
    setMaximumPoolSize(newPoolSize);
}
```

The `adjustThreadPoolSize()` method adjusts the size of the thread pool based on the current load. It takes two parameters: `ExecutorService` and an `integer  load`. The method calculates the optimal pool size using `calculateOptimalPoolSize(load)`. It then sets the core pool size and the maximum pool size of `ThreadPoolExecutor` to the new pool size. This adjustment ensures that the thread pool dynamically matches the current workload, optimizing resource utilization and maintaining application performance.

4.   Handle the tasks with `CompletableFuture`:

```
CompletableFuture.runAsync(() -> {
    // Task to be executed
}, executorService);
```

The `CompletableFuture.runAsync()` method runs a task asynchronously using the provided `ExecutorService`. It takes a lambda expression, which represents the task to be executed and `ExecutorService` as parameters. This allows the task to run in a separate thread managed by `ExecutorService`, enabling non-blocking execution.

By utilizing these concurrency tools, we can create a responsive and efficient auto-scaling solution that adapts to changing workloads in real-time. This approach not only optimizes resource utilization but also enhances the overall performance and reliability of cloud-based applications.

Having explored how concurrency tools empower cloud applications, let's now delve into strategies for optimizing Java applications to thrive in the dynamic cloud landscape.

Optimizing Java applications for cloud scalability – best practices

As cloud environments demand efficient and scalable applications, optimizing Java applications for cloud scalability is crucial. This section focuses on best practices, resource management techniques, and a practical code example to demonstrate how to optimize Java applications for auto-scaling.

To enhance Java application scalability in cloud environments, it is essential to follow best design practices, which are as follows:

- **Microservices architecture**: Break down applications into smaller, independently deployable services. This allows for better resource allocation and easier scaling.

- **Asynchronous processing**: Use asynchronous processing to improve responsiveness and scalability. Leveraging tools such as `CompletableFuture` can help manage tasks without blocking the main thread.

- **Stateless services**: Design services to be stateless wherever possible. Stateless services are easier to scale because they do not require session information to be shared between instances.

- **Load balancing**: Implement load balancing to distribute incoming requests evenly across multiple instances of your application. This prevents any single instance from becoming a bottleneck.

- **Caching**: Use caching mechanisms to reduce the load on backend services and databases. This can significantly improve response times and reduce latency.

- **Containerization**: Use containers (e.g., Docker) to package applications and their dependencies. Containers provide consistency across different environments and simplify scaling and deployment.

In the next section, we will look at an example to see how it works in real-world applications.

Code example – best practices in optimizing a Java application for auto-scaling with AWS services and Docker

To demonstrate best practices in optimizing a Java application for auto-scaling, we will create a real-world example using AWS services and Docker. This example will involve deploying a Java application to AWS using CloudFormation, Docker for containerization, and various AWS services to manage and scale the application.

Let's look at the following diagram:

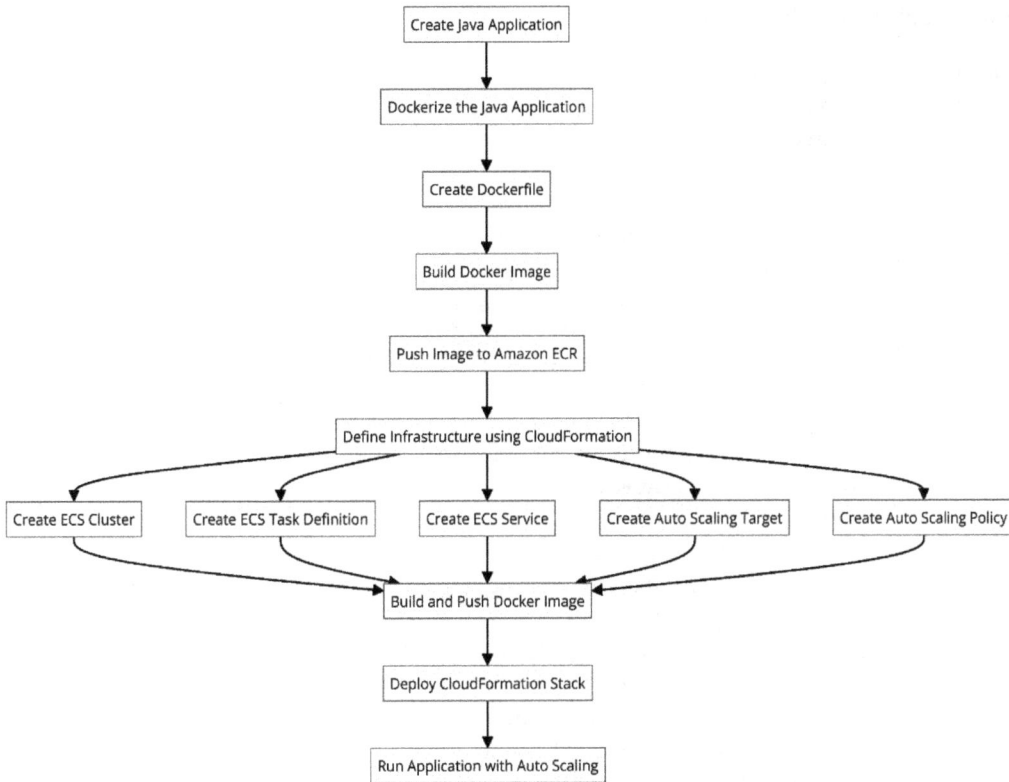

Figure 10.1: Deployment workflow for a Java application on AWS with auto-scaling

The diagram illustrates the architecture of an auto-scaling Java application deployed on AWS using Amazon **Elastic Container Service (ECS)**, Fargate, and CloudFormation. These steps and code blocks are designed to guide developers through the process of containerizing a Java application, deploying it on AWS, and ensuring it can scale automatically to handle varying loads. Before diving into the development process, it's essential to understand the purpose and sequence of these steps to ensure a smooth and efficient deployment. This preparation involves creating and configuring necessary AWS resources, building and pushing Docker images, and setting up infrastructure as code using CloudFormation to automate the entire process. Here are the steps to do so:

Step 1: **Dockerize the Java application**:

Create a simple Java application and package:

```java
public class App {
    private static final int initialPoolSize = 10;
    private static final int monitoringInterval = 5;
// in seconds

    public static void main(String[] args) {
        // Initialize ExecutorService with a fixed thread pool
        ExecutorService executorService = Executors.
newFixedThreadPool(initialPoolSize);
        ScheduledExecutorService scheduler = Executors.
newScheduledThreadPool(1);

        // Schedule load monitoring and adjustment task
        scheduler.scheduleAtFixedRate(() -> {
            int currentLoad = getCurrentLoad();
            adjustThreadPoolSize((
                ThreadPoolExecutor) executorService,
                    currentLoad);
        }, 0, monitoringInterval, TimeUnit.SECONDS);

        // Simulating task submission
        for (int i = 0; i < 100; i++) {
            CompletableFuture.runAsync(() -> performTask(),
                executorService);
        }
    }
    // Simulate getting current load
    private static int getCurrentLoad() {
        return (int) (Math.random() * 100);
    }
    // Adjust thread pool size based on load
    private static void adjustThreadPoolSize(
        ThreadPoolExecutor executorService, int load) {
            int newPoolSize = calculateOptimalPoolSize(
                load);
            executorService.setCorePoolSize(newPoolSize);
            executorService.setMaximumPoolSize(newPoolSize);
    }
    // Calculate optimal pool size
```

```
    private static int calculateOptimalPoolSize(int load) {
        return Math.max(1, load / 10);
    }
    // Simulate performing a task
    private static void performTask() {
        try {
            Thread.sleep((long) (Math.random() * 1000));
        } catch (InterruptedException e) {
            Thread.currentThread().interrupt();
        }
    }
}
```

Create a Dockerfile to containerize the Java application:

```
# Use an official OpenJDK runtime as a parent image
FROM openjdk:11-jre-slim

# Set the working directory
WORKDIR /usr/src/app

# Copy the current directory contents into the container at /usr/src/
app
COPY . .

# Compile the application
RUN javac App.java

# Run the application
CMD ["java", "App"]
```

Step 2: Define the infrastructure Using CloudFormation:

Create a CloudFormation template to define the required AWS resources. The CloudFormation template defines the necessary resources for deploying and managing a Dockerized Java application on AWS ECS with auto-scaling. The following is a breakdown of the template into sections, each explaining a specific resource.

First is the ECS cluster. This section creates an Amazon ECS cluster named `auto-scaling-cluster`. An ECS cluster is a logical grouping of tasks or services, providing the environment where your containerized application runs.

Next is the ECS task definition that specifies how Docker containers should be run on Amazon ECS. It defines various parameters and configurations for the containerized application, such as the required resources, network settings, and logging options. Here's a breakdown of the task definition:

```
Resources:
    ECSCluster:
        Type: AWS::ECS::Cluster
    Properties:
        ClusterName: auto-scaling-cluster
```

This section of the CloudFormation template is focused on creating an ECS cluster. Let's break down each component:

- `Resources`: This is the main section where we declare the different AWS resources we want to create.

- `ECSCluster`: This is the name we give to this particular resource within our template. We can reference it later using this name.

- `Type: AWS::ECS::Cluster`: This specifies that we're creating an ECS cluster resource, which is a logical grouping of container instances on which we can place tasks.

- `Properties`: This is where we define the configuration details of the cluster.

- `ClusterName: auto-scaling-cluster`: This is the name we're assigning to the ECS cluster. It's a good practice to choose a descriptive name that reflects the purpose of the cluster.

The ECS task definition specifies how Docker containers should be run on Amazon ECS. It defines various parameters and configurations for the containerized application, such as the required resources, network settings, and logging options. Here's a breakdown of the task definition:

```
TaskDefinition:
    Type: AWS::ECS::TaskDefinition
Properties:
    Family: auto-scaling-task
    NetworkMode: awsvpc
    RequiresCompatibilities:
        - FARGATE
    Cpu: 256
    Memory: 512
    ContainerDefinitions:
        - Name: auto-scaling-container
        Image: <your-docker-image-repo-url>
        Essential: true
    PortMappings:
        - ContainerPort: 8080
```

```
          LogConfiguration:
          LogDriver: awslogs
          Options:
                awslogs-group: /ecs/auto-scaling-logs
                awslogs-region: us-east-1
                awslogs-stream-prefix: ecs
```

This section defines the ECS task, which specifies how Docker containers should be run on ECS. The task definition auto-scaling-task includes the following:

- **Network mode**: awsvpc for Fargate tasks

- **CPU and memory**: It allocates 256 CPU units and 512 MB of memory

- **Container definitions**: It defines a container named auto-scaling-container that runs the Docker image specified by <your-docker-image-repo-url>, exposes port 8080, and configures logging to AWS CloudWatch

Now, we move on to ECS service. The ECS service is responsible for managing and running the ECS tasks defined in the task definition. It ensures that the specified number of tasks are maintained and running correctly. Here's a detailed breakdown of the ECS service definition:

```
ECSService:
    Type: AWS::ECS::Service
    Properties:
        Cluster: !Ref ECSCluster
        DesiredCount: 1
        LaunchType: FARGATE
        TaskDefinition: !Ref TaskDefinition
        NetworkConfiguration:
        AwsvpcConfiguration:
            AssignPublicIp: ENABLED
            Subnets:
                - subnet-12345678
# Replace with your subnet ID
                - subnet-87654321
# Replace with your subnet ID
            SecurityGroups:
                - sg-12345678 # Replace with your security group ID
```

This section creates an ECS service that manages and runs the ECS tasks defined in the task definition. The service does the following:

- It runs in the ECS cluster referenced by ECSCluster

- It has a desired count of 1 task

- It uses the Fargate launch type

- It uses the network configuration specified, including subnets and security groups

The auto-scaling target section defines how the ECS service scales in response to demand. This configuration ensures that the application can handle varying loads efficiently by automatically adjusting the number of running tasks. Here is the code for the auto-scaling target:

```
AutoScalingTarget:
    Type: AWS::ApplicationAutoScaling::ScalableTarget
    Properties:
        MaxCapacity: 10
        MinCapacity: 1
        ResourceId: !Join
        - /
        - - service
          - !Ref ECSCluster
          - !GetAtt ECSService.Name
        RoleARN: arn:aws:iam::<account-id>:role/aws-
service-role/ecs.application-autoscaling.amazonaws.com/
AWSServiceRoleForApplicationAutoScaling_ECSService
        ScalableDimension: ecs:service:DesiredCount
        ServiceNamespace: ecs
```

This section specifies the following:

- `MaxCapacity` and `MinCapacity` define the range for the number of tasks, with a minimum of 1 and a maximum of 10

- `ResourceId` identifies the ECS service to be scaled, constructed from the ECS cluster and service names

- `RoleARN` uses a specific **identity and access management** (**IAM**) role that grants permissions to Application auto-scaling to manage ECS service scaling

- `ScalableDimension` and `ServiceNamespace` indicate the ECS service's desired count as the scalable dimension within the ECS namespace

The auto-scaling policy section outlines the rules and conditions under which the ECS service will scale. This policy leverages AWS's Application auto-scaling to adjust the number of running tasks based on the specified metric. Here is the code for the auto-scaling policy:

```
AutoScalingPolicy:
    Type: AWS::ApplicationAutoScaling::ScalingPolicy
    Properties:
        PolicyName: ecs-auto-scaling-policy
        PolicyType: TargetTrackingScaling
```

```
    ScalingTargetId: !Ref AutoScalingTarget
    TargetTrackingScalingPolicyConfiguration:
        PredefinedMetricSpecification:
        PredefinedMetricType: ECSServiceAverageCPUUtilization
    TargetValue: 50.0
```

This section defines a scaling policy that adjusts the number of tasks based on CPU utilization. The policy does the following:

- It tracks the ECS service's average CPU utilization

- It scales the number of tasks to maintain the target CPU utilization at 50%

Step 3: Deploy the application:

Build and push the Docker image to Amazon **Elastic Container Registry (ECR)**:

    I.   **Install Docker**: Ensure Docker is installed on your local machine. You can download and install Docker from Docker's official website.

    II.   **Log in to AWS CLI**: Make sure you have the AWS CLI installed and configured with your AWS credentials:

```
aws configure
```

    III.   Create an ECR repository to store your Docker image:

```
aws ecr create-repository --repository-name auto-scaling-app
--region us-east-1
```

    IV.   Navigate to the directory containing your Dockerfile and build the Docker image:

```
docker build -t auto-scaling-app .
```

    V.   Tag the Docker image with the ECR repository URI:

```
docker tag auto-scaling-app:latest <aws-account-id>.dkr.ecr.
us-east-1.amazonaws.com/auto-scaling-app:latest
```

Replace <aws-account-id> with your actual AWS account ID.

    VI.   Retrieve an authentication token and authenticate your Docker client to your ECR registry:

```
aws ecr get-login-password --region us-east-1 | docker login
--username AWS --password-stdin <aws-account-id>.dkr.ecr.
us-east-1.amazonaws.com
```

Replace <aws-account-id> with your actual AWS account ID.

VII.   Push the tagged Docker image to your ECR repository:

```
docker push <aws-account-id>.dkr.ecr.us-east-1.amazonaws.com/
auto-scaling-app:latest
```

Step 4: Deploy the CloudFormation stack:

Once the Docker image is pushed to Amazon ECR, you can deploy the CloudFormation stack using the AWS CLI or AWS Management Console:

- **Deploy using AWS CLI:**

```
aws cloudformation create-stack --stack-name auto-scaling-
stack --template-body file://cloudformation-template.yml
--capabilities CAPABILITY_NAMED_IAM
```

- **Deploy using the AWS Management Console:**

    I.   Open the AWS CloudFormation console (`https://console.aws.amazon.com/cloudformation`).

    II.   Click on **Create Stack**.

    III.   Choose **Template is ready** and upload the `cloudformation-template.yaml` file.

    IV.   Follow the prompts to create the stack.

By following these steps, you will have your Dockerized Java application image stored in Amazon ECR and deployed using AWS CloudFormation, ECS, and Fargate, with auto-scaling capabilities configured.

Monitoring tools and techniques for Java applications

Effective monitoring is crucial for managing and optimizing Java applications during auto-scaling events. Various cloud providers offer comprehensive monitoring tools and services. Here, we discuss AWS CloudWatch, Google Cloud Monitoring, and Azure Monitor.

AWS CloudWatch

AWS CloudWatch is a monitoring and observability service that provides data and actionable insights to monitor applications, respond to system-wide performance changes, and optimize resource utilization. Key features include the following:

- **Metrics collection**: It collects and tracks metrics such as CPU usage, memory usage, and request counts

- **Logs management**: It collects and stores log files from AWS resources

- **Alarms**: It sets thresholds on metrics to automatically send notifications or trigger actions

- **Dashboards**: It creates custom dashboards to visualize and analyze metrics

Google Cloud Monitoring

Google Cloud Monitoring (formerly Stackdriver) provides visibility into the performance, uptime, and overall health of cloud-powered applications. Key features include the following:

- **Metrics and dashboards**: It visualizes key metrics with custom dashboards and tracks metrics such as CPU utilization, memory usage, and latency

- **Logs and traces**: It collects logs and traces to diagnose issues and understand application behavior

- **Alerting**: It configures alerts based on predefined or custom metrics to notify you of performance issues or anomalies

- **Integration**: It integrates with other Google Cloud services for seamless monitoring and management

Azure Monitor

Azure Monitor is a full-stack monitoring service that provides a comprehensive view of your application's performance and health. Key features include the following:

- **Metrics and logs**: It collects and analyzes performance metrics and logs from your applications and infrastructure.

- **Application insights**: It monitors live applications and automatically detects performance anomalies. It also provides deep insights into application performance and user behavior.

- **Alerts**: It sets up alerts to notify you of critical issues based on various conditions and thresholds.

- **Dashboards**: It has customizable dashboards to visualize and analyze application and infrastructure metrics.

By leveraging these tools, you can gain detailed insights into your Java application's performance, quickly detect and resolve issues, and ensure optimal resource utilization during auto-scaling events.

Code example – setting up monitoring and alerting for a Java-based cloud application

The following example demonstrates how to set up monitoring and alerting for a Java-based application using AWS CloudWatch. *Figure 10.2* illustrates the step-by-step process for configuring monitoring and alerting:

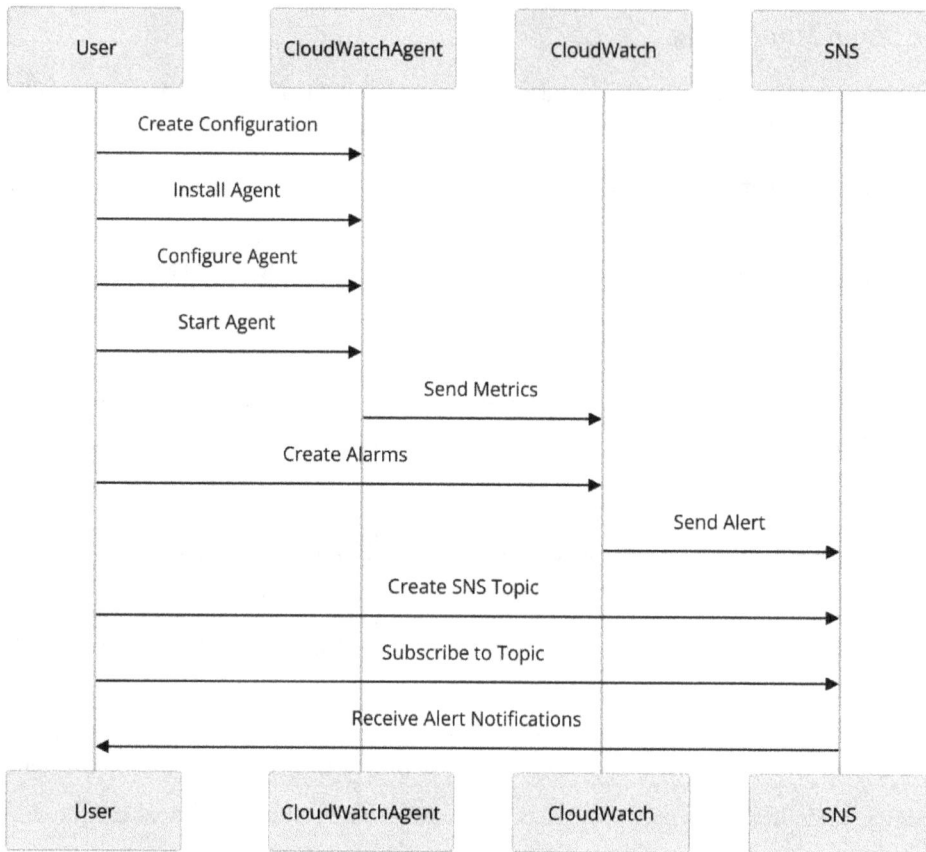

Figure 10.2: CloudWatch agent setup and alert notification workflow

Step 1: Configure CloudWatch agent on your java application:

- Create a CloudWatch agent configuration file (`cloudwatch-config.json`) to collect metrics and logs:

```
{
    "agent": {
        "metrics_collection_interval": 60,
        "logfile": "/opt/aws/amazon-cloudwatch-agent/logs/
amazon-cloudwatch-agent.log"
    },
    "metrics": {
        "append_dimensions": {
            "InstanceId": "${aws:InstanceId}"
        },
        "aggregation_dimensions": [["InstanceId"]],
```

```
    "metrics_collected": {
        "cpu": {
            "measurement": ["cpu_usage_idle",
                "cpu_usage_user", "cpu_usage_system"],
            "metrics_collection_interval": 60,
            "resources": ["*"]
            },
        "mem": {
            "measurement": ["mem_used_percent"],
            "metrics_collection_interval": 60
        }
    }
},
    "logs": {
    "logs_collected": {
        "files": {
        "collect_list": [
            {
                "file_path": "/var/log/java-app.log",
                "log_group_name": "/ecs/java-app",
                "log_stream_name": "{instance_id}"
            }
        ]
        }
    }
    }
}
```

This CloudWatch agent configuration script collects metrics and logs from a Java application running on an EC2 instance and sends them to AWS CloudWatch. It specifies the metrics to collect (CPU and memory usage) at 60-second intervals, along with the log file path. The collected metrics include the EC2 instance ID as a dimension, and the logs are stored in a specified CloudWatch Logs group and stream. This configuration enables real-time monitoring and analysis of the Java application's performance through CloudWatch.

Step 2: Install and start the CloudWatch agent:

- Install the CloudWatch agent:

```
sudo yum install amazon-cloudwatch-agent
```

- Configure the CloudWatch agent:

```
sudo /opt/aws/amazon-cloudwatch-agent/bin/amazon-CloudWatch-
agent-ctl -a fetch-config -m ec2 -c file:/path/to/cloudwatch-
config.json -s
```

This command installs and starts the AWS CloudWatch agent on an EC2 instance, fetching the configuration from the specified JSON file (`cloudwatch-config.json`) to begin collecting and reporting metrics and logs as defined.

Step 3: Create CloudWatch alarms:

- Create alarms to monitor CPU utilization and trigger alerts if it exceeds a certain threshold:

```
aws cloudwatch put-metric-alarm --alarm-name
"HighCPUUtilization" --metric-name "CPUUtilization" --namespace
"AWS/EC2" --statistic "Average" --period 60 --threshold 80
--comparison-operator "GreaterThanThreshold" --dimensions
"Name=InstanceId,Value=i-1234567890abcdef0" --evaluation-periods
2 --alarm-actions "arn:aws:sns:us-east-1:123456789012:my-sns-
topic" --unit "Percent"
```

This command creates a CloudWatch alarm named "`HighCPUUtilization`" that monitors the average CPU utilization of an EC2 instance and triggers a **Simple Notification Service (SNS)** notification if the CPU usage exceeds 80% for two consecutive 60-second periods.

Step 4: Set up SNS for alert notifications:

- Create an SNS topic:

```
aws sns create-topic --name my-sns-topic
```

- Subscribe to the SNS topic:

```
aws sns subscribe --topic-arn "arn:aws:sns:us-east-
1:123456789012:my-sns-topic" --protocol email --notification-
endpoint your-email@example.com
```

This command subscribes an email address (`your-email@example.com`) to an SNS topic (`my-sns-topic`), enabling the email recipient to receive notifications when the topic is triggered.

By following these steps, you can set up comprehensive monitoring and alerting for your Java-based cloud application, ensuring you are promptly notified of any scaling anomalies and can take immediate action to maintain stability and performance.

Having explored the technical aspects of monitoring and alerting, let's now turn our attention to real-world scenarios where companies such as Netflix and LinkedIn have successfully implemented Java-based auto-scaling solutions. These case studies will offer valuable insights into the practical application of the concepts we've discussed so far.

Real-world case studies and examples

In this section, we will explore real-world examples that demonstrate how Java can be effectively utilized in auto-scaling environments. We will delve into case studies of industry leaders such as Netflix and LinkedIn, highlighting their implementation of auto-scaling solutions and the benefits they have derived from such implementations. Additionally, we'll provide a few more practical examples from other companies to offer a broader perspective.

The following example relates to Netflix:

- **Background**: It experiences significant demand fluctuations due to varying viewer activity, especially during new show releases and peak viewing hours.

- **Solution**: It employs a Java-based microservices architecture using tools such as Eureka for service discovery, Ribbon for load balancing, and Hystrix for fault tolerance. This architecture allows Netflix to seamlessly scale services based on demand, ensuring high availability and cost efficiency.

- **Implementation**:

  - **Eureka**: This helps in service discovery, allowing services to find and communicate with each other dynamically

  - **Ribbon**: This provides client-side load balancing to distribute requests across multiple instances

  - **Hystrix**: This implements circuit breakers to handle failures gracefully, ensuring system resilience

- **Benefits**: During peak times, such as new show releases, Netflix's streaming services automatically scale up to handle increased traffic. This ensures a smooth and uninterrupted viewing experience for users while maintaining optimal resource utilization.

The following example relates to LinkedIn:

- **Background**: It needs to handle varying levels of user activity, such as job searches, profile updates, and messaging.

- **Solution**: It utilizes Java in its backend infrastructure for auto-scaling. They leverage Apache Samza for real-time data processing, Kafka for managing data pipelines, and Helix for cluster management.

- **Implementation**:

  - **Apache Samza**: This processes real-time data streams, enabling LinkedIn to provide timely insights and updates

  - **Kafka**: This manages data streams, ensuring reliable and scalable message brokering

  - **Helix**: This manages clusters, ensuring efficient resource utilization and failover mechanisms

- **Benefits**: LinkedIn can process and analyze data in real time, providing users with up-to-date information and insights. The scalable architecture ensures that their services can handle increased user activity without degradation in performance.

The following example relates to Airbnb:

- **Background**: It faces varying traffic loads, especially during holiday seasons and special events
- **Solution**: Airbnb uses a combination of Java-based microservices and Kubernetes for container orchestration to manage auto-scaling
- **Implementation**:

  - **Java microservices**: These decompose the application into smaller, manageable services that can be independently scaled

  - **Kubernetes**: It manages containerized applications, automatically scaling based on real-time demand

- **Benefits**: During peak booking periods, Airbnb's system can automatically scale up to handle the increased load, ensuring a seamless booking experience for users and optimal resource use

The following example relates to Spotify:

- **Background**: It needs to handle a large number of concurrent streams and user interactions
- **Solution**: It employs a combination of Java-based services and AWS auto-scaling to manage its infrastructure
- **Implementation**:

  - **Java services**: They handle core functionalities such as music streaming, playlist management, and user recommendations

  - **AWS auto-scaling**: It adjusts the number of instances based on CPU and memory usage to handle varying workloads

- **Benefits**: Spotify can ensure a high-quality streaming experience even during peak usage times, such as new album releases, by dynamically scaling its infrastructure to meet user demand

The following example relates to Pinterest:

- **Background**: It needs to manage fluctuating traffic loads, especially when users upload and share new content
- **Solution**: It uses Java-based backend services and integrates with KEDA for Kubernetes to manage auto-scaling based on event-driven metrics

- **Implementation**:
  - **Java backend services**: These manage core functionalities such as image processing, feed generation, and user interactions
  - **KEDA**: It scales the services based on event-driven metrics such as the number of uploads and API requests
- **Benefits**: Pinterest can efficiently handle spikes in user activity by automatically scaling its backend services based on real-time events, ensuring a responsive and reliable user experience

These practical examples from Netflix, LinkedIn, Airbnb, Spotify, and Pinterest demonstrate the versatility and effectiveness of using Java for auto-scaling in diverse environments. By leveraging modern tools and techniques, these companies ensure their applications can handle varying workloads efficiently, providing optimal performance and resource utilization.

To solidify our understanding, we will embark on a hands-on journey by constructing a real-world simulation project that showcases how Java and cloud services can be harnessed to create an auto-scaling solution, with detailed steps and visual aids to guide us through the process.

Practical application – building scalable Java-based solutions for real-time analytics and event-driven auto-scaling

In this section, we will explore two practical examples that demonstrate how Java can be effectively leveraged in building scalable solutions for real-time analytics and event-driven auto-scaling. The first example will showcase how to implement a Kubernetes-based auto-scaling solution using Java, while the second application focuses on developing a Java-based real-time analytics platform using AWS services. These two applications highlight the versatility of Java in addressing the diverse challenges presented by modern cloud-native architectures, showcasing its ability to integrate seamlessly with a range of cloud-based tools and services.

Auto-scaling a Java application with Kubernetes

In this example, we will create a realistic Spring Boot application that simulates a simple e-commerce order processing service. This service will have an endpoint to place an order, which will simulate some CPU-intensive processing to better demonstrate auto-scaling in a real-world scenario. We will use Kubernetes to manage the deployment and **Horizontal Pod Autoscaler** (HPA) to handle auto-scaling:

Step 1: Create the Spring Boot application:

pom.xml: Enter the dependency:

```
<dependencies>
    <dependency>
        <groupId>org.springframework.boot</groupId>
```

```
            <artifactId>spring-boot-starter-web</artifactId>
        </dependency>
    </dependencies>
```

`Application.java`: This is the entry point of the Spring Boot application. It initializes and runs the Spring Boot application:

```
import org.springframework.boot.SpringApplication;
import org.springframework.boot.autoconfigure.SpringBootApplication;

@SpringBootApplication
public class Application {

    public static void main(String[] args) {
        SpringApplication.run(Application.class, args);
    }
}
```

`OrderController.java`: This is a REST controller for handling order requests. It provides an endpoint to place orders and simulates CPU-intensive processing:

```
@RestController
public class OrderController {

    @PostMapping("/order")
    public String placeOrder(@RequestBody Order order) {
        // Simulate CPU-intensive processing
        processOrder(order);
        return "Order for " + order.getItem() + " placed
        successfully!";
    }
    private void processOrder(Order order) {
        // Simulate some CPU work
        for (int i = 0; i < 1000000; i++) {
            Math.sqrt(order.getQuantity() * Math.random());
        }
    }
}
class Order {
    private String item;
    private int quantity;

    // Getters and setters
    public String getItem() {
```

```
        return item;
    }
    public void setItem(String item) {
        this.item = item;
    }
    public int getQuantity() {
        return quantity;
    }

    public void setQuantity(int quantity) {
        this.quantity = quantity;
    }
}
```

Step 2: Dockerize the Application:

Dockerfile: This defines the Docker image for the application. Specifies the base image, sets the working directory, copies the application JAR, and defines the entry point to run the application:

```
FROM openjdk:11-jre-slim
WORKDIR /app
COPY target/application.jar /app/application.jar
ENTRYPOINT ["java", "-jar", "/app/application.jar"]
```

Build and push the Docker image: Run these Docker commands:

```
# Build the Docker image
docker build -t myrepo/auto-scaling-demo .

# Tag the image
docker tag myrepo/auto-scaling-demo:latest myrepo/auto-scaling-demo:latest

# Push the image to Docker Hub (or your preferred container registry)
docker push myrepo/auto-scaling-demo:latest
```

Step 3: Deploy to Kubernetes:

deployment.yaml: This defines the deployment of the application in Kubernetes. It specifies the application image, the number of replicas, and resource requests and limits:

```
apiVersion: apps/v1
kind: Deployment
metadata:
    name: auto-scaling-demo
spec:
```

```
        replicas: 2
        selector:
            matchLabels:
                app: auto-scaling-demo
        template:
            metadata:
                labels:
                    app: auto-scaling-demo
        spec:
            containers:
                - name: auto-scaling-demo
            image: myrepo/auto-scaling-demo:latest
            ports:
                - containerPort: 8080
            resources:
                requests:
                    cpu: "200m"
                limits:
                    cpu: "500m"
```

`service.yaml`: This defines the service to expose the application. It creates a `LoadBalancer` service to expose the application on port 80:

```
apiVersion: v1
kind: Service
metadata:
    name: auto-scaling-demo
spec:
    selector:
        app: auto-scaling-demo
    ports:
        - protocol: TCP
          port: 80
          targetPort: 8080
    type: LoadBalancer
```

Apply the deployment and service: Run the following command:

```
kubectl apply -f deployment.yaml
kubectl apply -f service.yaml
```

Step 4: Configure the HPA:

`hpa.yaml`: This defines the HPA configuration. It specifies the target deployment, minimum and maximum replicas, and CPU utilization threshold for scaling:

```
apiVersion: autoscaling/v2beta2
kind: HorizontalPodAutoscaler
metadata:
    name: auto-scaling-demo-hpa
spec:
    scaleTargetRef:
        apiVersion: apps/v1
        kind: Deployment
        name: auto-scaling-demo
    minReplicas: 2
    maxReplicas: 10
    metrics:
        - type: Resource
    resource:
        name: cpu
        target:
        type: Utilization
        averageUtilization: 50
```

Apply the HPA configuration and run the following command:

```
kubectl apply -f hpa.yaml
```

Step 5: Testing auto-scaling:

To test the auto-scaling, you can generate load on the application to increase CPU usage. Use a tool such as hey or ab (Apache Benchmark) to send a large number of requests to the application. Run these commands:

```
# Install hey (if not already installed)
brew install hey
# Generate load on the application
hey -z 1m -c 100 http://<your-load-balancer-dns>/order -m POST -H
"Content-Type: application/json" -d '{"item": "book", "quantity": 10}'
```

Monitor the Kubernetes pods to see the auto-scaling in action:

```
kubectl get pods -w
```

You should see that the number of pods increases as the load increases and then decrease once the load is reduced.

In this example, we demonstrated how to containerize the application using Docker and deploy it to Kubernetes. Building upon this foundation, the next practical application we will explore is developing a serverless real-time analytics pipeline using Java and AWS.

Developing a serverless real-time analytics pipeline using Java and AWS

We will simulate a real-time analytics platform that processes streaming data to generate insights. This platform will use Java-based AWS Lambda functions to handle various tasks, including data ingestion, processing, storage, and notification. AWS Step Functions will orchestrate these tasks, DynamoDB will be used for data storage, AWS SNS for notifications, and API Gateway to expose endpoints:

Step 1: Set up the environment:

- **Install Docker**: Ensure Docker is installed on your local machine.

- **Install AWS CLI**: Follow the instructions here: https://docs.aws.amazon.com/cli/latest/userguide/getting-started-install.html.

- **Install AWS SAM CLI**: Follow the instructions here: https://docs.aws.amazon.com/serverless-application-model/latest/developerguide/install-sam-cli.html.

Step 2: Create Java Lambda functions:

We will need to create several Lambda functions to handle different tasks. Let's use DataIngestionFunction as an example:

```
@SpringBootApplication
public class DataIngestionFunctionApplication {

    public static void main(String[] args) {
        SpringApplication.run(
            DataIngestionFunctionApplication.class, args);
    }

    @Bean
    public Function<Map<String, Object>,
        Map<String, Object>> dataIngestion() {
        return input -> {
            Map<String, Object> response = new HashMap<>();
            if (validateData(input)) {
                storeData(input);
                response.put("status",
                    "Data Ingested Successfully");
            } else {
                response.put("status", "Invalid Data");
```

```
        }
        return response;
    };
}

private boolean validateData(Map<String, Object> input) {
    return input.containsKey(
        "timestamp") && input.containsKey("value");
}

private void storeData(Map<String, Object> input) {
    // Code to store data in DynamoDB
}
}
```

Create a Dockerfile:

```
FROM openjdk:11-jre-slim
WORKDIR /app
COPY target/DataIngestionFunction-1.0.jar app.jar
ENTRYPOINT ["java", "-jar", "app.jar"]
```

This Dockerfile sets up a container based on the OpenJDK 11 JRE slim image, copies the `DataIngestionFunction-1.0.jar` file into the container, and sets the entry point to run the JAR file using `java -jar`.

Step 3: Create the CloudFormation template:

Create a `cloudformation-template.yaml` file to define the infrastructure. In this crucial step, we'll define our real-time analytics infrastructure using AWS CloudFormation. CloudFormation allows us to describe and provision all the infrastructure resources in a declarative way, ensuring consistency and ease of deployment.

Our template will encompass various AWS services essential for our real-time analytics platform, including the following:

- DynamoDB for data storage
- **Simple Storage Service (S3)** for processed data
- SNS for notifications
- Lambda functions for data processing
- Step Functions for workflow orchestration
- API Gateway for exposing endpoints

We'll break down each resource in the template, explaining its purpose and configuration. This approach will give you a clear understanding of how each component fits into the overall architecture and how they interact with each other.

By using CloudFormation, we ensure that our infrastructure is version-controlled, easily replicable, and can be updated or rolled back as needed. Let's dive into the details of each resource in our CloudFormation template.

First, let's look at the DynamoDB table:

```
Resources:
    DataIngestionTable:
        Type: AWS::DynamoDB::Table
        Properties:
            TableName: DataIngestionTable
            AttributeDefinitions:
                - AttributeName: DataId
                  AttributeType: S
            KeySchema:
                - AttributeName: DataId
                  KeyType: HASH
            ProvisionedThroughput:
                ReadCapacityUnits: 5
                WriteCapacityUnits: 5
```

Explanation: This resource defines a DynamoDB table named `DataIngestionTable` with a primary key of `DataId` to store ingested data. It includes provisions for read and write capacities.

Let's look at the S3 bucket:

```
ProcessedDataBucket:
    Type: AWS::S3::Bucket
    Properties:
        BucketName: processed-data-bucket
```

Explanation: This resource creates an S3 bucket named processed-data-bucket to store processed data files.

Let's look at an SNS topic:

```
DataNotificationTopic:
    Type: AWS::SNS::Topic
    Properties:
        TopicName: DataNotificationTopic
```

Explanation: This resource sets up an SNS topic named `DataNotificationTopic` to send notifications related to data processing events.

Let's look at an IAM role for Lambda execution:

```
LambdaExecutionRole:
    Type: AWS::IAM::Role
    Properties:
        AssumeRolePolicyDocument:
            Version: '2012-10-17'
        Statement:
            - Effect: Allow
            Principal:
                Service: lambda.amazonaws.com
            Action: sts:AssumeRole
        Policies:
            - PolicyName: LambdaPolicy
            PolicyDocument:
                Version: '2012-10-17'
            Statement:
                - Effect: Allow
                Action:
                    - dynamodb:PutItem
                    - dynamodb:GetItem
                    - dynamodb:UpdateItem
                    - s3:PutObject
                    - sns:Publish
                Resource: '*'
```

Explanation: This resource defines an IAM role that grants Lambda functions permissions to interact with DynamoDB, S3, and SNS.

```
Lambda Function for data ingestion:
DataIngestionFunction:
    Type: AWS::Lambda::Function
    Properties:
        FunctionName: DataIngestionFunction
        Handler: org.springframework.cloud.function.adapter.aws.
FunctionInvoker
        Role: !GetAtt LambdaExecutionRole.Arn
        Runtime: java11
        Timeout: 30 # Set the timeout to 30 seconds or more if needed
        MemorySize: 1024 # Increase memory size to 1024 MB
        Code:
            S3Bucket: !Ref LambdaCodeBucket
```

```
        S3Key: data-ingestion-1.0-SNAPSHOT-aws.jar
    Environment:
        Variables:
          SPRING_CLOUD_FUNCTION_DEFINITION:
              dataIngestion
```

Explanation: This resource creates a Lambda function named `DataIngestionFunction` to handle data ingestion tasks. The function is associated with the previously defined IAM role for necessary permissions.

Let's look at a Step Functions state machine:

```
RealTimeAnalyticsStateMachine:
    Type: AWS::StepFunctions::StateMachine
    Properties:
        StateMachineName: RealTimeAnalyticsStateMachine
        RoleArn: !GetAtt LambdaExecutionRole.Arn
        DefinitionString: !Sub |
        {
            "Comment": "Real-Time Analytics Workflow",
            "StartAt": "DataIngestion",
            "States": {
                "DataIngestion": {
                "Type": "Task",
                "Resource": "${DataIngestionFunction.Arn}",
                "End": true
                }
            }
        }
```

Explanation: This resource defines a Step Functions state machine named `RealTimeAnalyticsStateMachine`, which orchestrates the data ingestion process using `DataIngestionFunction`.

Let's look at API Gateway:

```
RealTimeAnalyticsApi:
    Type: AWS::ApiGateway::RestApi
    Properties:
        Name: RealTimeAnalyticsApi
```

Explanation: This resource sets up an API Gateway named `RealTimeAnalyticsApi` to expose endpoints for the real-time analytics platform.

Let's look at a real-time analytics resource:

```
Type: AWS::ApiGateway::Resource
Properties:
    ParentId: !GetAtt RealTimeAnalyticsApi.RootResourceId
    PathPart: ingest
    RestApiId: !Ref RealTimeAnalyticsApiPI Gateway Resource:
```

Explanation: This resource defines an API Gateway resource under `RealTimeAnalyticsApi`, creating a path segment/ingest for the data ingestion endpoint.

Let's look at an API Gateway method:

```
RealTimeAnalyticsMethod:
    Type: AWS::ApiGateway::Method
    Properties:
        AuthorizationType: NONE
        HttpMethod: POST
        ResourceId: !Ref RealTimeAnalyticsResource
        RestApiId: !Ref RealTimeAnalyticsApi
        Integration:
            IntegrationHttpMethod: POST
        Type: AWS
        Uri: !Sub "arn:aws:apigateway:${AWS::Region}:
            states:action/StartExecution"
        IntegrationResponses:
          - StatusCode: 200
        RequestTemplates:
          application/json: |
            {
                "input": "$util.escapeJavaScript(
                    $input.json('$'))",
                "stateMachineArn": "arn:aws:states:${
                    AWS::Region}:${AWS::AccountId}:
                        stateMachine:${
                            RealTimeAnalyticsStateMachine}"
            }
        PassthroughBehavior: WHEN_NO_TEMPLATES
        Credentials: !GetAtt LambdaExecutionRole.Arn
        MethodResponses:
          - StatusCode: 200
```

Explanation: This resource creates an HTTP POST method for the /ingest endpoint, integrating it with the Step Functions state machine to start the data ingestion process.

Step 4: Deploy the application:

Build and push the Docker image:

```
docker build -t data-ingestion-function .
docker tag data-ingestion-function:latest <aws-account-id>.dkr.ecr.
us-east-1.amazonaws.com/data-ingestion-function:latest
aws ecr get-login-password --region us-east-1 | docker login
--username AWS --password-stdin <aws-account-id>.dkr.ecr.us-east-1.
amazonaws.com
docker push <aws-account-id>.dkr.ecr.us-east-1.amazonaws.com/data-
ingestion-function:latest
```

Deploy the CloudFormation stack:

```
aws cloudformation create-stack --stack-name RealTimeAnalyticsStack
--template-body file://cloudformation-template.yml --capabilities
CAPABILITY_NAMED_IAM
```

Monitor the stack creation:

```
aws cloudformation describe-stacks --stack-name RealTimeAnalyticsStack
```

Verify resources: Ensure all resources (DynamoDB tables, API Gateway, Lambda functions, IAM roles, etc.) are created successfully.

Test the API Gateway:

```
curl -X POST https://YOUR_API_GATEWAY_URL/ingest -d '{"dataId":
"12345", "timestamp": "2024-06-09T12:34:56Z", "value": 42.0}'
```

Step 5: Cleanup:

To delete the resources created by the CloudFormation stack:

```
aws cloudformation delete-stack --stack-name RealTimeAnalyticsStack
```

This practical project demonstrates how to use Java and cloud services to implement an auto-scaling solution for a real-time analytics platform. By following these steps, readers will gain hands-on experience in deploying and managing scalable Java applications in a cloud environment.

Advanced topics

This section will explore advanced techniques such as predictive auto-scaling using ML algorithms and integration with cloud-native tools and services, which provide more efficient and intelligent scaling solutions for optimal performance and cost efficiency.

Predictive auto-scaling using ML algorithms

Predictive auto-scaling, a more proactive approach than traditional reactive methods, harnesses ML algorithms to forecast future demand based on historical data and relevant metrics. This allows for optimized resource allocation and improved application performance.

To implement predictive auto-scaling, follow these steps:

1. **Collect and preprocess data**: Gather historical metrics such as CPU usage, memory usage, network traffic, and request rates using monitoring tools (e.g., AWS CloudWatch, Google Cloud Monitoring, or Azure Monitor). Cleanse and preprocess this data to handle any missing values, and outliers and ensure consistency.

2. **Train ML models**: Utilize ML algorithms such as linear regression, **autoregressive integrated moving average** (**ARIMA**), or more sophisticated techniques such as **long short-term memory** (**LSTM**) networks to train models on historical data. Cloud-based platforms such as Amazon SageMaker, Google Cloud AI Platform, or Azure ML can facilitate this process.

3. **Deploy and integrate**: Deploy the trained models as services, using either serverless functions (e.g., AWS Lambda, Google Cloud Functions, Azure Functions) or containerized applications. Integrate these models with your auto-scaling policies, enabling them to dynamically adjust resource allocation based on the predictions.

To demonstrate how these steps can be implemented in practice, let's take a look at a Spring Boot application that integrates with Amazon SageMaker for predictive scaling.

This code snippet demonstrates a Spring Boot application that integrates with Amazon SageMaker to perform predictive scaling. It defines a bean that invokes a trained linear regression model endpoint in SageMaker and adjusts the auto-scaling policies based on the predictions.

The Spring Boot application is `PredictiveScalingApplication.java`. Train the model in SageMaker:

```
@SpringBootApplication
public class PredictiveScalingApplication {

    public static void main(String[] args) {
        SpringApplication.run(
            PredictiveScalingApplication.class, args);
    }
```

```java
    @Bean
    public AmazonSageMakerRuntime sageMakerRuntime() {
        return AmazonSageMakerRuntimeClientBuilder.defaultClient();
    }


    @Bean
    public Function<Message<String>, String>
predictiveScaling(AmazonSageMakerRuntime sageMakerRuntime) {
        return input -> {
            String inputData = input.getPayload();

            InvokeEndpointRequest invokeEndpointRequest = new
            InvokeEndpointRequest()
                .withEndpointName("linear-endpoint")
                .withContentType("text/csv")
                .withBody(ByteBuffer.wrap(inputData.
                getBytes(StandardCharsets.UTF_8)));

            InvokeEndpointResult result = sageMakerRuntime.
            invokeEndpoint(invokeEndpointRequest);
            String prediction = StandardCharsets.UTF_8.decode(result.
            getBody()).toString();
            adjustAutoScalingBasedOnPrediction(prediction);
            return "Auto-scaling adjusted based on prediction: " +
            prediction;
        };
    }
    private void adjustAutoScalingBasedOnPrediction(String prediction)
    {
        // Logic to adjust auto-scaling policies based on prediction
    }
}
```

In this Java-based Spring Boot application, we define a `predictiveScaling()` function that takes input data, sends it to a designated SageMaker endpoint for prediction, and then adjusts auto-scaling policies based on the returned prediction. Remember to replace the placeholder endpoint name (`"linear-endpoint"`) with your actual SageMaker endpoint. While this example focuses on the integration with an existing endpoint, typically, you would first train a model in SageMaker using appropriate algorithms such as linear regression or time series forecasting to generate these predictions. The choice of algorithm will depend on your specific use case. The `adjustAutoScalingBasedOnPrediction()` method is where you would implement the logic for adjusting auto-scaling policies using the AWS auto-scaling API or other relevant services.

The `application.yaml` file is a crucial configuration component in Spring Boot applications, serving as a central place to define various application settings. In the context of our predictive scaling function for AWS Lambda, this file plays a particularly important role.

Let's examine the key configuration:

```
spring:
    cloud:
        function:
            definition: predictiveScaling
```

This concise yet powerful configuration does several important things:

- It utilizes **Spring Cloud Function**, a project that simplifies the development of serverless applications.
- The `predictiveScaling` definition line is especially significant. It tells Spring Cloud Function that our `predictiveScaling` function (which we'll define in our `PredictiveScalingApplication` class) should be the primary entry point for our serverless application.
- This configuration ensures that when our Spring Boot application is built and packaged, the `predictiveScaling` function is properly included and set up as the main executable component.

Understanding this configuration is crucial as it bridges the gap between our Spring Boot application and the serverless environment of AWS Lambda. It enables our Java code to seamlessly integrate with the cloud infrastructure, allowing us to focus on the business logic of predictive scaling rather than the intricacies of serverless deployment.

Let's take a look at the Dockerfile:

```
FROM openjdk:11-jre-slim
WORKDIR /app
COPY target/predictive-scaling-1.0.jar app.jar
ENTRYPOINT ["java", "-jar", "app.jar"]
```

This Dockerfile sets up a container based on the OpenJDK 11 JRE slim image, copies the predictive scaling JAR file into the container, and sets the entry point to run the JAR file.

Build and push the Docker image:

```
docker build -t predictive-scaling-app .
docker tag predictive-scaling-app:latest <aws-account-id>.dkr.ecr.
us-east-1.amazonaws.com/predictive-scaling-app:latest
aws ecr get-login-password --region us-east-1 | docker login
--username AWS --password-stdin <aws-account-id>.dkr.ecr.us-east-1.
amazonaws.com
```

```
docker push <aws-account-id>.dkr.ecr.us-east-1.amazonaws.com/
predictive-scaling-app:latest
```

Replace <aws-account-id> with your actual AWS account ID. Open your terminal and navigate to the project directory containing your Dockerfile. Run the preceding command.

This script builds a Docker image tagged predictive-scaling-app, then tags it for an Amazon ECR repository in the us-east-1 region. It then logs into that ECR repository using AWS credentials, preparing the image for deployment to a cloud environment.

Integration with cloud-native tools and services

To enhance the deployment and management of our predictive scaling application, we can integrate it with popular cloud-native tools and services. Let's explore how we can leverage Kubernetes, Istio, and AWS SAM to improve our application's scalability, observability, and infrastructure management.

Kubernetes

Kubernetes is a powerful container orchestration platform that enables automated deployment, scaling, and management of containerized applications. One of the key features of Kubernetes is the HPA, which allows us to automatically scale the number of pods based on CPU utilization or other custom metrics.

Here's an example of an HPA configuration:

```
apiVersion: autoscaling/v2beta2
kind: HorizontalPodAutoscaler
metadata:
name: java-app-autoscaler
spec:
    scaleTargetRef:
    apiVersion: apps/v1
    kind: Deployment
    name: java-app
    minReplicas: 2
    maxReplicas: 10
    metrics:
        - type: Resource
resource:
    name: cpu
    target:
        type: Utilization
        averageUtilization: 50
```

This configuration defines an HPA that targets a deployment named java-app. It specifies the minimum and maximum number of replicas and sets the target CPU utilization to 50%. Kubernetes will automatically scale the number of pods based on the observed CPU utilization, ensuring that our application can handle varying levels of traffic.

To apply this HPA configuration to your Kubernetes cluster, save the configuration as a YAML file (e.g., hpa.yaml) and run the following command in your terminal:

```
kubectl apply -f hpa.yaml
```

Istio service mesh

Istio is a powerful service mesh that provides a wide range of features for managing microservices in a distributed environment. It enables fine-grained traffic control, observability, and security for our application.

Here's an example of an Istio VirtualService configuration:

```
apiVersion: networking.istio.io/v1alpha3
kind: VirtualService
metadata:
    name: java-app
spec:
    hosts:
        - "*"
    http:
        - route:
            - destination:
                host: java-app
          subset: v1
          weight: 50
        - destination:
            host: java-app
            subset: v2
            weight: 50
```

This VirtualService configuration defines routing rules for our Java application. It specifies that 50% of the traffic should be routed to subset v1 and the other 50% to subset v2. This allows us to implement advanced deployment strategies such as canary releases or A/B testing.

To apply this Istio `VirtualService` configuration in your Kubernetes cluster, follow these steps:

1. **Save configuration**: Save the provided Istio `VirtualService` configuration as a YAML file (e.g., `virtual-service.yaml`).

2. **Apply configuration**: Open your terminal and run the following command:

```
kubectl apply -f virtual-service.yaml
```

AWS SAM

AWS SAM is a framework that extends AWS CloudFormation to define and manage serverless applications. It provides a simplified syntax for defining AWS Lambda functions, API Gateway endpoints, and other serverless resources.

Here's an example of a SAM template defining a Lambda function:

```
Resources:
    PredictiveScalingFunction:
        Type: AWS::Lambda::Function
    Properties:
        FunctionName: PredictiveScalingFunction
        Handler: com.example.
PredictiveScalingApplication::predictiveScaling
        Role: !GetAtt LambdaExecutionRole.Arn
        Runtime: java11
        Timeout: 30 # Set the timeout to 30 seconds or more if needed
        MemorySize: 1024 # Increase memory size to 1024 MB
        Code:
            S3Bucket: !Ref LambdaCodeBucket
            S3Key: predictive-scaling-1.0-SNAPSHOT-aws.jar
        Environment:
            Variables:
                SPRING_CLOUD_FUNCTION_DEFINITION:
                    predictiveScaling
```

This template defines a Lambda function resource named `PredictiveScalingFunction` with properties such as the function name, the fully qualified name of the Java method serving as the entry point, the IAM role granting permissions, the runtime environment (Java 11), the maximum allowed execution time (30 seconds), the allocated memory (1,024 MB), the location of the function code in an S3 bucket, and an environment variable indicating the name of the function to be invoked.

To implement this, do the following:

1. **Save as template**: Save this code as a YAML file (e.g., `template.yaml`).

2. **Install SAM CLI**: If you haven't already, install the AWS SAM CLI (`https://docs.aws.amazon.com/serverless-application-model/latest/developerguide/install-sam-cli.html`).

3. **Build**: From your terminal, navigate to the directory containing `template.yaml` and run `sam build`. This will build your function code and prepare it for deployment.

4. **Deploy**: Run `sam deploy –guided` and follow the prompts to deploy your function to AWS Lambda.

Now, your Java Lambda function is running in the cloud, ready to be triggered by events or invoked directly.

By leveraging these cloud-native tools and services, we can enhance the scalability, observability, and management of our predictive scaling application. Kubernetes enables automated scaling based on resource utilization, Istio provides advanced traffic management and observability features, and AWS SAM simplifies the definition and deployment of serverless components.

Summary

In this chapter, we explored the synchronization of Java's concurrency models with cloud auto-scaling dynamics. We delved into the fundamentals of cloud auto-scaling, examining how Java's concurrency tools can be leveraged to optimize applications for scalability. Key discussions included best practices for enhancing Java application performance, monitoring and managing Java processes during auto-scaling events, and real-world case studies from industry leaders such as Netflix and LinkedIn.

We also walked through a practical project that demonstrated the deployment and management of a scalable Java-based real-time analytics platform using AWS services and Docker. Advanced topics such as predictive auto-scaling using ML and the integration of Java applications with cloud-native tools such as Kubernetes, Istio, and AWS SAM were covered to provide a comprehensive understanding of modern scaling solutions.

The skills and knowledge gained from this chapter are essential for building robust, scalable, and cost-effective Java applications in cloud environments. By mastering these techniques, readers can ensure optimal performance, efficient resource utilization, and seamless adaptability to the demands of contemporary cloud-based systems.

In the next chapter, *Advanced Java Concurrency Practices in Cloud Computing*, we will delve deeper into the intricacies of concurrent Java applications optimized for cloud environments. We will explore powerful techniques such as leveraging GPU computing, utilizing **Compute Unified Device Architecture** (**CUDA**) and OpenCL libraries, and integrating Java with native libraries for unparalleled parallel execution. This chapter will equip readers with a robust toolkit to ensure their Java applications remain resilient and ultra-performant in any cloud setting, advancing their skills to the next level.

Questions

1. What is the primary advantage of cloud auto-scaling in Java applications?

    A. Manual monitoring and scaling

    B. Fixed resource allocation

    C. Dynamic resource allocation based on demand

    D. Increased operational overhead

2. Which Java concurrency tool is essential for managing asynchronous tasks in cloud auto-scaling environments?

    A. `ThreadLocal`

    B. `CompletableFuture`

    C. `StringBuilder`

    D. `InputStream`

3. What is the role of `ExecutorService` in Java's concurrency model for cloud auto-scaling?

    A. Managing a fixed number of threads

    B. Encrypting data

    C. Handling single-threaded tasks only

    D. Directly handling HTTP requests

4. Which practice is recommended for optimizing Java applications for cloud scalability?

    A. Using synchronous processing

    B. Avoiding the use of caching

    C. Implementing stateless services

    D. Designing monolithic applications

5. What benefit do parallel streams provide in Java applications with respect to cloud auto-scaling?

    A. Simplifying error handling

    B. Blocking the main thread

    C. Improving performance through concurrent data processing

    D. Reducing the need for load balancing

11

Advanced Java Concurrency Practices in Cloud Computing

In today's rapidly evolving technological landscape, **cloud computing** has become an integral part of modern software architecture. As Java continues to be a dominant language in enterprise applications, understanding how to leverage its concurrency capabilities in cloud environments is crucial for developers and architects alike. This chapter delves into advanced Java concurrency practices specifically tailored for cloud computing scenarios.

Throughout this chapter, you'll gain practical knowledge on implementing robust, scalable, and efficient concurrent Java applications in the cloud. We'll explore cutting-edge techniques for enhancing redundancy and failover mechanisms, leveraging **graphics processing unit** (**GPU**) acceleration for computational tasks, and implementing specialized monitoring solutions for cloud-based Java applications.

By the end of this chapter, you'll be equipped with the skills to design and optimize Java applications that can fully harness the power of cloud infrastructure. You'll learn how to implement cloud-specific redundancies, utilize GPU acceleration through **Compute Unified Device Architecture** (**CUDA**) and **Open Computing Language** (**OpenCL**), and set up comprehensive monitoring systems that integrate both cloud-native and Java-centric tools.

These advanced practices will enable you to create high-performance, resilient Java applications that can scale effortlessly in cloud environments. Whether you're working on data-intensive applications, real-time processing systems, or complex distributed architectures, the techniques covered in this chapter will help you maximize the potential of Java concurrency in the cloud.

In this chapter, we're going to cover the following main topics:

- Enhancing cloud-specific redundancies and failovers in Java applications
- GPU acceleration in Java: leveraging CUDA, OpenCL, and native libraries
- Specialized monitoring for Java concurrency in the cloud

Let's embark on this journey to master advanced Java concurrency practices in cloud computing!

Technical requirements

To fully engage with *Chapter 11*'s content and examples, ensure the following are installed and configured:

- **CUDA Toolkit**: This provides the environment for building and running GPU-accelerated applications. Download and install from the NVIDIA developer website: `https://developer.nvidia.com/cuda-downloads`.

- **Java bindings for CUDA (JCuda) library**: This enables CUDA integration into Java. Download from `http://www.jcuda.org/downloads/downloads.html` and add the JAR files to your project's classpath.

- **Amazon Web Services (AWS) command-line interface (CLI) and AWS SDK for Java**: These facilitate interaction with AWS services.

  - Install the AWS CLI (`https://aws.amazon.com/cli/`).

  - Include the AWS **software development kit (SDK)** dependencies in your project.

  - Configure your AWS credentials using `aws configure`.

- **Java virtual machine (JVM) monitoring tool (JConsole or VisualVM)**: Monitor JVM performance during CUDA execution. Launch and connect to your running application.

Here are some additional notes:

- **GPU hardware**: A CUDA-capable NVIDIA GPU is required for running the examples

- **Operating system (OS) compatibility**: Ensure your OS is compatible with the CUDA Toolkit and JCuda versions

Refer to each tool's documentation for installation instructions and troubleshooting.

The code in this chapter can be found on GitHub:

`https://github.com/PacktPublishing/Java-Concurrency-and-Parallelism`

Enhancing cloud-specific redundancies and failovers in Java applications

In the realm of cloud computing, redundancy and failover mechanisms are paramount to ensure the uninterrupted availability and resilience of applications. **Redundancy** involves duplicating critical components or resources, while **failover** refers to the automatic switchover to a backup system in case of a primary system failure. These mechanisms are essential for mitigating the impact of hardware failures, network outages, or other unforeseen disruptions that can occur in cloud environments. By implementing redundancy and failover strategies, developers can minimize downtime, prevent data loss, and maintain the overall reliability of their applications.

Java offers a robust toolkit for building resilient cloud applications, enabling developers to implement redundancy, replication, and failover mechanisms even when leveraging managed cloud services.

Leveraging Java libraries and frameworks

By leveraging Java libraries and frameworks, Java developers can seamlessly integrate with cloud providers' managed services through their respective SDKs (such as the AWS SDK for Java) or utilize cloud-agnostic frameworks such as Spring Cloud. These tools abstract away much of the underlying infrastructure complexity, simplifying the implementation of redundancy and failover strategies.

For load balancing, Java applications interact with cloud-based load balancers (e.g., AWS **Elastic Load Balancing** (**ELB**), Azure Load Balancer) using provider-specific SDKs or frameworks. The Java code can dynamically discover healthy instances and update load balancer configurations, ensuring traffic is routed efficiently. Additionally, in scenarios where direct control is desired, Java applications can implement client-side load-balancing algorithms.

Regarding data replication, Java libraries simplify interaction with cloud storage services (e.g., Amazon **Simple Storage Service** (**S3**), DynamoDB), abstracting the complexities of replication. Java code handles data consistency challenges by implementing strategies such as eventual consistency, conflict resolution, or leveraging the consistency levels offered by the cloud service. Developers can also utilize cloud provider APIs or SDKs to manage backup and recovery processes programmatically.

For failover mechanisms, Java applications can actively monitor the health of cloud resources using provider APIs, enabling prompt failover actions when necessary. By integrating with services such as Amazon Route 53 or Eureka, Java applications can dynamically locate healthy instances and adjust configurations in response to failures. Moreover, Java's built-in exception-handling mechanisms and retry libraries enable graceful recovery from failures and seamless switchover to backup resources.

Writing correct test scenarios for failover and advanced mechanisms

When implementing failover and other advanced mechanisms in Java applications for cloud environments, it is crucial to write comprehensive and correct test scenarios to ensure the reliability and effectiveness of these mechanisms. Here are some key considerations and best practices for testing failover and advanced mechanisms:

- **Simulate network failures**:
  - Use tools such as **Traffic Control** (**TC**) in Linux to introduce network delays, drops, or partitions
  - Ensure your application can handle partial network failures and still route traffic to healthy instances

- **Test resource unavailability**:

  - Simulate unavailability of critical resources such as databases, message brokers, or external APIs

  - Verify that your application can switch to backup resources or enter a degraded mode without crashing

- **Automate failover testing**:

  - Use automation tools such as Chaos Monkey or Gremlin to randomly terminate instances or induce failures

  - Automate the validation of failover processes and check for successful switchover to backup systems

- **Monitor failover performance**:

  - Measure the time it takes for your application to detect a failure and switch to the backup system

  - Ensure that performance metrics remain within acceptable limits during and after the failover process

By incorporating these testing practices and continuously refining the test scenarios based on real-world observations, developers can ensure the robustness and reliability of their Java applications in cloud environments.

Let's develop a practical exercise demonstrating Java techniques for cloud redundancy and failover in an AWS environment. We'll create a sample application that showcases load balancing, data replication, and failover mechanisms.

Practical exercise – resilient cloud-native Java application

Before we begin the practical exercise, it's important to note that it assumes some familiarity with Spring Boot and Spring Cloud. Spring Boot is a popular Java framework that simplifies the development of standalone, production-grade Spring applications. It provides a streamlined way to configure and run Spring applications with minimal setup. Spring Cloud, on the other hand, is a collection of tools and libraries that enhance Spring Boot applications with cloud-specific features, such as service discovery, configuration management, and circuit breakers.

If you're new to Spring Boot and Spring Cloud, don't worry! While a deep understanding of these technologies is beneficial, the exercise will focus on the key concepts and components relevant to building a resilient cloud-native Java application. To get started with Spring Boot, you can refer to the official documentation and guides at `https://spring.io/projects/spring-boot`. For an introduction to Spring Cloud and its various modules, check out the Spring Cloud documentation at `https://spring.io/projects/spring-cloud`.

In this exercise, we will create a comprehensive Java-based application that demonstrates cloud redundancy, failover mechanisms, and data replication with consistency and conflict resolution in an AWS environment. We will use AWS services such as ELB, Amazon DynamoDB, Amazon S3, and Amazon Route 53. We'll also leverage the AWS SDK for Java and Spring Cloud for cloud-agnostic implementations.

Figure 11.1 illustrates the resilient cloud-native Java application:

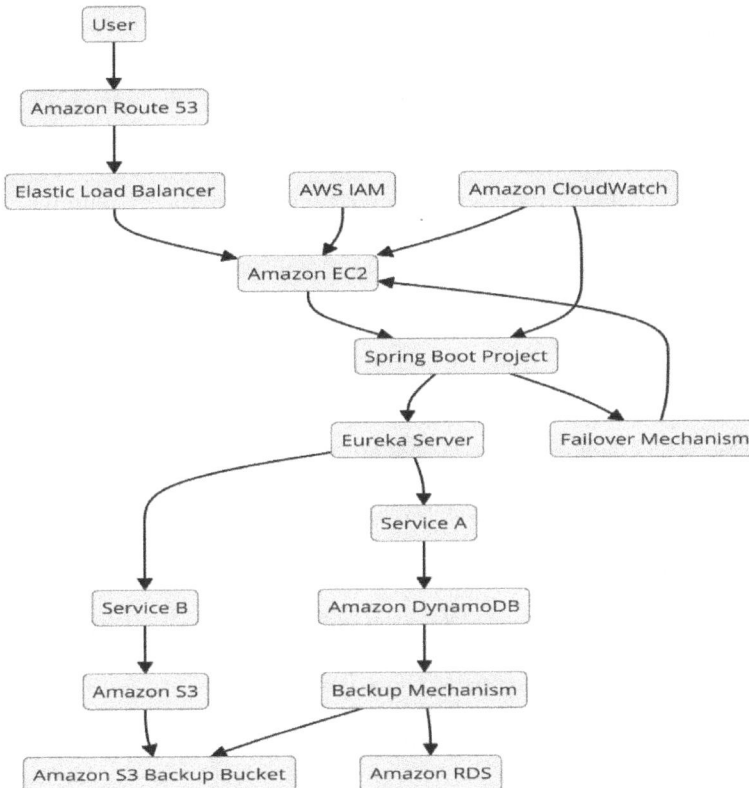

Figure 11.1: AWS-based Java application architecture with backup and failover mechanisms

This diagram illustrates a comprehensive architecture for a Java-based application deployed in an AWS environment, featuring cloud redundancy, failover mechanisms, and data replication with consistency and conflict resolution. Key components include Amazon Route 53 for **domain name system (DNS)** routing, ELB for distributing traffic across multiple **Elastic Compute Cloud (EC2)** instances, and the Spring Boot project hosting service instances managed by the Eureka server. Service A interacts with Amazon DynamoDB, while Service B interacts with Amazon S3, with a backup mechanism ensuring data replication to a dedicated S3 backup bucket. Amazon **Relational Database Service (RDS)** is integrated for relational database management, **Identity and Access Management (IAM)** for secure access management, and CloudWatch for monitoring and performance insights. A failover mechanism is in place to ensure high availability and reliability.

Here's a summary of the steps involved in this application:

- **Step 1: Set up the Spring Boot project**:

  - Create a new Spring Boot project using Spring Initializer or your preferred method.

  - Add the necessary dependencies to your project's configuration file (e.g., `pom.xml`).

- **Step 2: Implement load balancing**:

  - Create a REST controller with endpoints to simulate load-balanced services.

  - Configure a client-side load balancing mechanism using Ribbon.

- **Step 3: Implement data replication with consistency and conflict resolution**:

  - Create a service to interact with Amazon S3 and DynamoDB for data replication.

  - Implement methods for replicating data to S3 and DynamoDB, handling eventual consistency, and resolving conflicts.

  - Implement backup and recovery mechanisms for DynamoDB data using S3.

- **Step 4: Create REST endpoints for data operations**:

  - Create a REST controller to expose endpoints for data operations, including backup and restore.

- **Step 5: Implement failover mechanisms**:

  - Create a health check endpoint and integrate with Eureka for service discovery and failover.

- **Step 6: Configure AWS resources using CloudFormation**:

  - Update the CloudFormation template to include the necessary AWS resources, such as S3 buckets and DynamoDB tables.

- **Step 7: Deploy and test**:

  - Deploy the CloudFormation stack to provision the required AWS resources.

  - Deploy the Spring Boot application to AWS EC2 instances behind a load balancer.

  - Test the application's load balancing, data replication, consistency handling, backup, and failover mechanisms.

- **Step 8: Additional considerations**: (Detailed implementation will not be covered)

This exercise provides a comprehensive hands-on experience in building a resilient cloud-native Java application using Spring Boot, AWS services, and various architectural patterns. By following these steps, readers will gain practical knowledge of implementing load balancing, data replication, consistency management, failover mechanisms, and other essential aspects of building robust applications in the cloud.

Step 1: Set up the Spring Boot project:

Create a new Spring Boot project using Spring Initializer or your preferred method. Add the following dependencies to your pom.xml file:

```
<dependencies>
    <dependency>
        <groupId>org.springframework.boot</groupId>
        <artifactId>spring-boot-starter-web</artifactId>
    </dependency>
    <dependency>
        <groupId>com.amazonaws</groupId>
        <artifactId>aws-java-sdk</artifactId>
        <version>2.17.102</version>
    </dependency>
    <dependency>
        <groupId>org.springframework.cloud</groupId>
        <artifactId>spring-cloud-starter-aws</artifactId>
    </dependency>
    <dependency>
        <groupId>org.springframework.cloud</groupId>
        <artifactId>spring-cloud-starter-netflix-eureka-client</artifactId>
    </dependency>
</dependencies>
```

Step 2: Implement load balancing:

Create a REST controller with endpoints to simulate load-balanced services:

```
@RestController
public class LoadBalancedController {

    @GetMapping("/serviceA")
    public String serviceA() {
        return "Service A Response";
    }

    @GetMapping("/serviceB")
```

```
    public String serviceB() {
        return "Service B Response";
    }
}
```

Create a configuration class to enable Ribbon for client-side load balancing:

```
@Configuration
@RibbonClient(name = "serviceA")
public class RibbonConfiguration {
    // Custom Ribbon configuration can be added here
}
```

Step 3: Implement data replication with consistency and conflict resolution:

Create a service to interact with Amazon S3 and DynamoDB for data replication, handling eventual consistency, conflict resolution, and backup/recovery. The following are the essential parts of the `DataReplicationService` class. For the complete implementation, please refer to the book's accompanying GitHub repository:

```
@Service
public class DataReplicationService {
    private final S3Client s3Client;
    private final DynamoDbClient dynamoDbClient;
    private final String tableName = "MyTable";

    public DataReplicationService() {
        this.s3Client = S3Client.builder().build();
        this.dynamoDbClient = DynamoDbClient.builder().build();
    }

    public void replicateToS3(String key, String content) {
        PutObjectRequest putObjectRequest = PutObjectRequest.builder()
                    .bucket("my-bucket")
                    .key(key)
                    .build();
        s3Client.putObject(putObjectRequest,
            RequestBody.fromString(content));
    }

    public void replicateToDynamoDB(String key,
        String value) {
        PutItemRequest putItemRequest = PutItemRequest.builder()
            .tableName(tableName)
            .item(Map.of("Key",
```

```
                    AttributeValue.builder().s(key).build(),
                        "Value", AttributeValue.builder().s(value).
                        build()))
                .build();
        dynamoDbClient.putItem(putItemRequest);
    }

    public Optional<String> retrieveFromDynamoDB(
        String key) {
            GetItemRequest getItemRequest = GetItemRequest.builder()
                .tableName(tableName)
                .key(Map.of("Key", AttributeValue.builder().s(key).
                build()))
                .build();
        try {
            GetItemResponse response = dynamoDbClient.
            getItem(getItemRequest);
            return Optional.ofNullable(response.item().get(
                "Value")).map(AttributeValue::s);
        } catch (DynamoDbException e) {
            throw new RuntimeException(
                "Failed to retrieve item from DynamoDB",e);
        }
    }


    // For complete implementation, refer to the book's accompanying
GitHub repository.
}
```

Step 4: Create REST endpoints for data operations:

Create a REST controller to expose endpoints for data operations, including backup and restore:

```
@RestController
@RequestMapping("/data")
public class DataController {
    private final DataReplicationService dataService;
    public DataController(
        DataReplicationService dataService) {
            this.dataService = dataService;
    }


    @PostMapping("/s3")
    public String replicateToS3(@RequestParam String key,
    @RequestParam String content) {
```

```
        dataService.replicateToS3(key, content);
        return "Data replicated to S3";
    }


    @PostMapping("/dynamo")
    public String replicateToDynamoDB(@RequestParam String key,
    @RequestParam String value) {
        dataService.replicateToDynamoDB(key, value);
        return "Data replicated to DynamoDB";
    }


    @GetMapping("/dynamo/{key}")
    public String retrieveFromDynamoDB(@PathVariable String key) {
    return dataService.retrieveFromDynamoDB(
        key).orElse("No data found");
    }


    @PostMapping("/dynamo/conflict")
    public String resolveConflict(@RequestParam String key,
    @RequestParam String newValue) {
        dataService.resolveConflict(key, newValue);
        return "Conflict resolved in DynamoDB";
    }


    @PostMapping("/backup/{key}")
    public String backupToS3(@PathVariable String key){
        dataService.backupDynamoDBToS3(key);
        return "Data backed up to S3";
    }


    @PostMapping("/restore/{key}")
    public String restoreFromS3(@PathVariable String key) {
        dataService.restoreFromS3(key);
        return "Data restored from S3 to DynamoDB";
    }
}
```

Step 5: Implement failover mechanisms:

Create a health check endpoint and integrate with Eureka for service discovery and failover:

```
@RestController
public class FailoverController {
    private final EurekaClient eurekaClient;
    public FailoverController(EurekaClient eurekaClient) {
```

```
            this.eurekaClient = eurekaClient;
    }
    @GetMapping("/health")
    public String health() {
        return "OK";
    }
    @GetMapping("/failover")
    public String failover() {
        InstanceInfo instance = eurekaClient.
        getNextServerFromEureka("serviceB", false);
        return "Failing over to " + instance.getHomePageUrl();
    }
}
```

Step 6: Configure AWS resources using CloudFormation:

Update your CloudFormation template to include the backup S3 bucket and other necessary resources:

```
AWSTemplateFormatVersion: '2010-09-09'
Resources:
    MyBucket:
        Type: 'AWS::S3::Bucket'
        Properties:
            BucketName: 'my-bucket'

    BackupBucket:
        Type: 'AWS::S3::Bucket'
        Properties:
            BucketName: 'my-bucket-backup'

    MyTable:
        Type: 'AWS::DynamoDB::Table'
        Properties:
            TableName: 'MyTable'
            AttributeDefinitions:
            - AttributeName: 'Key'
            AttributeType: 'S'
        KeySchema:
            - AttributeName: 'Key'
            KeyType: 'HASH'
        ProvisionedThroughput:
            ReadCapacityUnits: 5
            WriteCapacityUnits: 5
```

```
MyLoadBalancer:
    Type: 'AWS::ElasticLoadBalancing::LoadBalancer'
    Properties:
    AvailabilityZones: !GetAZs ''
    Listeners:
        - LoadBalancerPort: '80'
        InstancePort: '8080'
        Protocol: 'HTTP'
    HealthCheck:
        Target: 'HTTP:8080/health'
        Interval: '30'
        Timeout: '5'
        UnhealthyThreshold: '2'
        HealthyThreshold: '10'


MyRoute53:
    Type: 'AWS::Route53::RecordSet'
    Properties:
        HostedZoneName: 'example.com.'
        Name: 'myapp.example.com.'
    Type: 'A'
        AliasTarget:
        HostedZoneId: !GetAtt MyLoadBalancer.
CanonicalHostedZoneNameID
        DNSName: !GetAtt MyLoadBalancer.DNSName
```

Step 7: Deploy and test:

Deploy the CloudFormation stack: Open a terminal and ensure that you have the AWS CLI installed and configured with the appropriate credentials. Run the following command to create the CloudFormation stack:

```
aws cloudformation create-stack --stack-name ResilientJavaApp
--template-body file://template.yaml --parameters
ParameterKey=UniqueSuffix,ParameterValue=youruniquesuffix,
ParameterKey=HostedZoneName,ParameterValue=yourHostedZoneName.
ParameterKey=DomainName,ParameterValue=yourDomainName
```

Wait for the stack creation to complete. You can check the status using the following command:

```
aws cloudformation describe-stacks --stack-name <your-stack-name>
```

Deploy the Spring Boot application: Package your Spring Boot application into a JAR file using the following command:

```
mvn clean package
```

Upload the JAR file to your EC2 instances using SCP:

```
scp -i /path/to/key-pair.pem target/your-application.jar
ec2-user@<EC2-Instance-Public-IP>:/home/ec2-user/
```

Run application on EC2: SSH into your EC2 instance and run the Spring Boot application:

```
ssh -i /path/to/key-pair.pem ec2-user@<EC2-Instance-Public-IP>
java -jar /home/ec2-user/your-application.jar
```

Test the application:

- **Test load balancing**:

  - Access the load balancer URL in your browser.

  - Ensure traffic is distributed across instances by refreshing the page multiple times and checking responses.

- **Test data replication and consistency**:

  - Use the REST endpoints to replicate data, handle conflicts, and test backup and restore functionality.

  - Here are some example API calls:

```
curl -X POST "http://<Load-Balancer-URL>/data/
s3?key=testKey&content=testContent"
curl -X POST "http://<Load-Balancer-URL>/data/
dynamo?key=testKey&value=testValue
```

Test failover:

- Simulate instance failure by stopping one of the EC2 instances from the AWS Management Console.
- Ensure the failover mechanism directs traffic to healthy instances.

Step 8: Additional considerations:

While this book focuses on building Java applications for the cloud, it is important to note that there are several additional considerations to be aware of when working with AWS for this application. Due to the scope of this book, we will not delve into the details of these AWS technologies, but here are a few key points:

- **Implementing proper authentication and authorization**: Secure your endpoints and AWS resources

- **Adding metrics and monitoring**: Set up AWS CloudWatch alarms and dashboards

- **Implementing circuit breakers for resilience**: Use tools such as Hystrix or Resilience4j

- **Adding caching mechanisms to reduce database load**: Integrate with AWS ElastiCache
- **Implementing proper testing**: Ensure thorough testing coverage with unit tests and integration tests
- **Setting up CI/CD pipelines for automated deployment**: Use AWS CodePipeline or Jenkins

For further details and reference links to the related AWS technologies, please refer to Appendix A.

This practical exercise has demonstrated how to build a resilient, cloud-native Java application leveraging AWS services. We've implemented key concepts such as load balancing, data replication with consistency management, and failover mechanisms. By utilizing Spring Boot, AWS SDK, and various AWS services such as S3, DynamoDB, and ELB, we've created a robust architecture capable of handling high availability and fault tolerance in cloud environments.

As we transition to the next section, we shift our focus from cloud resilience to computational performance. While cloud computing provides scalability and reliability, GPU acceleration offers the potential for massively parallel processing, opening new horizons for computationally intensive tasks in Java applications. This next section will explore how Java developers can harness the power of GPUs to significantly boost performance in suitable scenarios, complementing the resilience strategies we've just discussed.

GPU acceleration in Java – leveraging CUDA, OpenCL, and native libraries

To harness the immense computational power of GPUs within Java applications, developers have several options at their disposal. This section explores how Java developers can leverage CUDA, OpenCL, and native libraries to accelerate computations and tap into the parallel processing capabilities of GPUs. We'll delve into the strengths and weaknesses of each approach, guiding you toward the most suitable solution for your specific use case.

Fundamentals of GPU computing

GPUs have evolved from their original purpose of rendering graphics to becoming powerful tools for general-purpose computation. This shift, known as **general-purpose computing on graphics processing units (GPGPU)**, leverages the parallel processing capabilities of GPUs to perform computations more efficiently than traditional CPUs in certain tasks.

Unlike CPUs, which have a few cores optimized for sequential processing, GPUs have many smaller cores optimized for parallel tasks. This architecture allows for significant speedups in tasks that can be divided into smaller, concurrent operations.

Let's look at *Figure 11.2*:

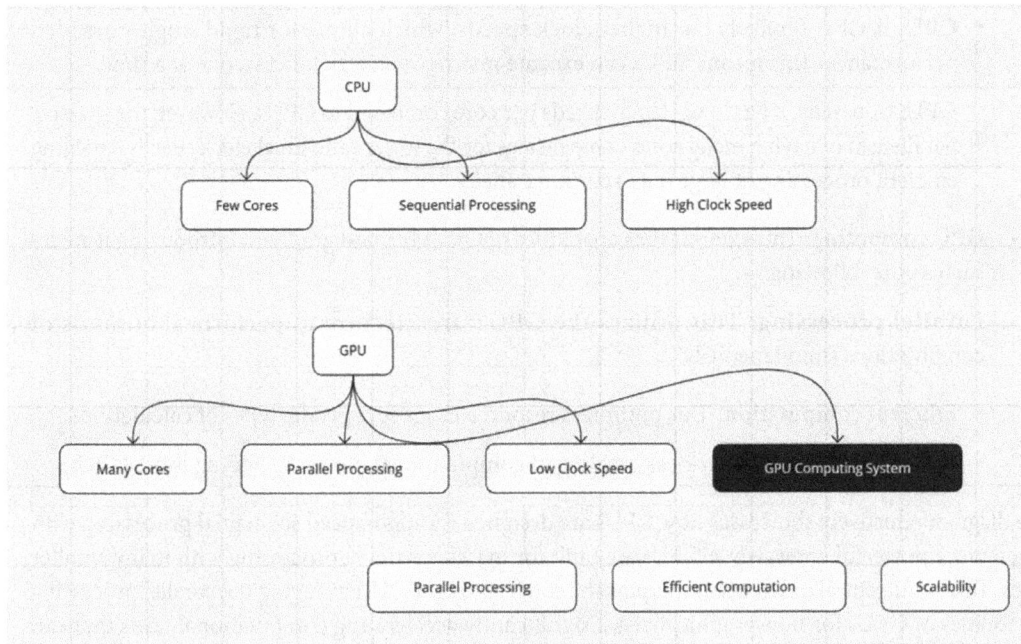

Figure 11.2: GPU versus CPU architecture

This diagram shows the fundamental architectural differences between a **central processing unit** (**CPU**) and a GPU, and the concept of GPU computing, which are explained further here:

- **Number of cores**:

  - **CPU**: This is characterized by a relatively small number of cores. These cores are powerful and designed for handling sequential processing tasks efficiently.

  - **GPU**: This features a large number of smaller cores. These cores are optimized for handling parallel processing tasks, allowing the GPU to perform many computations simultaneously.

- **Processing style**:

  - **CPU**: This is optimized for sequential task execution. This means CPUs are designed to handle a series of instructions in a specific order, making them ideal for tasks that require high single-threaded performance.

  - **GPU**: This is designed for parallel task execution. GPUs excel at dividing tasks into smaller, concurrent operations, which makes them ideal for tasks that can be parallelized, such as graphics rendering and scientific computations.

- **Clock speed**:

  - **CPU**: A CPU typically has higher clock speeds, which allows for rapid single-threaded performance. This means CPUs can execute instructions very quickly, one at a time.

  - **GPU**: Generally, it has lower clock speeds per core compared to CPUs. However, the massive parallelism of having many cores compensates for the lower individual clock speeds, enabling efficient processing of large data sets in parallel.

- **GPU computing**: This extends the capabilities of GPUs beyond graphics, introducing features such as the following:

- **Parallel processing**: This utilizes the GPU's architecture to perform thousands of computations simultaneously.

  - **Efficient computation**: This optimizes resource usage for specific types of calculations.

  - **Scalability**: This allows for easy scaling of computational power by adding more GPUs.

The diagram effectively showcases how CPUs are designed for high-speed sequential processing with fewer, more powerful cores, while GPUs are built for massive parallel processing with many smaller cores. This architectural difference underpins the concept of GPGPU, leveraging the parallel processing capabilities of GPUs for non-graphical tasks, significantly accelerating computational tasks that can be parallelized.

For readers interested in delving deeper into GPU architecture and its intricacies, several excellent resources are available online. The NVIDIA Developer website provides detailed documentation on CUDA and GPU architecture, including the CUDA C++ Programming Guide (`https://docs.nvidia.com/cuda/cuda-c-programming-guide/`) and the CUDA Runtime API (`https://docs.nvidia.com/cuda/cuda-runtime-api/`). These resources offer in-depth explanations of the CUDA programming model, memory hierarchy, and optimization techniques. For a more visual representation of GPU architecture, NVIDIA's *GPU Gems* series (`https://developer.nvidia.com/gpugems`) presents a collection of articles and tutorials on advanced GPU programming techniques and case studies.

CUDA and OpenCL overview – differences and uses in Java applications

CUDA and OpenCL are two prominent frameworks for GPU computing. They serve similar purposes but have distinct differences and use cases, particularly in Java applications.

CUDA is a proprietary parallel computing platform and API designed specifically for NVIDIA GPUs. It offers excellent performance optimizations and fine-grained control over NVIDIA GPU hardware, making it ideal for computationally intensive tasks. CUDA comes with a comprehensive suite of libraries, development tools, and debuggers for efficient GPU programming. It provides access to NVIDIA-specific libraries such as the **CUDA Deep Neural Network (cuDNN)** library for deep learning, the **CUDA Fast Fourier Transform (cuFFT)** library for fast Fourier transforms, and the **CUDA Basic Linear Algebra Subprograms(cuBLAS)** library for linear algebra operations. However, CUDA is limited to NVIDIA GPUs, which restricts its portability to other hardware. While Java bindings exist (e.g., JCuda, JCublas), the integration and ease of use may not be as seamless as with C/C++.

OpenCL is an open standard for cross-platform parallel programming maintained by the Khronos Group. It runs on a wide range of hardware from different vendors, including NVIDIA, AMD, and Intel. OpenCL code can run on various GPUs and CPUs, making it more versatile across different platforms. It is widely adopted and supported by multiple vendors, offering a broader range of applications. In Java, OpenCL is well supported through libraries such as JOCL, providing a convenient way to leverage OpenCL in Java applications. However, OpenCL may not achieve the same level of performance optimization as CUDA on NVIDIA GPUs due to its more generic nature, and its tooling and ecosystem might not be as extensive as CUDA's.

Figure 11.3 presents a table that provides a concise comparison between CUDA and OpenCL. It highlights key differences across several important aspects, including supported hardware, programming languages, performance characteristics, ecosystem, and typical use cases.

Feature	CUDA	OpenCL
Supported Hardware	NVIDIA GPUs	Multiple vendors (AMD, Intel, NVIDIA)
Programming Language	C/C++ with NVIDIA extensions	C with extensions for various platforms
Performance	Optimized for NVIDIA hardware	Portable across different hardware
Ecosystem	Robust with extensive libraries and tools	Diverse but less cohesive than CUDA
Use Cases	High-performance computing, AI, deep learning	General-purpose parallel computing

Description: This table compares CUDA and OpenCL, highlighting their key features, use cases, and performance metrics in Java applications.

Figure 11.3: A comparison between CUDA and OpenCL

Some examples of Java applications using CUDA and OpenCL are as follows:

- **Image and video processing**: Accelerating tasks such as image filtering, video encoding/decoding, and computer vision algorithms

- **Scientific computing**: Speeding up simulations, numerical calculations, and data analysis

- **Machine learning and deep learning**: Training and inference of neural networks on GPUs

- **Financial modeling**: Accelerating complex calculations in quantitative finance

Choosing between CUDA and OpenCL for Java

The choice between CUDA and OpenCL in Java depends on specific requirements:

- **Target hardware**: For NVIDIA GPUs and maximum performance, CUDA is likely the better choice. For cross-platform compatibility, OpenCL is preferred.

- **Performance versus portability**: Consider the trade-off between absolute performance (CUDA on NVIDIA GPUs) and portability across different hardware (OpenCL).

- **Ease of use and tooling**: CUDA offers a more mature ecosystem for NVIDIA GPUs, while OpenCL might require more manual setup and optimization.

- **Application-specific needs**: Specialized libraries or features available only in CUDA or OpenCL can also guide the decision.

For a comprehensive understanding of CUDA, refer to the CUDA Toolkit documentation (`https://docs.nvidia.com/cuda/`), which covers everything from installation to programming guides and API references. The OpenCL specification and documentation can be found on the Khronos Group website (`https://www.khronos.org/opencl/`), providing detailed insights into the OpenCL programming model and API. Additionally, the *OpenCL Programming Guide* by Aaftab Munshi, Benedict R. Gaster, Timothy G. Mattson, and Dan Ginsburg (`https://www.amazon.com/OpenCL-Programming-Guide-Aaftab-Munshi/dp/0321749642`) is a highly recommended resource for mastering OpenCL programming concepts and best practices.

TornadoVM – GraalVM-based GPU Acceleration

In addition to CUDA and OpenCL, another option for accelerating Java applications with GPUs is **TornadoVM**. TornadoVM is a plugin for **GraalVM**, a high-performance runtime for Java, that enables seamless execution of Java code on GPUs and other accelerators.

TornadoVM leverages the Graal compiler to automatically translate Java bytecode into OpenCL or PTX (CUDA) code, allowing developers to take advantage of GPU acceleration without the need for extensive code modifications or low-level programming. It supports a wide range of GPU architectures, including NVIDIA, AMD, and Intel GPUs.

One of the key advantages of TornadoVM is its ability to optimize code execution based on the specific characteristics of the target GPU architecture. It employs advanced compiler optimizations and runtime techniques to maximize performance and resource utilization.

To use TornadoVM, developers need to install GraalVM and the TornadoVM plugin. They can then annotate their Java code with TornadoVM-specific annotations to mark the methods or loops that should be offloaded to the GPU. TornadoVM takes care of the rest, automatically compiling and executing the annotated code on the GPU.

For more information on TornadoVM and its usage, readers can refer to the official TornadoVM documentation: `https://github.com/beehive-lab/TornadoVM`.

In the next section, we will create a practical exercise demonstrating how to leverage GPU for computational tasks in a Java application. Using CUDA, we will create a simple matrix multiplication application to showcase GPU acceleration.

Practical exercise – GPU-accelerated matrix multiplication in Java

Objective: Implement a matrix multiplication algorithm using Java and CUDA, comparing its performance with a CPU-based implementation.

Here is a step-by-step guide:

Step 1: Set up the development environment and add these dependencies to your Maven `pom.xml` file to include JCuda libraries:

```
<dependencies>
    <dependency>
        <groupId>org.jcuda</groupId>
        <artifactId>jcuda</artifactId>
        <version>12.0.0</version> </dependency>

    <dependency>
        <groupId>org.jcuda</groupId>
        <artifactId>jcublas</artifactId>
        <version>12.0.0</version>
    </dependency>
</dependencies>
```

Step 2: Implement the CPU version of matrix multiplication: Here's a standard CPU implementation of matrix multiplication:

```
public class MatrixMultiplication {
    public static float[][] multiplyMatricesCPU(
        float[][] a, float[][] b) {
            int m = a.length;
```

```
                int n = a[0].length;
                int p = b[0].length;
                float[][] result = new float[m][p];

                for (int i = 0; i < m; i++) {
                    for (int j = 0; j < p; j++) {
                        for (int k = 0; k < n; k++) {
                            result[i][j] += a[i][k] * b[k][j];
                        }
                    }
                }
            return result;
        }
    }
```

This `multiplyMatricesCPU()` method performs matrix multiplication using nested loops, suitable for CPU execution.

Step 3: Implement the GPU version using JCuda: Let's create a GPU-accelerated version using JCuda:

```
public class MatrixMultiplicationGPU {
    public static float[][] multiplyMatricesGPU(
        float[][] a, float[][] b) {
            int m = a.length;
            int n = a[0].length;
            int p = b[0].length;

        // Initialize JCublas
        JCublas.cublasInit();

        // Allocate memory on GPU
            Pointer d_A = new Pointer();
            Pointer d_B = new Pointer();
            Pointer d_C = new Pointer();

            JCublas.cublasAlloc(m * n, Sizeof.FLOAT, d_A);
            JCublas.cublasAlloc(n * p, Sizeof.FLOAT, d_B);
            JCublas.cublasAlloc(m * p, Sizeof.FLOAT, d_C);

        // Copy data to GPU
            JCublas.cublasSetVector(
                m * n, Sizeof.FLOAT, Pointer.to(
                    flattenMatrix(a)), 1, d_A, 1);
            JCublas.cublasSetVector(n * p, Sizeof.FLOAT,
```

```
                    Pointer.to(flattenMatrix(b)), 1, d_B, 1);

        // Perform matrix multiplication
            JCublas.cublasSgemm('n', 'n', m, p, n, 1.0f,
                d_A, m, d_B, n, 0.0f, d_C, m);

        // Copy result back to CPU
            float[] resultFlat = new float[m * p];
            JCublas.cublasGetVector(m * p, Sizeof.FLOAT,
                d_C, 1, Pointer.to(resultFlat), 1);

        // Free GPU memory
            JCublas.cublasFree(d_A);
            JCublas.cublasFree(d_B);
            JCublas.cublasFree(d_C);

        // Shutdown JCublas
            JCublas.cublasShutdown();

            return unflattenMatrix(resultFlat, m, p);
}
private static float[] flattenMatrix(float[][] matrix) {
    int m = matrix.length;
    int n = matrix[0].length;
    float[] flattened = new float[m * n];
    for (int i = 0; i < m; i++) {
        System.arraycopy(matrix[i], 0, flattened, i * n,n);
    }
    return flattened;
}


    private static float[][] unflattenMatrix(
        float[] flattened, int m, int p) {
            float[][] result = new float[m][p];
            for (int i = 0; i < m; i++) {
                System.arraycopy(flattened, i * p,
                    result[i], 0, p);
            }
            return result;
        }
}
```

This class uses the Java cuBLAS library (**JCublas**) to perform matrix multiplication on the GPU, including memory allocation, data transfer, and computation.

Step 4: Create a main class to compare CPU and GPU performance: So, let's create a comparison class:

```java
public class MatrixMultiplicationComparison {
    public static void main(String[] args) {
        int size = 1000; // Size of the square matrices
        float[][] a = generateRandomMatrix(size, size);
        float[][] b = generateRandomMatrix(size, size);

        // CPU multiplication
        long startTimeCPU = System.currentTimeMillis();
        float[][] resultCPU = MatrixMultiplication.
        multiplyMatricesCPU(a, b);
        long endTimeCPU = System.currentTimeMillis();
        System.out.println("CPU time: " + (
            endTimeCPU - startTimeCPU) + " ms");

        // GPU multiplication
        long startTimeGPU = System.currentTimeMillis();
        float[][] resultGPU = MatrixMultiplicationGPU.
        multiplyMatricesGPU(a, b);
        long endTimeGPU = System.currentTimeMillis();
        System.out.println("GPU time: " + (
            endTimeGPU - startTimeGPU) + " ms");

        // Verify results
        boolean correct = verifyResults(resultCPU,
            resultGPU);
        System.out.println(
            "Results are correct: " + correct);
    }

    private static float[][] generateRandomMatrix(int rows,
        int cols) {
        float[][] matrix = new float[rows][cols];
        for (int i = 0; i < rows; i++) {
            for (int j = 0; j < cols; j++) {
                matrix[i][j] = (float) Math.random();
            }
        }
        return matrix;
    }
    private static boolean verifyResults(float[][] a,
```

```
            float[][] b) {
            if (a.length != b.length || a[0].length != b[0].length) {
                return false;
            }
            for (int i = 0; i < a.length; i++) {
                for (int j = 0; j < a[0].length; j++) {
                    if (Math.abs(a[i][j] - b[i][j]) > 1e-5) {
                        return false;
                    }
                }
            }
            return true;
        }
    }
```

This class generates random matrices, performs multiplication using both CPU and GPU methods, and compares their performance and accuracy.

Step 5: Run the application: Execute the `MatrixMultiplicationComparison` class. It will perform matrix multiplication using both CPU and GPU implementations and compare their execution times.

Step 6: Analyze the results: Compare the execution times of CPU and GPU implementations. For large matrices, you should see a significant speedup with the GPU version.

This practical exercise demonstrates how to leverage GPU acceleration for a common computational task in Java. It showcases the integration of CUDA through JCuda, providing a real-world example of how GPU computing can significantly improve performance for suitable tasks.

Remember to handle potential exceptions and edge cases in a production environment. Also, for even better performance, consider using more advanced CUDA features such as shared memory and optimized memory access patterns in your kernels.

For those eager to explore more advanced topics in GPU acceleration and its applications in Java, the following resources are highly recommended:

- *GPU Computing Gems Emerald Edition* by Wen-mei W. Hwu provides a collection of techniques and algorithms for GPU computing across various domains
- The *Java on GPUs* website (`https://xperti.io/blogs/accelerating-java-with-gpu/`) offers tutorials, articles, and case studies specifically focused on leveraging GPUs in Java applications

These resources will help you deepen your understanding of GPU acceleration and its practical applications in Java-based projects.

As we conclude our exploration of GPU acceleration in Java, we've seen how leveraging CUDA or OpenCL can significantly boost performance for parallel processing tasks. This knowledge sets the stage for our next critical topic, *Specialized monitoring for Java concurrency in the cloud*. Here, we'll examine how to effectively monitor and optimize these high-performance Java applications in distributed cloud environments.

Specialized monitoring for Java concurrency in the cloud

Monitoring concurrent operations in cloud environments is crucial for several key reasons. Performance optimization stands at the forefront, as it allows developers to identify bottlenecks and inefficiencies in parallel execution. Effective monitoring ensures efficient resource management across distributed systems, a critical aspect of cloud computing. It also plays a vital role in error detection, enabling quick identification and diagnosis of issues related to race conditions or deadlocks. Furthermore, monitoring provides valuable scalability insights, helping teams understand how applications perform under varying loads, which in turn informs scaling decisions. Lastly, it contributes to cost control by optimizing resource usage, an essential factor in managing cloud computing expenses effectively.

Challenges in monitoring

Monitoring Java concurrency in cloud environments presents a unique set of challenges. The distributed nature of cloud systems makes it difficult to obtain a cohesive view of concurrent operations spread across multiple instances or services. Dynamic scaling, a hallmark of cloud computing, requires monitoring systems to adapt rapidly to changing infrastructure as resources scale up or down. The sheer volume of monitoring data generated in cloud environments poses significant management and analysis challenges. Latency and network issues can lead to delays and inconsistencies in data collection, complicating real-time monitoring efforts. Security and compliance concerns necessitate careful consideration to ensure monitoring practices adhere to cloud security standards and data protection regulations. Finding monitoring tools that are compatible with both Java concurrency constructs and cloud-native technologies can be challenging. Lastly, there's a delicate balance to strike between the need for detailed monitoring and the performance impact of the monitoring tools themselves.

These challenges underscore the need for specialized approaches to effectively monitor Java concurrency in cloud environments, a topic we will explore in depth in the following sections.

Monitoring tools and techniques

Cloud-native monitoring tools are essential for maintaining the performance and reliability of applications running in the cloud. Leading cloud providers offer robust solutions to help monitor, troubleshoot, and optimize your cloud infrastructure.

Here's a brief introduction to some of the popular cloud-native monitoring tools:

- **AWS CloudWatch**: A comprehensive monitoring and observability service from AWS. It enables you to collect metrics, monitor log files, set alarms for specific thresholds, and even react automatically to changes in your AWS resources. With custom metrics, you can track application-specific data points, providing a deeper understanding of your application's behavior.

- **Google Cloud Monitoring**: This powerful monitoring solution from **Google Cloud Platform (GCP)** offers a unified view of your entire cloud environment. It automatically collects metrics from various GCP resources and provides insights into the health, performance, and availability of your applications and services. Google Cloud Monitoring also integrates with other GCP services, such as Cloud Logging and Cloud Trace, for a complete observability solution.

- **Azure Monitor**: Microsoft Azure's comprehensive monitoring solution, **Azure Monitor**, collects and analyzes telemetry data from your cloud and on-premises environments. It allows you to monitor various aspects, including application performance, infrastructure health, and platform logs. Azure Monitor's customizable dashboards and alerts help you proactively identify and address issues before they impact your users.

- **Other cloud-native tools**: Several other cloud-native tools such as Datadog, New Relic, and Prometheus provide advanced monitoring capabilities and integrations with various cloud providers. These tools offer features such as distributed tracing, **application performance monitoring (APM)**, and infrastructure monitoring, giving you a holistic view of your cloud environment.

By leveraging these cloud-native monitoring tools, you can gain valuable insights into your application's performance, identify potential bottlenecks, and proactively optimize your cloud infrastructure. This leads to improved reliability, reduced downtime, and a better overall user experience.

Let us now explore Java-centric tools. Java applications deployed in the cloud often demand specialized monitoring tools to manage their unique complexities, particularly around concurrency and performance. Popular Java-centric tools include the following:

- **Java Management Extensions (JMX)** is a Java technology that supplies tools for managing and monitoring applications, system objects, devices, and service-oriented networks. It allows monitoring of JVM health, such as memory usage, garbage collection, and thread states. Custom MBeans can be created to expose application-specific metrics.

- **VisualVM** is a visual tool integrating several command-line **Java development kit (JDK)** tools and lightweight profiling capabilities. It provides detailed insights into the JVM performance and supports heap dump analysis, thread analysis, and profiling.

- **Custom monitoring solutions** can be built to address specific needs using libraries, such as Dropwizard Metrics or Micrometer. These solutions offer flexibility to define and collect metrics specific to the application and integrate with various backends such as Prometheus, Graphite, or AWS CloudWatch.

Utilizing these specialized monitoring tools and techniques ensures that Java concurrency in cloud environments is managed effectively, leading to enhanced performance, reliability, and cost efficiency.

Integrating cloud-native and Java-centric monitoring for optimal performance

Effectively monitoring cloud-native Java applications often involves a combination of cloud-native and Java-centric tools. Cloud-native tools such as AWS CloudWatch, Google Cloud Monitoring, and Azure Monitor provide a high-level overview of the entire cloud infrastructure, including resource utilization, network traffic, and overall system health. They offer valuable insights into how your Java application interacts with the cloud environment.

Java-centric tools such as JMX, VisualVM, and custom monitoring solutions dive deeper into the internals of the Java application itself. They monitor JVM metrics such as garbage collection, thread states, and memory usage, along with application-specific metrics exposed through custom MBeans or libraries such as Dropwizard Metrics. These tools are essential for understanding the performance and behavior of your Java code.

In practice, it's often most convenient and effective to use both types of tools in conjunction. Cloud-native tools provide the big picture of how your application fits into the cloud ecosystem, while Java-centric tools give you granular insights into the application's internal workings. Integrating these tools can help you correlate cloud-level events with Java-level metrics, leading to a more comprehensive understanding of your application's performance and easier troubleshooting.

For example, you might use CloudWatch to monitor CPU and memory utilization across your entire AWS infrastructure, while simultaneously using JMX to track garbage collection frequency and duration within your Java application. If CloudWatch reveals a spike in CPU usage, you can then use JMX to determine whether it's related to excessive garbage collection or some other Java-specific issue. This integrated approach enables you to quickly identify and address performance bottlenecks and other problems before they impact your users.

Use case – monitoring a Java web application in the Cloud

In this use case, we will monitor a Java-based web application deployed on AWS using both AWS CloudWatch (a cloud-native tool) and JMX (a Java-centric tool). The goal is to achieve comprehensive monitoring that covers both the cloud infrastructure and application-specific metrics.

Let's look at *Figure 11.4*:

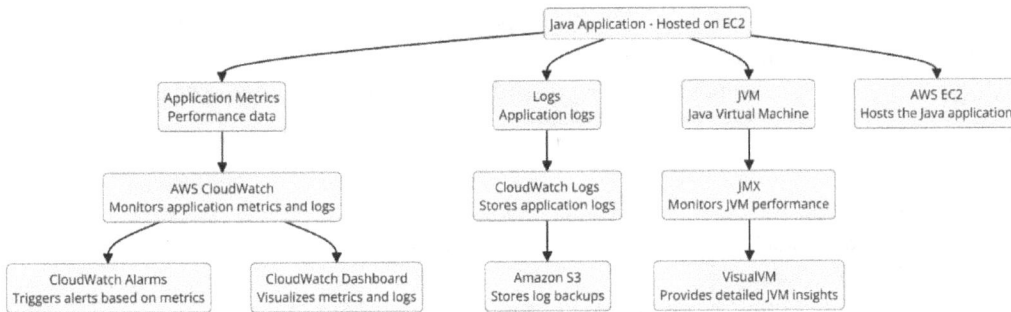

Figure 11.4: Enhanced monitoring setup for Java application

This diagram illustrates the monitoring setup for a Java application hosted on AWS EC2. It details how application metrics and logs are collected, monitored, and visualized using various tools. Application metrics are sent to AWS CloudWatch, which also stores logs and triggers alarms for alerts. JMX is used for monitoring JVM performance, providing detailed insights through VisualVM. Logs are backed up to Amazon S3 for additional storage and retrieval. This setup ensures comprehensive monitoring and alerting for optimal application performance and reliability.

Step 1: Set up AWS CloudWatch:

First, ensure that your AWS SDK for Java is included in your project dependencies. Then, create a simple Java application that publishes custom metrics to CloudWatch.

Add to your pom.xml file if using Maven:

```
<!-- Add to your pom.xml -->
<dependency>
    <groupId>com.amazonaws</groupId>
    <artifactId>aws-java-sdk-cloudwatch</artifactId>
        <version> 2.17.102 </version>
</dependency>
```

Create the CloudWatchMonitoring class. Provide a simplified interface for publishing custom metrics to AWS CloudWatch using the AWS SDK for Java:

```
public class CloudWatchMonitoring {

    private final CloudWatchClient cloudWatch;

    public CloudWatchMonitoring(String accessKey,
        String secretKey) {
```

```
                AwsBasicCredentials awsCredentials = AwsBasicCredentials.
                create(accessKey, secretKey);
                this.cloudWatch = CloudWatchClient.builder()
                    .region(Region.US_EAST_1)
                    .credentialsProvider(StaticCredentialsProvider.
                    create(awsCredentials))
                    .build();
        }

        public void publishCustomMetric(String metricName,
            double value) {
                MetricDatum datum = MetricDatum.builder()
                    .metricName(metricName)
                    .unit(StandardUnit.COUNT)
                    .value(value)
                    .build();

            PutMetricDataRequest request = PutMetricDataRequest.builder()
                    .namespace("MyAppNamespace")
                    .metricData(datum)
                    .build();

            cloudWatch.putMetricData(request);
            System.out.println("Metric '" + metricName + "' published to
            CloudWatch.");
        }
    }
```

Step 2: Set up JMX:

Enable JMX in your Java application to monitor JVM performance metrics such as memory usage, garbage collection, and thread states. Here is a sample code:

```
public class JMXMonitoring {
    public interface CustomMBean {
        int getMetric();
        void setMetric(int metric);
    }
    public static class Custom implements CustomMBean {
        private int metric = 0;
        @Override
        public int getMetric() {
            return metric;
        }
        @Override
        public void setMetric(int metric) {
```

```
                this.metric = metric;
        }
    }
    public static CustomMBean createAndRegisterMBean(
        MBeanServer mbs) throws Exception {
            CustomMBean customMBean = new Custom();
            ObjectName name = new ObjectName(
                "com.example:type=CustomMBean");
            mbs.registerMBean(customMBean, name);
            return customMBean;
        }
    }
}
```

Step 3: Publish Metrics to AWS CloudWatch:

After setting up AWS CloudWatch, the next step is to publish custom metrics from your Java application to CloudWatch. Follow these steps:

First, publish custom metrics:

- Use the provided Java code to send custom metrics to CloudWatch. This involves implementing a class that interacts with the AWS SDK to publish metrics data.

- Ensure that the metrics are meaningful and relevant to your application's performance and health.

After that, monitor metrics in the AWS Management Console:

- Once your metrics are being published, monitor them via the AWS Management Console under the CloudWatch service. Set up necessary dashboards and alarms to track key metrics and receive notifications on potential issues.

- Create the MonitoringApplication class, which we will do next.

Create a new class called `MonitoringApplication` that will serve as the main entry point for your application and integrate both CloudWatch and JMX monitoring:

```
public class MonitoringApplication {
    private static JMXMonitoring.CustomMBean customMBean;
    private static CloudWatchMonitoring cloudWatchMonitor;

    public static void main(String[] args) {
        // Initialize CloudWatch monitoring
        cloudWatchMonitor = new CloudWatchMonitoring(
            "your-access-key", "your-secret-key");

        // Initialize JMX monitoring
```

```
        setupJMXMonitoring();

        // Start periodic monitoring
        startPeriodicMonitoring();

        System.out.println("Monitoring systems initialized.
        Application running...");

        // Keep the application running
        try {
            Thread.sleep(Long.MAX_VALUE);
        } catch (InterruptedException e) {
            e.printStackTrace();
        }
    }

    private static void setupJMXMonitoring() {
        try {
            MBeanServer mbs = ManagementFactory.
            getPlatformMBeanServer();
            customMBean = JMXMonitoring.createAndRegisterMBean(mbs);
            customMBean.setMetric(0); // Set initial metric value
            System.out.println("JMX Monitoring setup complete. Initial
            metric value: " + customMBean.getMetric());
        } catch (Exception e) {
            e.printStackTrace();
        }
    }

    private static void startPeriodicMonitoring() {
        ScheduledExecutorService executor = Executors.
        newSingleThreadScheduledExecutor();
        executor.scheduleAtFixedRate(() -> {
            try {
                // Simulate metric change
                int currentMetric = customMBean.getMetric();
                int newMetric = currentMetric + 1;
                customMBean.setMetric(newMetric);

                // Publish to CloudWatch
                cloudWatchMonitor.publishCustomMetric("JMXMetric",
                newMetric);

                System.out.println("Updated JMX metric: " +
                newMetric);
```

```
        } catch (Exception e) {
            e.printStackTrace();
        }
    }, 0, 60, TimeUnit.SECONDS); // Run every 60 seconds
    }
}
```

Note that this `MonitoringApplication` class does the following:

- Initializes both CloudWatch and JMX monitoring.

- Sets up periodic monitoring that updates the JMX metric and publishes it to CloudWatch every 60 seconds.

- Keeps the application running indefinitely.

Remember to replace `"your-access-key"` and `"your-secret-key"` with your actual AWS credentials and consider using more secure methods to manage these credentials in a production environment.

Step 4: Run the application:

To run the application with JMX enabled, use the following command:

```
mvn exec:java -Dexec.mainClass="com.example.MonitoringApplication"
-Dcom.sun.management.jmxremote -Dcom.sun.management.jmxremote.
port=9090 -Dcom.sun.management.jmxremote.authenticate=false -Dcom.sun.
management.jmxremote.ssl=false
```

Step 5: Connect to JMX: Once your application is running with JMX enabled, you can connect to it using a JMX client:

- Use a JMX client such as JConsole or VisualVM to connect to your application.

- If using JConsole, do the following:

    i.   Open a terminal and type: `jconsole`.

    ii. Select **Remote Process**.

    iii. Enter `localhost:9090` for the connection.

    iv. Click **Connect**.

- If using VisualVM, do the following:

    i.   Open VisualVM.

    ii. Right-click on **Local** in the left panel and select **Add JMX Connection**.

    iii. Enter `localhost:9090` and give the connection a name.

    iv.   Double-click on the new connection to open it.

    v.   Once connected, you should be able to see your custom MBean under the MBeans tab.

By integrating both AWS CloudWatch and JMX, you can gain a comprehensive view of your application's performance, combining the strengths of cloud-native monitoring with detailed JVM insights. This approach ensures optimal performance and reliability for your Java applications running in the cloud.

Effective monitoring is crucial for maintaining high-performance, reliable Java applications in cloud environments. By combining cloud-native tools such as AWS CloudWatch with Java-centric solutions such as JMX, developers can gain a comprehensive view of their application's behavior, from infrastructure-level metrics to JVM-specific insights. This integrated approach allows for quick identification and resolution of performance bottlenecks, efficient resource management, and proactive optimization of concurrent operations. As cloud technologies continue to evolve, mastering these monitoring techniques will be essential for Java developers aiming to leverage the full potential of cloud computing while ensuring their applications remain robust and scalable.

Summary

In this chapter, we've explored advanced Java concurrency practices tailored for cloud computing environments, equipping you with powerful tools and techniques to optimize your Java applications in the cloud.

We began by delving into cloud-specific redundancies and failover mechanisms, learning how to enhance the resilience of Java applications in distributed systems. You've gained practical knowledge on implementing load balancing, data replication with consistency management, and robust failover strategies using AWS services and Spring Boot. This foundation ensures your applications can maintain high availability and fault tolerance in dynamic cloud environments.

Next, we ventured into the realm of GPU acceleration, discovering how to leverage CUDA and OpenCL in Java applications. You've learned the fundamentals of GPU computing and how it differs from traditional CPU processing. Through practical exercises, such as implementing GPU-accelerated matrix multiplication, you've seen firsthand how to significantly boost performance for computationally intensive tasks.

Finally, we tackled the crucial aspect of monitoring Java concurrency in the cloud. You've learned about the importance of specialized monitoring, the challenges it presents, and how to overcome them using a combination of cloud-native and Java-centric tools. The practical examples of integrating AWS CloudWatch with JMX have given you a comprehensive approach to monitoring that spans both infrastructure and application-level metrics.

By mastering these advanced practices, you're now well equipped to design, implement, and maintain high-performance, scalable Java applications in cloud environments. You can confidently handle complex concurrency scenarios, optimize computational tasks using GPUs, and maintain visibility into your application's performance across distributed systems.

As cloud computing continues to evolve, the skills you've acquired in this chapter will prove invaluable in leveraging Java's concurrency capabilities to their fullest potential in the cloud. Remember to keep exploring new tools and techniques as they emerge, and always consider the unique requirements of your specific applications when applying these advanced practices.

As we conclude our journey through advanced Java concurrency in cloud computing, it's natural to wonder what lies ahead. In our final chapter, *The Horizon Ahead*, we'll explore emerging trends in cloud technologies and Java's evolving role in this dynamic landscape. We'll uncover how Java continues to adapt and shape the future of cloud computing, ensuring you're prepared for the next wave of innovations in this ever-changing field.

Questions

1. Which of the following is *not* a key reason for monitoring concurrent operations in cloud environments?

    A. Performance optimization

    B. Resource management

    C. Error detection

    D. User interface design

2. What is the primary advantage of using GPU acceleration in Java applications for cloud computing?

    A. Simplified code structure

    B. Reduced power consumption

    C. Improved performance for parallel tasks

    D. Enhanced network security

3. Which tool is specifically designed for monitoring Java applications and provides detailed insights into JVM performance?

    A. AWS CloudWatch

    B. VisualVM

    C. Google Cloud Monitoring

    D. Azure Monitor

4. In the context of cloud-specific redundancies, what does data replication primarily aim to achieve?

    A. Reduce network latency

    B. Minimize data loss and improve availability

    C. Enhance user interface responsiveness

    D. Decrease cloud storage costs

5. What is a key challenge in monitoring Java concurrency in cloud environments?

    A. Limited availability of monitoring tools

    B. Excessive simplicity of cloud architectures

    C. Difficulty in obtaining a cohesive view of distributed operations

    D. Lack of support for Java applications in cloud platforms

12

The Horizon Ahead

As cloud technologies continue to evolve at a rapid pace, it is crucial for developers and organizations to stay ahead of the curve and prepare for the next wave of innovations. This chapter will explore the emerging trends and advancements in the cloud computing landscape, with a particular focus on Java's role in shaping these future developments.

We will begin by examining the evolution of serverless Java, where frameworks such as Quarkus and Micronaut are redefining the boundaries of **functions as a service**. These tools leverage innovative techniques, such as native image compilation, to deliver unprecedented performance and efficiency in serverless environments. Additionally, we will delve into the concept of serverless containers, which allow for the deployment of entire Java applications in a serverless fashion, harnessing the benefits of container orchestration platforms such as Kubernetes and **Amazon Web Services (AWS)** Fargate.

Next, we will explore the role of Java in the emerging paradigm of edge computing. As data processing and decision-making move closer to the source, Java's platform independence, performance, and extensive ecosystem make it an ideal candidate for building edge applications. We will discuss the key frameworks and tools that enable Java developers to leverage the power of edge computing architectures.

Furthermore, we will investigate Java's evolving position in the integration of **artificial intelligence (AI)** and **machine learning (ML)** within cloud-based ecosystems. From serverless AI/ML workflows to the seamless integration of Java with cloud-based AI services, we will explore the opportunities and challenges that this convergence presents.

Finally, we will delve into the captivating realm of **quantum computing**, a field that promises to revolutionize various industries. While still in its early stages, understanding the fundamental principles of quantum computing, such as qubits, quantum gates, and algorithms, can prepare developers for future advancements and their potential integration with Java-based applications.

By the end of this chapter, you will have a comprehensive understanding of the emerging trends in cloud computing and Java's pivotal role in shaping these innovations. You will be equipped with the knowledge and practical examples to position your applications and infrastructure for success in the rapidly evolving cloud landscape.

The following are the key topics that will be covered in this chapter:

- Future trends in cloud computing and Java's role

- Edge computing and Java

- AI and ML integration

- Emerging concurrency and parallel processing tools in Java

- Preparing for the next wave of cloud innovations

So, let's get started!

Technical requirements

To fully engage with *Chapter 12*'s content and examples, ensure the following are installed and configured:

- **Java Development Kit or JDK**:

  - Quarkus requires a JDK to run. If you don't have one, download and install a recent version (JDK 17 or newer is recommended) from the official source:

    - **AdoptOpenJDK**: `https://adoptium.net/`

    - **OpenJDK**: `https://openjdk.org/`

- **Quarkus Command Line Interface (CLI)**:

  - Use package managers such as Chocolatey (`choco install quarkus`) or Scoop (`scoop install quarkus`)

  - Alternatively, use JBang (`jbang app install --fresh quarkus@quarkusio`)

  - **Quarkus CLI installation guide**: `https://quarkus.io/guides/cli-tooling`

- **GraalVM**:

  I.   Download the GraalVM Community Edition for Windows from `https://www.graalvm.org/downloads/`.

  II.   Follow the installation instructions provided.

  III. Set the `GRAALVM_HOME` environment variable to the GraalVM installation directory.

  IV.   Add `%GRAALVM_HOME%\bin` to your PATH environment variable.

- **Docker Desktop**:

  I.   Download and install Docker Desktop for Windows from `https://www.docker.com/products/docker-desktop/`.

  II.   Follow the installation wizard and configure Docker as needed.

The code in this chapter can be found on GitHub:

`https://github.com/PacktPublishing/Java-Concurrency-and-Parallelism`

Future trends in cloud computing and Java's role

As cloud computing continues to evolve, several emerging trends are shaping the future of this technology landscape. Innovations such as edge computing, AI and ML integration, and serverless architectures are at the forefront, driving new possibilities and efficiencies. Java, with its robust ecosystem and continuous advancements, is playing a pivotal role in these developments. This section will explore the latest trends in cloud computing, how Java is adapting to and facilitating these changes, and provide real-world examples of Java's adoption in cutting-edge cloud technologies.

Emerging trends in cloud computing – serverless Java beyond function as a service

Emerging trends in cloud computing are reshaping the landscape of serverless Java, extending beyond the traditional functions-as-a-service model. Innovations in serverless Java frameworks such as Quarkus and Micronaut are driving this evolution.

Quarkus

Quarkus, recognized for its strengths in microservices, is now making a substantial impact in serverless environments. It empowers developers to build serverless functions that adhere to microservice principles, seamlessly merging these two architectural approaches. A standout feature is Quarkus' native integration with GraalVM, enabling the compilation of Java applications into native executables. This is a game-changer for serverless computing, as it tackles the long-standing issue of cold start latency. By harnessing GraalVM, Quarkus dramatically reduces startup times for Java applications, often from seconds to mere milliseconds, compared to traditional **Java virtual machine** (**JVM**) based alternatives. Moreover, the resulting native binaries are more memory efficient, facilitating optimized scaling and resource utilization in the dynamic world of serverless environments. These advancements are revolutionizing serverless Java, providing developers with a powerful toolkit to create high-performance, cloud-native applications that are both efficient and responsive.

Micronaut

Micronaut is another innovative framework making significant progress in the serverless Java space. It is designed to optimize the performance of microservices and serverless applications through several key features:

- **Compile-time dependency injection**: Unlike traditional frameworks that resolve dependencies at runtime, Micronaut performs this task during compilation. This approach eliminates the need for runtime reflection, resulting in faster startup times and reduced memory consumption.

- **Aspect-oriented programming (AOP)**: AOP is a programming paradigm that increases modularity by allowing the separation of cross-cutting concerns. In Micronaut, AOP is implemented at compile time rather than runtime. This means that features such as transaction management, security, and caching are woven into the bytecode during compilation, eliminating the need for runtime proxies and further reducing memory usage and startup time.

These compile-time techniques make Micronaut an ideal choice for building lightweight, fast, and efficient serverless applications. The framework's design is particularly well suited to environments where rapid startup and low resource consumption are crucial.

Additionally, Micronaut supports the creation of GraalVM native images. This feature further enhances its suitability for serverless environments by minimizing cold start times and resource usage, as native images can start almost instantaneously and consume less memory compared to traditional JVM-based applications.

Serverless containers and Java applications

Serverless containers represent another dimension of serverless computing, enabling the deployment of entire Java applications rather than individual functions. This approach leverages container orchestration platforms such as Kubernetes and AWS Fargate to run containers in a serverless fashion. Java applications packaged as containers benefit from the same serverless advantages of automatic scaling and pay-per-use pricing, but with more control over the runtime environment compared to traditional serverless functions. Developers can ensure consistency across different environments by packaging the application with its dependencies. Full control over the runtime environment allows for the inclusion of necessary libraries and tools, providing flexibility that is sometimes lacking in traditional serverless functions. Additionally, serverless containers can scale automatically based on demand, offering the benefits of serverless computing while maintaining the robustness of containerized applications.

By combining the innovations in serverless Java frameworks such as Quarkus and Micronaut with the flexibility of serverless containers, developers can create highly scalable, efficient, and responsive Java applications that meet the demands of modern cloud-native environments. These advancements are paving the way for the next generation of serverless Java, moving beyond simple functions to encompass full-fledged applications and services.

Example use-case – building a serverless REST API with Quarkus and GraalVM

Objective: Create a serverless REST API for product management and deploy it on AWS Lambda using Quarkus, demonstrating key Quarkus features and integration with AWS services.

This example covers key concepts and elements of Quarkus. The full application will be available in the GitHub repository.

1. **Set up the project**: Use Quarkus CLI or Maven to bootstrap a new project. For this example, we'll use Maven. Run the following Maven command to create the Quarkus project:

```
mvn io.quarkus:quarkus-maven-plugin:2.7.5.Final:create \
    -DprojectGroupId=com.example \
    -DprojectArtifactId=quarkus-serverless \
    -DclassName="com.example.ProductResource" \
    -Dpath="/api/products"
```

2. **Key components**:

   - **Product resource (REST API)**: The `ProductResource` class is a RESTful resource that defines the endpoints for managing products within the application. Using JAX-RS annotations, it provides methods for retrieving all products, fetching the count of products, and getting details of individual products by ID. This class serves as the primary interface for client interactions with the product-related data in the application. It demonstrates Quarkus features such as dependency injection, metrics, and OpenAPI documentation:

```
@Path("/api/products")
@Produces(MediaType.APPLICATION_JSON)
@Consumes(MediaType.APPLICATION_JSON)
@Tag(name = "Product",
    description = "Product management operations")
public class ProductResource {

    @Inject
    ProductRepository productRepository;

    @GET
    @Counted(name = "getAllProductsCount",
        description = "How many times getAllProducts has been
        invoked")
    @Timed(name = "getAllProductsTimer",
        description = "A measure of how long it takes to perform
        getAllProducts",
        unit = MetricUnits.MILLISECONDS)
    @Operation(summary = "Get all products",
```

```
        description = "Returns a list of all products with
        pagination and sorting")
    public Response getAllProducts(@QueryParam(
        "page") @DefaultValue("0") int page,
            @QueryParam("size") @DefaultValue("20") int size,
            @QueryParam("sort") @DefaultValue("name") String
            sort) {
        // Implementation omitted for brevity
    }


    @POST
    @Operation(summary = "Create a new product",
        description = "Creates a new product and returns the
        created product")
    public Response createProduct(Product product) {
        // Implementation omitted for brevity
    }


    // Additional CRUD methods omitted for brevity
}
```

- ProductRepository: The ProductRepository class acts as the data access layer, managing interactions with *AWS DynamoDB* for product data persistence. It demonstrates Quarkus' seamless integration with AWS **Software Development Kit (SDK)** v2, specifically for DynamoDB operations. The class uses dependency injection to obtain DynamoDbClient, showcasing how Quarkus simplifies cloud service integration. It implements methods for **create, read, update, and delete (CRUD)** operations, translating between Java objects and DynamoDB item representations, thus demonstrating how Quarkus applications can efficiently work with NoSQL databases in a cloud environment:

```
@ApplicationScoped
public class ProductRepository {

    @Inject
    DynamoDbClient dynamoDbClient;

    private static final String TABLE_NAME = "Products";

    public void persist(Product product) {
        Map<String, AttributeValue> item = new HashMap<>();
        item.put("id", AttributeValue.builder().s(
            product.getId()).build());
        item.put("name", AttributeValue.builder().s(
            product.getName()).build());
```

```
        // Add other attributes

        PutItemRequest request = PutItemRequest.builder()
                .tableName(TABLE_NAME)
                .item(item)
                .build();

        dynamoDbClient.putItem(request);
    }

    // Additional methods omitted for brevity
}
```

- ImageAnalysisCoordinator: The ImageAnalysisCoordinator class showcases Quarkus' ability to create AWS Lambda functions that interact with multiple AWS services. It demonstrates handling **Simple Storage Service (S3)** events and triggering **Elastic Container Service (ECS)** tasks, illustrating how Quarkus can be used to build complex, event-driven architectures. This class uses dependency injection for AWS clients (ECS and S3), showing how Quarkus simplifies working with multiple cloud services in a single component. It's an excellent example of using Quarkus for serverless applications that orchestrate other AWS services:

```
@ApplicationScoped
public class ImageAnalysisCoordinator implements
RequestHandler<S3Event, String> {

    @Inject
    EcsClient ecsClient;

    @Inject
    S3Client s3Client;

    @Override
    public String handleRequest(S3Event s3Event,
        Context context) {
            String bucket = s3Event.getRecords().get(
                0).getS3().getBucket().getName();
            String key = s3Event.getRecords().get(
                0).getS3().getObject().getKey();

        RunTaskRequest runTaskRequest = RunTaskRequest.builder()
            .cluster("your-fargate-cluster")
            .taskDefinition("your-task-definition")
            .launchType("FARGATE")
```

```
            .overrides(TaskOverride.builder()
          .containerOverrides(
             ContainerOverride.builder()
               .name("your-container-name")
               .environment(
                   KeyValuePair.builder()
                       .name("BUCKET")
                       .value(bucket)
                       .build(),
                   KeyValuePair.builder()
                       .name("KEY")
                       .value(key)
                       .build()))
                 .build())
               .build())
             .build();

        // Implementation omitted for brevity
      }
   }
```

- ProductHealthCheck: The ProductHealthCheck class implements Quarkus'
  health check mechanism, which is crucial for maintaining application reliability in cloud
  environments. It demonstrates the use of Microprofile Health, allowing the application to
  report its status to orchestration systems such as Kubernetes. The class checks the accessibility
  of the DynamoDB table, showcasing how Quarkus applications can provide meaningful health
  information about external dependencies. This component is essential for implementing
  robust microservices that can self-report their operational status:

```
@Readiness
@ApplicationScoped
public class ProductHealthCheck implements HealthCheck {

    @Inject
    DynamoDbClient dynamoDbClient;

    private static final String TABLE_NAME = "Products";

    @Override
    public HealthCheckResponse call() {
        HealthCheckResponseBuilder responseBuilder =
        HealthCheckResponse.named(
            "Product service health check");
```

```
        try {
            dynamoDbClient.describeTable(DescribeTableRequest.
            builder()
                    .tableName(TABLE_NAME)
                    .build());

            return responseBuilder.up()
                    .withData("table", TABLE_NAME)
                    .withData("status", "accessible")
                    .build();
        } catch (DynamoDbException e) {
            return responseBuilder.down()
                    .withData("table", TABLE_NAME)
                    .withData("status",
                        "inaccessible")
                    .withData("error", e.getMessage())
                    .build();
        }
    }
}
```

3. **Configure a native build:**

- **Maven profile:** Ensure your pom.xml file includes a profile for native builds. This will specify the necessary dependencies and plugins for GraalVM:

```
<profiles>
    <profile>
        <id>native</id>
        <activation>
            <property>
                <name>native</name>
            </property>
        </activation>
            <properties>
                <skipITs>false</skipITs>
                <quarkus.package.type>native</quarkus.package.
type>
                <quarkus.native.enabled>true</quarkus.native.
enabled>
            </properties>
        </profile>
    </profiles>
```

- `Dockerfile.native`: The provided Dockerfile is essential for building and packaging a Quarkus application with GraalVM for deployment on AWS Lambda. It starts by using a GraalVM image to compile the application into a native executable, ensuring optimal performance and minimal startup time. The build stage includes copying the project files and running the Maven build process. Subsequently, the runtime stage uses a minimal base image to keep the final image lightweight. The compiled native executable is copied from the build stage to the runtime stage, where it is set as the entry point for the container. This setup guarantees a streamlined and efficient deployment process for serverless environments:

```
# Start with a GraalVM image for native building
FROM quay.io/quarkus/ubi-quarkus-native-image:21.0.0-java17 AS
build
COPY src /usr/src/app/src
COPY pom.xml /usr/src/app
USER root
RUN chown -R quarkus /usr/src/app
USER quarkus
RUN mvn -f /usr/src/app/pom.xml -Pnative clean package

FROM registry.access.redhat.com/ubi8/ubi-minimal
WORKDIR /work/
COPY --from=build /usr/src/app/target/*-runner /work/application
RUN chmod 775 /work/application
EXPOSE 8080
CMD ["./application", "-Dquarkus.http.host=0.0.0.0"]
```

This Dockerfile describes a two-stage build process for a Quarkus native application:

1. **Build stage**:

   - Uses a GraalVM-based image to compile the application

   - Copies project files and builds a native executable

2. **Runtime stage**:

   - Uses a minimal Red Hat UBI as the base image

   - Copies the native executable from the build stage

   - Sets the executable as the entry point

Multi-stage builds have the following benefits:

- **Smaller image size**: The final image is lean, containing only the necessary runtime dependencies
- **Improved security**: Reduces the attack surface by including fewer tools and packages
- **Clear separation**: Simplifies maintenance by separating the build environment from the runtime environment

> **Note for Apple Silicon Users**
>
> When building Docker images on Apple Silicon (M1 or M2) devices, you might encounter compatibility issues due to the default **Advance RISC Machine** (**ARM**) architecture. Most cloud environments, including AWS, Azure, and Google Cloud, use AMD64 (x86_64) architecture. To avoid these issues, specify the target platform when building Docker images to ensure compatibility.
>
> Specify the `--platform` argument when building Docker images on Apple Silicon devices to ensure compatibility with cloud environments.

For example, use the following command to build an image compatible with AMD64 architecture:

```
docker build --platform linux/amd64 -t myapp:latest .
```

While the `application.properties` file is not directly used for enabling native builds, you can include properties to optimize the application for running as a native image. Here is a sample `application.properties` file:

```
# you can include properties to optimize the application for running
as a native image:
# Disable reflection if not needed
quarkus.native.enable-http-url-handler=true
# Native image optimization
quarkus.native.additional-build-args=-H:+ReportExceptionStackTraces
# Example logging configuration for production
%prod.quarkus.log.console.level=INFO
```

Deploy to AWS Lambda:

`Template.yaml`: An AWS **Serverless Application Model** (**SAM**) template that defines the infrastructure for our Quarkus-based Lambda function, specifying its runtime environment, handler, resource allocations, and necessary permissions:

```
AWSTemplateFormatVersion: '2010-09-09'
Transform: AWS::Serverless-2016-10-31
Description: >
```

```
    quarkus-serverless

Resources:
    QuarkusFunction:
        Type: AWS::Serverless::Function
        Properties:
            Handler: com.example.LambdaHandler
            Runtime: provided.al2
            CodeUri: s3://your-s3-bucket-name/your-code.zip
            MemorySize: 128
            Timeout: 15
            Policies:
                - AWSLambdaBasicExecutionRole
```

Build and package: Run this Maven command to create a native `.jar` file:

```
mvn clean package -Pnative -Dquarkus.native.container-build=true
```

Deploy with SAM CLI: Use the AWS SAM CLI to package and deploy our Quarkus-based Lambda function: the first command packages the application and uploads it to an Amazon S3 bucket, while the second command deploys the packaged application to AWS, creating or updating a CloudFormation stack with the necessary resources and permissions:

```
sam package --output-template-file packaged.yaml --s3-bucket your-s3-
bucket-name
sam deploy --template-file packaged.yaml --stack-name quarkus-
serverless --capabilities CAPABILITY_IAM
```

By following these steps, you will have successfully built a serverless REST API using Quarkus, packaged it as a native image with GraalVM, and deployed it to AWS Lambda. This setup ensures optimal performance and reduces cold start times for your serverless application.

The serverless paradigm continues to evolve, with Java frameworks such as Quarkus leading the charge in optimizing for cloud-native, serverless environments. As we've seen, modern serverless Java applications can leverage advanced features such as rapid startup times, low memory footprints, and seamless integration with cloud services. This enables developers to build complex, scalable applications that go far beyond simple function executions, encompassing full-fledged microservices architectures.

As the cloud computing landscape continues to evolve, another emerging trend is gaining significant traction: edge computing. Let's explore how Java is adapting to meet the unique challenges and opportunities presented by edge computing environments.

Edge computing and Java

Edge computing represents a paradigm shift in how data is processed, with computation occurring at or near the data source instead of relying solely on centralized cloud data centers. This approach reduces latency, optimizes bandwidth usage, and improves response times for critical applications.

Java's role in edge computing architectures

Java, with its mature ecosystem and robust performance, is increasingly becoming a pivotal player in edge computing architectures.

Java's versatility and platform independence make it an ideal candidate for edge computing environments, which often consist of heterogeneous hardware and **operating systems (OSs)**. Java's ability to run on various devices, from powerful servers to constrained **internet of things (IoT)** devices, ensures that developers can leverage a consistent programming model across the entire edge-to-cloud continuum. Additionally, the extensive set of libraries and frameworks available in the Java ecosystem enables rapid development and deployment of edge applications.

Key benefits of using Java in edge computing include the following:

- **Cross-platform compatibility**: Java's "write once, run anywhere" philosophy allows edge applications to be deployed across diverse hardware platforms without modification
- **Performance and scalability**: Java's robust performance and efficient memory management are critical for handling the resource-constrained environments often found in edge devices
- **Security**: Java provides a strong security model, which is essential for safeguarding sensitive data processed at the edge

These advantages make Java a compelling choice for edge computing. To further empower developers, several frameworks and tools have been developed to streamline Java-based edge application development and deployment.

Frameworks and tools for Java-based edge applications

To effectively leverage Java in edge computing, developers can utilize a variety of frameworks and tools specifically designed for building and managing edge applications. Some of the prominent frameworks and tools include the following:

- **Eclipse Foundation's IoT initiative**:
  - **Eclipse Kura**: An open-source framework for building IoT gateways. It provides a set of Java APIs for accessing hardware interfaces, managing network configurations, and interacting with cloud services.

- **Eclipse Kapua**: A modular IoT cloud platform that works in conjunction with Eclipse Kura to provide end-to-end IoT solutions. It offers features such as device management, data management, and application integration.

- **Apache Edgent**: Apache Edgent (formerly known as Quarks) is a lightweight, embeddable programming model and runtime for edge devices. It allows developers to create analytics applications that can run on small-footprint devices and integrate with central data systems.

- **Vert.x**: Vert.x is a toolkit for building reactive applications on the JVM. Its event-driven architecture and lightweight nature make it well suited for edge-computing scenarios where low latency and high concurrency are essential.

- **AWS IoT Greengrass**: AWS IoT Greengrass extends AWS capabilities to edge devices, enabling them to act locally on the data they generate while still using the cloud for management, analytics, and durable storage. Java developers can create Greengrass Lambda functions to process and respond to local events.

- **Azure IoT Edge**: Azure IoT Edge allows developers to deploy and run containerized applications at the edge. Java applications can be packaged in Docker containers and deployed using Azure IoT Edge runtime, enabling seamless integration with Azure cloud services.

- **Google Cloud IoT Edge**: Google Cloud IoT Edge brings Google Cloud's ML and data processing capabilities to edge devices. Java developers can utilize TensorFlow Lite and other Google Cloud services to create intelligent edge applications.

Java's robust ecosystem, platform independence, and extensive library support make it a strong contender for edge computing. By leveraging frameworks and tools designed for edge environments, Java developers can build efficient, scalable, and secure edge applications that harness the full potential of edge computing architectures. As edge computing continues to evolve, Java is well positioned to play a critical role in shaping the future of distributed and decentralized data processing.

AI and ML integration

As we look toward the future of Java in cloud computing, the integration of AI and ML presents exciting opportunities and challenges. While *Chapter 7* focused on Java's concurrency mechanisms for ML workflows, this section explores Java's evolving role in cloud-based AI/ML ecosystems and its integration with advanced cloud AI services.

Java's position in cloud-based AI/ML workflows

Here are some of Java's evolving roles in cloud-based AI/ML ecosystems:

- **Serverless AI/ML with Java**: The future of Java in cloud-based AI/ML workflows is increasingly serverless. Frameworks such as AWS Lambda and Google Cloud Functions allow developers to deploy AI/ML models as serverless functions. This trend is expected to grow, enabling more efficient and scalable AI/ML operations without the need for managing infrastructure.

- **Java as an orchestrator**: Java is positioning itself as a powerful orchestrator for complex AI/ML pipelines in the cloud. Its robustness and extensive ecosystem make it ideal for managing workflows that involve multiple AI/ML services, data sources, and processing steps. Expect to see more Java-based tools and frameworks designed specifically for AI/ML pipeline orchestration in cloud environments.

- **Edge AI with Java**: As edge computing gains prominence, Java's *write once, run anywhere* philosophy becomes increasingly valuable. Java is being adapted for edge AI applications, allowing models trained in the cloud to be deployed and run on edge devices. This trend will likely accelerate, with Java serving as a bridge between cloud-based training and edge-based inference.

Next, let's explore Java's integration with advanced cloud-based AI services.

Integration of Java with cloud AI services

Integrating Java applications with cloud-based AI services opens up a world of possibilities for developers, enabling the creation of intelligent and adaptive software solutions. Cloud AI services offer pre-trained models, scalable infrastructure, and APIs that make it easier to implement advanced ML and AI capabilities without the need for extensive in-house expertise. The following is a list of popular cloud AI services that can be integrated with Java applications:

- **Native Java SDKs for cloud AI services**: Major cloud providers are investing in developing a robust **Java Development Kit** (**JDK**) for their AI services. For example, AWS has released the AWS SDK for Java 2.0, which provides streamlined access to services such as Amazon SageMaker. Google Cloud has also enhanced its Java client libraries for AI and ML services. This trend is expected to continue, making it easier for Java developers to integrate cloud AI services into their applications.

- **Java-friendly AutoML platforms**: Cloud providers are developing AutoML platforms that are increasingly Java friendly. For instance, Google Cloud AutoML now offers Java client libraries, allowing Java applications to easily train and deploy custom ML models without extensive ML expertise. This trend is likely to expand, making advanced AI capabilities more accessible to Java developers.

- **Containerized Java AI/ML deployments**: The future of Java in cloud AI/ML workflows is closely tied to containerization. Platforms such as Kubernetes are becoming the de facto standard for deploying and managing AI/ML workloads in the cloud. Java's compatibility with containerization technologies positions it well for this trend. Expect to see more tools and best practices emerge for deploying Java-based AI/ML applications in containerized environments.

- **Java in federated learning**: Federated learning is an ML technique that trains algorithms across multiple decentralized edge devices or servers holding local data samples, without exchanging them. This approach addresses growing privacy concerns by allowing model training on distributed datasets without centrally pooling the data.

  As privacy concerns grow, federated learning is gaining traction. Java's robust security features and its wide adoption in enterprise environments make it a strong candidate for implementing federated learning systems. Cloud providers are likely to offer more support for Java in their federated learning offerings, enabling models to be trained across decentralized data sources without compromising privacy.

- **Java for Machine Learning Operations (MLOps)**: The emerging field of MLOps is seeing increased adoption of Java. Its stability and extensive tooling make Java well suited for building robust MLOps pipelines in the cloud. Expect to see more Java-based MLOps tools and integrations with cloud CI/CD services specifically designed for AI/ML workflows.

In conclusion, Java's role in cloud-based AI/ML is evolving beyond just a language for implementing algorithms. It's becoming a crucial part of the broader AI/ML ecosystem in the cloud, from serverless deployments to edge computing, and from AutoML to MLOps. As cloud AI services continue to mature, Java's integration with these services will deepen, offering developers powerful new ways to build intelligent, scalable applications in the cloud.

Use case – serverless AI image analysis with AWS Lambda and Fargate

This use case demonstrates a scalable, serverless architecture for AI-powered image analysis using AWS Lambda and Fargate. AWS Fargate is AWS's implementation of serverless containers. This technology allows for the deployment of entire Java applications in a serverless fashion, leveraging container orchestration platforms such as Kubernetes and AWS Fargate. By packaging Java applications as containers, developers can enjoy the benefits of serverless computing – such as automatic scaling and pay-per-use pricing – while maintaining control over the runtime environment. This approach ensures consistency across different environments, provides flexibility with the inclusion of necessary libraries and tools, and offers robust scalability.

The system consists of two main components, each built as a separate Quarkus project:

- `ImageAnalysisCoordinator`:

  - Built as a native executable for optimal performance in a serverless environment

  - Triggered when an image is uploaded to an S3 bucket

  - Performs quick analysis using Amazon Rekognition

  - Initiates a more detailed analysis by launching an AWS Fargate task

- `FargateImageAnalyzer`:

  - Built as a JVM-based application and containerized using Docker

  - Runs as a task in AWS Fargate when triggered by the Lambda function

  - Performs in-depth image processing using advanced AI techniques

  - Stores detailed analysis results back in S3

This two-component architecture allows for efficient resource utilization: the lightweight Lambda function handles the initial processing and orchestration, while the Fargate container manages the more intensive computational tasks. Together, they form a robust, scalable solution for serverless AI-powered image analysis.

Step 1: Create a Fargate container:

`Dockerfile.jvm`: The `Dockerfile.jvm` is used to build the Docker image for the Fargate container component of the serverless AI image analysis architecture. Unlike the Lambda function, which is built as a native executable, the Fargate container runs the `FargateImageAnalyzer` application as a JVM-based Quarkus application. This choice is due to the Fargate container being responsible for the more computationally intensive image processing tasks, where the benefits of the Quarkus framework can outweigh the potential performance advantages of a native executable.

This Dockerfile defines the steps to build a Docker image for a Quarkus application. The image is designed to run within a Red Hat **Universal Base Image** (**UBI**) environment with OpenJDK 17:

```
FROM registry.access.redhat.com/ubi8/openjdk-17:1.14
ENV LANGUAGE='en_US:en'
# We make four distinct layers so if there are application changes the
library layers can be re-used
COPY --chown=185 target/quarkus-app/lib/ /deployments/lib/
COPY --chown=185 target/quarkus-app/*.jar /deployments/
COPY --chown=185 target/quarkus-app/app/ /deployments/app/
COPY --chown=185 target/quarkus-app/quarkus/ /deployments/quarkus/
EXPOSE 8080
USER 185
```

```
ENV JAVA_OPTS="-Dquarkus.http.host=0.0.0.0 -Djava.util.logging.
manager=org.jboss.logmanager.LogManager"
ENV JAVA_APP_JAR="/deployments/quarkus-run.jar"
ENTRYPOINT [ "/opt/jboss/container/java/run/run-java.sh" ]
```

`FargateImageAnalyzer.java` performs in-depth image processing:

```java
@QuarkusMain
@ApplicationScoped
public class FargateImageAnalyzer implements QuarkusApplication {

    @Inject
    S3Client s3Client;

    @Inject
    RekognitionClient rekognitionClient;

    @Override
    public int run(String... args) throws Exception {
        String bucket = System.getenv("IMAGE_BUCKET");
        String key = System.getenv("IMAGE_KEY");

        try {
            DetectLabelsRequest labelsRequest = DetectLabelsRequest.
            builder()
                    .image(Image.builder().s3Object(
                        S3Object.builder().bucket(
                            bucket).name(key).build()
                            ).build())
                    .maxLabels(10)
                    .minConfidence(75F)
                    .build();

            DetectLabelsResponse labelsResult = rekognitionClient.
            detectLabels(labelsRequest);

            DetectFacesRequest facesRequest = DetectFacesRequest.
            builder()
                    .image(Image.builder().s3Object(
                        S3Object.builder().bucket(
                            bucket).name(key).build()
                            ).build())
                    .attributes(Attribute.ALL)
                    .build();
```

```
                    DetectFacesResponse facesResult = rekognitionClient.
                    detectFaces(facesRequest);

                    String analysisResult =
                    generateAnalysisResult(labelsResult, facesResult);

                    s3Client.putObject(builder -> builder
                            .bucket(bucket)
                            .key(key + "_detailed_analysis.json")
                            .build(),
                            RequestBody.fromString(analysisResult));

            } catch (Exception e) {
                System.err.println("Error processing image: " +
                e.getMessage());
                return 1;
            }

            return 0;
        }
    private String generateAnalysisResult(
        DetectLabelsResponse labelsResult, DetectFacesResponse
        facesResult) {
            // Implement result generation logic
            return "Analysis result";
        }
    }
}
```

The FargateImageAnalyzer class is the main application that runs inside the Fargate container as part of the serverless AI image analysis architecture. It is designed as a Quarkus application and implements the QuarkusApplication interface. The class is responsible for extracting the S3 bucket and object key information, using the AWS Rekognition client to perform image analysis, generating a detailed analysis result, and storing it back in the same S3 bucket. It is designed to run as a standalone Quarkus application within the Fargate task, leveraging the benefits of running in a containerized environment and the ease of deployment and scaling that Fargate provides.

Step 2: Create a Lambda function:

`Dockerfile.native`: This Dockerfile is used to build the Docker image for the native executable of the Lambda function component in the serverless AI image analysis architecture. This Dockerfile follows the Quarkus convention for building native executables by using the `quay.io/quarkus/ubi-quarkus-native-image` base image and performing the necessary build steps. By using `Dockerfile.native`, the Lambda function can be packaged as a native executable, which provides improved performance and reduced cold start times compared to a JVM-based deployment. This is particularly beneficial for serverless applications where rapid response times are crucial:

```
FROM quay.io/quarkus/ubi-quarkus-native-image:21.0.0-java17 AS build
COPY src /usr/src/app/src
COPY pom.xml /usr/src/app
USER root
RUN chown -R quarkus /usr/src/app
USER quarkus
RUN mvn -f /usr/src/app/pom.xml -Pnative clean package

FROM registry.access.redhat.com/ubi8/ubi-minimal
WORKDIR /work/
COPY --from=build /usr/src/app/target/*-runner /work/application
RUN chmod 775 /work/application
EXPOSE 8080
CMD ["./application", "-Dquarkus.http.host=0.0.0.0"]
```

`ImageAnalysisCoordinator.java`: This is an AWS Lambda function that gets triggered when a new image is uploaded to an S3 bucket:

```
@ApplicationScoped
public class ImageAnalysisCoordinator implements
RequestHandler<S3Event, String> {

    @Inject
    EcsClient ecsClient;

    @Inject
    S3Client s3Client;

    @Override
    public String handleRequest(S3Event s3Event,
        Context context) {
            String bucket = s3Event.getRecords().get(
                0).getS3().getBucket().getName();
            String key = s3Event.getRecords().get(
                0).getS3().getObject().getKey();
```

```java
RunTaskRequest runTaskRequest = RunTaskRequest.builder()
    .cluster("your-fargate-cluster")
    .taskDefinition("your-task-definition")
    .launchType("FARGATE")
    .overrides(TaskOverride.builder()
        .containerOverrides(
            ContainerOverride.builder()
            .name("your-container-name")
// Replace with your actual container name
            .environment(KeyValuePair.builder()
                .name("BUCKET")
                .value(bucket)
                .build(),
            KeyValuePair.builder()
                .name("KEY")
                .value(key)
                .build())
            .build())
        .build())
    .build();

try {
    ecsClient.runTask(runTaskRequest);
    return "Fargate task launched for image analysis:
    " + bucket + "/" + key;
} catch (Exception e) {
    context.getLogger().log(
        "Error launching Fargate task: " +
        e.getMessage());
        return "Error launching Fargate task";
    }
  }
}
```

The `ImageAnalysisCoordinator` class is an AWS Lambda function that serves as the entry point for the serverless AI image analysis architecture. Its primary responsibilities are as follows:

- Extracting the S3 bucket and object key information from the incoming S3 event that triggers the Lambda function

- Initiating a Fargate task to perform the computationally intensive image analysis by launching an ECS task and passing the necessary environment variables (bucket and key)

- Handling any errors that occur during the Fargate task launch process and returning appropriate status messages

This Lambda function acts as a lightweight coordinator, responsible for orchestrating the overall image analysis workflow. It triggers the more resource-intensive processing to be performed by the Fargate container, which runs the `FargateImageAnalyzer` application. By separating the responsibilities in this way, the architecture achieves efficient resource utilization and scalability.

Step 3: Build the projects:

For the Lambda function, run the following command to package the function:

```
mvn package -Pnative -Dquarkus.native.container-build=true
```

For the Fargate container, run the following command to build the Docker image:

```
docker build -f src/main/docker/Dockerfile.jvm -t quarkus-ai-image-
analysis .
```

Step 4: Deploy:

To streamline the deployment of the serverless AI infrastructure, an AWS CloudFormation template has been prepared. This template automates the entire deployment process, including the following steps:

1. Create the necessary AWS resources, such as the following:

   - S3 bucket for storing the images and analysis results

   - ECS cluster and task definition for the Fargate container

   - Lambda function for the `ImageAnalysisCoordinator` class

2. Upload the built artifacts (Lambda function `.jar` file and Docker image) to the appropriate locations.

3. Configure the necessary permissions and triggers for the Lambda function to be invoked when an image is uploaded to the S3 bucket.

4. Deploy the Fargate task definition and set up the necessary network configurations.

To use the CloudFormation template, you can find it in the book's accompanying GitHub repository alongside the source code. Simply download the template, fill in any necessary parameters, and deploy it using the AWS CloudFormation service. This will set up the entire serverless AI infrastructure for you, streamlining the deployment process and ensuring consistency across different environments.

Emerging concurrency and parallel processing tools in Java

As Java continues to evolve, new tools and frameworks are being developed to address the growing demands of concurrent and parallel programming. These advancements aim to simplify development, improve performance, and enhance scalability in modern applications.

Introduction to Project Loom – virtual threads for efficient concurrency

Project Loom is an ambitious initiative by the OpenJDK community to enhance Java's concurrency model. The primary goal is to simplify writing, maintaining, and observing high-throughput concurrent applications by introducing virtual threads (also known as **fibers**).

Virtual threads are lightweight and are managed by the Java runtime rather than the OS. Unlike traditional threads, which are limited by the number of OS threads, virtual threads can scale to handle millions of concurrent operations without overwhelming system resources. They allow developers to write code in a synchronous style while achieving the scalability of asynchronous models.

Its key features include the following:

- **Lightweight nature**: Virtual threads are much lighter than traditional OS threads, reducing memory and context-switching overhead

- **Blocking calls**: They handle blocking calls efficiently, suspending only the virtual thread while keeping the underlying OS thread available for other tasks

- **Simplicity**: Developers can write straightforward, readable code using familiar constructs such as loops and conditionals without resorting to complex asynchronous paradigms

To illustrate the practical application of Project Loom and virtual threads, let's explore a code example that demonstrates implementing a high-concurrency microservice using Project Loom and Akka within an AWS cloud environment.

Code example – implementing a high-concurrency microservice using Project Loom and Akka for the AWS cloud environment

In this section, we will demonstrate how to implement a high-concurrency microservice using Project Loom and Akka, designed to run in an AWS cloud environment. This example will showcase how to leverage virtual threads from Project Loom and the actor model provided by Akka to build a scalable and efficient microservice:

Step 1: Project setup: Enter pom.xml dependencies:

```
<!-- Akka Dependencies -->
    <dependency>
        <groupId>com.typesafe.akka</groupId>
        <artifactId>akka-actor-typed_${
            scala.binary.version}</artifactId>
        <version>${akka.version}</version>
    </dependency>
    <dependency>
        <groupId>com.typesafe.akka</groupId>
```

```
        <artifactId>akka-stream_${scala.binary.version}</artifactId>
        <version>${akka.version}</version>
    </dependency>
<!-- AWS SDK -->
    <dependency>
        <groupId>software.amazon.awssdk</groupId>
        <artifactId>s3</artifactId>
        <version>2.17.100</version>
    </dependency>
```

Step 2: Code implementation:

`HighConcurrencyService.java`: The main entry point for the service, which sets up `ActorSystem` and uses `ExecutorService` to manage virtual threads:

```
public class HighConcurrencyService {

    public static void main(String[] args) {
        ActorSystem<Void> actorSystem = ActorSystem.create(
            Behaviors.empty(), "high-concurrency-system");
        S3Client s3Client = S3Client.create();
        ExecutorService executorService = Executors.
        newCachedThreadPool();
// Use a compatible thread pool

        for (int i = 0; i < 1000; i++) {
            final int index = i;
            executorService.submit(() -> {
                // Create and start the actor
                Behavior<RequestHandlerActor.HandleRequest> behavior =
                RequestHandlerActor.create(s3Client);
                var requestHandlerActor = actorSystem.systemActorOf(
                    behavior, "request-handler-" + index,
                    Props.empty());

                // Send a request to the actor
                requestHandlerActor.tell(
                    new RequestHandlerActor.HandleRequest(
                        "example-bucket",
                        "example-key-" + index,
                        "example-content"));
            });
        }

        // Clean up
```

```
            executorService.shutdown();
            actorSystem.terminate();
        }
    }
```

The `HighConcurrencyService` class serves as the entry point for a high-concurrency microservice application designed to handle numerous requests efficiently. Utilizing Akka's actor model and Java's concurrency features, this class demonstrates how to manage thousands of concurrent tasks effectively. The main function initializes `ActorSystem` for creating and managing actors, sets up an S3 client for interacting with AWS S3 services, and employs an executor service to submit multiple tasks. Each task involves creating a new actor instance to handle a specific request, showcasing how to leverage virtual threads and actors for scalable and concurrent processing in a cloud environment.

`RequestHandlerActor.java`: This actor handles individual requests to process data and interact with AWS S3:

```java
public class RequestHandlerActor {
    public static Behavior<HandleRequest> create(
        S3Client s3Client) {
            return Behaviors.setup(context ->
                Behaviors.receiveMessage(message -> {
            processRequest(s3Client, message.bucket,
                message.key, message.content);
            return Behaviors.same();
        }));
    }
    private static void processRequest(S3Client s3Client,
        String bucket, String key, String content) {
            PutObjectRequest putObjectRequest = PutObjectRequest.
            builder()
                .bucket(bucket)
                .key(key)
                .build();
        PutObjectResponse response = s3Client.putObject(
            putObjectRequest,
            RequestBody.fromString(content));
        System.out.println(
            "PutObjectResponse: " + response);
    }
    public static class HandleRequest {
        public final String bucket;
        public final String key;
        public final String content;
```

```
        public HandleRequest(String bucket, String key,
            String content) {
                this.bucket = bucket;
                this.key = key;
                this.content = content;
        }
    }
}
```

The `RequestHandlerActor` class defines the behavior of an actor responsible for handling individual requests in the high-concurrency microservice. It processes requests to store data in AWS S3 by utilizing the S3 client. The `HandleRequest` inner class encapsulates the details of a request, including the S3 bucket name, key, and content to be stored. The actor's behavior is defined as receiving these `HandleRequest` messages, processing the request by interacting with the S3 service, and logging the result. This class exemplifies the use of Akka's actor model to manage and process concurrent tasks efficiently, ensuring scalability and robustness in cloud-based applications.

Step 3: Deployment to AWS:

Dockerfile: The Dockerfile should be created and saved in the root directory of your application project. This is the standard location for the Dockerfile, as it allows the Docker build process to access all the necessary files and resources without requiring additional context switches:

```
FROM amazoncorretto:17-alpine as builder
WORKDIR /workspace
COPY pom.xml .
COPY src ./src
RUN ./mvnw package -DskipTests

FROM amazoncorretto:17-alpine
WORKDIR /app
COPY --from=builder /workspace/target/high-concurrency-microservice-
1.0.0-SNAPSHOT.jar /app/app.jar
ENTRYPOINT ["java", "-jar", "/app/app.jar"]
```

The key points about this Dockerfile are as follows:

- It uses the `amazoncorretto: 17-alpine` base image, which provides the Java 17 runtime environment based on the Alpine Linux distribution

- It follows a two-stage build process:

  - The *builder* stage compiles the application and packages it into a JAR file

  - The final stage copies the packaged JAR file and sets the entry point to run the application

Deploy using AWS ECS/Fargate:

We have also prepared a CloudFormation template for these processes, which can be found in the code repository. Follow these steps to deploy:

1. **Create an ECS cluster and task definition in AWS**: Set up your ECS cluster and define the task that will run your Docker container.

2. **Upload the Docker image to Amazon Elastic Container Registry (ECR)**: Push the Docker image to Amazon ECR for easy deployment.

3. **Configure ECS service to use Fargate and run the container**: Configure your ECS service to use AWS Fargate, a serverless compute engine, to run the containerized application.

This streamlined process ensures that your high-concurrency microservice is efficiently deployed in a scalable cloud environment.

This high-concurrency microservice example demonstrates the power of leveraging Project Loom's virtual threads and Akka's actor model to build scalable, efficient, and cloud-ready applications. By harnessing these advanced concurrency tools, developers can simplify their code, improve resource utilization, and enhance the overall performance and responsiveness of their services, particularly in the context of the AWS cloud environment. This lays the foundation for exploring the next wave of cloud innovations, where emerging technologies such as AWS Graviton processors and Google Cloud Spanner can further enhance the scalability and capabilities of cloud-based applications.

Preparing for the next wave of cloud innovations

As cloud technologies continue to evolve rapidly, developers and organizations must stay ahead of the curve. Anticipating advancements in cloud services, here's how you can prepare for upcoming advancements in cloud services:

- **AWS Graviton**: AWS Graviton is a family of ARM-based processors designed by AWS to offer improved price performance compared to traditional x86-based processors, particularly for workloads that can take advantage of the parallel processing capabilities of ARM architecture. The latest **Graviton3** iteration can provide up to 25% better performance and 60% better price performance than previous-generation Intel-based EC2 instances.

- **Amazon Corretto**: On the other hand, Amazon Corretto is a no-cost, multiplatform, production-ready distribution of the OpenJDK, a free and open-source implementation of the Java platform. Corretto is available for both x86-based and ARM-based (including Graviton) architectures, providing a certified, tested, and supported version of the JDK for AWS customers. The ARM-based Corretto JDK is optimized to run on AWS Graviton-powered instances.

Consider using the Amazon Corretto JDK. Here is a code snippet to build a Docker image:

```
FROM --platform=$BUILDPLATFORM amazoncorretto:17
COPY . /app
WORKDIR /app
RUN ./gradlew build
CMD ["java", "-jar", "app.jar"]
```

Build and push run the following command:

```
docker buildx build --platform linux/amd64,linux/arm64 -t myapp:latest
--push .
```

Google Cloud Spanner: Cloud Spanner is a fully managed, scalable, relational database service offering strong consistency and high availability:

- **Global distribution**: Spanner supports multi-regional and global deployment, providing high availability and low-latency access to data

- **Strong consistency**: Unlike many NoSQL databases, Spanner maintains strong consistency, making it suitable for applications that require transactional integrity

- **Seamless scaling**: Spanner automatically handles horizontal scaling, allowing applications to grow without compromising performance or availability

Example: Using Java with Cloud Spanner:

```java
import com.google.cloud.spanner.*;
try (Spanner spanner = SpannerOptions.newBuilder(
    ).build().getService()) {
    DatabaseClient dbClient = spanner.getDatabaseClient(
        DatabaseId.of(projectId, instanceId, databaseId));
    try (ResultSet resultSet = dbClient
        .singleUse() // Create a single-use read-only //transaction
        .executeQuery(Statement.of("SELECT * FROM Users"))){
    while (resultSet.next()) {
        System.out.printf("User ID: %d, Name: %s\n",
            resultSet.getLong("UserId"),
            resultSet.getString("Name"));
        }
    }}
```

This code snippet demonstrates the use of the Java client library to interact with Google Cloud Spanner. The code first creates a Spanner client using the `SpannerOptions` builder and retrieves the service instance. It then gets a `DatabaseClient` instance, which is used to interact with a specific Spanner database identified by the `projectId`, `instanceId`, and `databaseId` parameters.

Within a try-with-resources block, the code creates a single-use, read-only transaction using the `singleUse()` method and executes a `SQL SELECT` query to retrieve all records from the `Users` table. The results are then iterated through, and the `UserId` and `Name` columns are printed for each user record.

This example showcases the basic usage of the Google Cloud Spanner Java client library, including establishing a connection to the database, executing a query, and processing the results, while ensuring proper resource management and cleanup.

Quantum computing

Quantum computing, though still in its early stages, promises to revolutionize various industries by solving complex problems that are infeasible for classical computers. Quantum computers leverage the principles of quantum mechanics, such as superposition and entanglement, to perform computations in parallel.

While not immediately practical for most applications, it's beneficial to start learning about quantum computing principles and how they might apply to your domain. Key concepts to explore include qubits, quantum gates, and quantum algorithms such as Shor's algorithm for factoring large numbers and Grover's algorithm for search problems.

Understanding these principles will prepare you for future advancements and potential integration of quantum computing into your workflows. By familiarizing yourself with the foundational concepts now, you'll be better positioned to take advantage of quantum computing as it becomes more accessible and applicable to real-world problems.

Staying informed and exploring these technologies, even at an introductory level, will help ensure your organization is ready to adapt and thrive in the rapidly evolving cloud landscape.

Summary

As the final chapter of this book, we now stand at the precipice of the future, where cloud technologies continue to evolve at a breathtaking pace. In this concluding section, we explored the emerging trends and advancements that are poised to reshape the way we develop and deploy applications in the cloud, with a particular emphasis on Java's pivotal role in shaping these innovations.

We began by delving into the evolution of serverless Java, where we saw how frameworks such as Quarkus and Micronaut are redefining the boundaries of function as a service. These cutting-edge tools leverage techniques such as native image compilation to deliver unprecedented performance and efficiency in serverless environments, while also enabling the deployment of full-fledged Java applications as serverless containers. This represents a significant shift, empowering developers to create highly scalable, responsive, and cloud-native applications that go beyond simple function executions.

Next, we turned our attention to the edge computing landscape, where data processing and decision-making are moving closer to the source. Java's platform independence, performance, and extensive ecosystem make it an ideal choice for building edge applications. We introduced the key frameworks and tools that enable Java developers to leverage the power of edge computing, ensuring their applications can seamlessly integrate with this rapidly advancing paradigm.

Furthermore, we explored Java's evolving role in the integration of AI and ML within cloud-based ecosystems. From serverless AI/ML workflows to the seamless integration of Java with cloud-based AI services, we uncovered the opportunities and challenges that this convergence presents, equipping you with the knowledge to harness the power of these technologies in your Java-based applications.

Finally, we ventured into the captivating realm of quantum computing, a field that promises to revolutionize various industries. While still in its early stages, understanding the fundamental principles of quantum computing, such as qubits, quantum gates, and algorithms, can prepare developers for future advancements and their potential integration with Java-based applications.

As we conclude this book, you now possess a comprehensive understanding of the emerging trends in cloud computing and Java's pivotal role in shaping these innovations. Armed with this knowledge, you are poised to position your applications and infrastructure for success in the rapidly evolving cloud landscape, ensuring your organization can adapt and thrive in the years to come.

Questions

1. What is a key benefit of using Quarkus and GraalVM for building serverless Java applications?

    A. Improved startup time and reduced memory usage

    B. Easier integration with cloud-based AI/ML services

    C. Seamless deployment across multiple cloud providers

    D. All of the above

2. Which of the following is a key advantage of using Java in edge computing environments?

    A. Platform independence

    B. Extensive library support

    C. Robust security model

    D. All of the above

3. Which cloud AI service allows Java developers to easily train and deploy custom ML models without extensive ML expertise?

   A. AWS SageMaker

   B. Google Cloud AutoML

   C. Microsoft Azure Cognitive Services

   D. IBM Watson Studio

4. Which quantum computing concept is demonstrated in the provided code example that puts a qubit into superposition and measures the outcome?

   A. Quantum entanglement

   B. Quantum teleportation

   C. Quantum superposition

   D. Quantum tunneling

5. What is a key benefit of using serverless containers for Java applications in the cloud?

   A. Reduced operational overhead for managing infrastructure

   B. Increased cold start times for serverless functions

   C. Inability to include custom libraries and dependencies

   D. Limited control over the runtime environment

Appendix A
Setting up a Cloud-Native Java Environment

In *Appendix A*, you will learn how to set up a cloud-native environment for Java applications. This comprehensive guide covers everything from building and packaging Java applications to deploying them on popular cloud platforms like **Amazon Web Services (AWS)**, **Microsoft Azure**, and **Google Cloud Platform (GCP)**. Key topics include:

- **Building and packaging**: Step-by-step instructions on using build tools like Maven and Gradle to create and manage Java projects.

- **Ensuring cloud-readiness**: Best practices for making your Java applications stateless and configurable to thrive in cloud environments.

- **Containerization**: How to create Docker images for your Java applications and deploy them using Docker.

- **Cloud deployments**: Detailed procedures for deploying Java applications on AWS, Azure, and GCP, including setting up the necessary cloud environments, creating and managing cloud resources, and using specific cloud services like Elastic Beanstalk, Kubernetes, and serverless functions.

By the end of this appendix, you will have a solid understanding of how to effectively build, package, containerize, and deploy Java applications in a cloud-native environment.

General approach – build and package Java applications

This section provides a detailed guide on the essential steps required to build and package your Java applications, ensuring they are ready for deployment in a cloud environment.

1. Ensure your app is cloud-ready

    I.   **Stateless:** Design your application to be stateless. This means that each request should be independent and not rely on previous requests. Store session data in a distributed cache like Redis instead of in memory.

    II. **Configurable:** Use external configuration files or environment variables to manage configuration. This can be done using Spring Boot's `application.properties` or `application.yaml` files, or by using a configuration management tool. Here is an example:

    ```
    server.port=8080
    spring.datasource.url=jdbc:mysql://localhost:3306/mydb
    spring.datasource.username=root
    spring.datasource.password=password
    ```

2. Use a Build tool like Maven or Gradle

    • **Maven:** Create a `pom.xml` file in your project root directory if it doesn't already exist. Add the necessary dependencies. Here is an example of `pom.xml`:

    ```
    <project xmlns="http://maven.apache.org/POM/4.0.0"
    xmlns:xsi="http://www.w3.org/2001/XMLSchema-instance"
        xsi:schemaLocation="http://maven.apache.org/POM/4.0.0
    http://maven.apache.org/xsd/maven-4.0.0.xsd">
        <modelVersion>4.0.0</modelVersion>
        <groupId>com.example</groupId>
        <artifactId>myapp</artifactId>
        <version>1.0-SNAPSHOT</version>
        <dependencies>
            <dependency>
                <groupId>org.springframework.boot</groupId>
                <artifactId>spring-boot-starter-web</artifactId>
            </dependency>
            <!-- Add other dependencies here -->
        </dependencies>
        <build>
            <plugins>
                <plugin>
                    <groupId>org.springframework.boot</groupId>
                    <artifactId>spring-boot-maven-plugin</
    ```

```
         artifactId>
                 </plugin>
             </plugins>
         </build>
</project>
```

- **Gradle**: Create a `build.gradle` file in your project root directory if it doesn't already exist. Add the necessary dependencies, here is an example of `build.gradle`:

```
plugins {
    id 'org.springframework.boot' version '2.5.4'
    id 'io.spring.dependency-management' version '1.0.11.
RELEASE'
    id 'java'
}
group = 'com.example'
version = '1.0-SNAPSHOT'
sourceCompatibility = '11'
repositories {
    mavenCentral()
}
dependencies {
    implementation 'org.springframework.boot:spring-boot-
starter-web'
    // Add other dependencies here
}
test {
    useJUnitPlatform()
}
```

3. **Build the JAR file**: Build the JAR file using Maven or Gradle.

   - **Maven**: Run the following command:

   ```
   mvn clean package
   ```

   This command will generate a JAR file in the target directory, typically named `myapp-1.0-SNAPSHOT.jar`.

   - **Gradle**: Run the following command:

   ```
   gradle clean build
   ```

   This command will generate a JAR file in the build/libs directory, typically named `myapp-1.0-SNAPSHOT.jar`.

> **Note:**
>
> If you are not using containers, you can stop here. The JAR file located in the target or build/ libs directory can now be used to run your application directly.

4. Containerize your application using Docker

    I.   **Create a Dockerfile**: Create a Dockerfile in your project root directory with the following content:

    ```
    # Use an official OpenJDK runtime as a parent image (Java 21)
    FROM openjdk:21-jre-slim

    # Set the working directory
    WORKDIR /app

    # Copy the executable JAR file to the container
    COPY target/myapp-1.0-SNAPSHOT.jar /app/myapp.jar

    # Expose the port the app runs on
    EXPOSE 8080

    # Run the JAR file
    ENTRYPOINT ["java", "-jar", "myapp.jar"]
    ```

    Make sure to adjust the COPY instruction if your JAR file is located in a different directory or has a different name.

    II. **Build the Docker Image**:

    Build the Docker image using the Docker build command. Run this command in the directory where your Dockerfile is located:

    ```
    docker build -t myapp:1.0 .
    ```

    This command will create a Docker image named myapp with the tag 1.0.

    III. **Run the Docker Container**: Run the Docker container using the docker run command:

    ```
    docker run -p 8080:8080 myapp:1.0
    ```

    This command will start a container from the myapp:1.0 image and map port 8080 of the container to port 8080 on your host machine.

This section provides a detailed guide on the essential steps required to build and package your Java applications, ensuring they are ready for deployment in a cloud environment.

After learning how to build and package your Java applications, the next step is to explore the specific procedures for deploying these applications on popular cloud platforms.

Step-by-step guides for deploying Java applications on popular cloud platforms:

1. Setting Up the AWS environment

    I.   **Create an AWS Account**: Sign up for an AWS account if you don't already have one.

    II.   **Install AWS CLI**: Download and install the AWS **Command Line Interface** (CLI) to manage your AWS services.

    III. **Configure AWS CLI**: Run `aws configure` and enter your AWS Access Key, Secret Key, region, and output format.

    IV.   **Create an IAM role policy**: `trust-policy.json`

```
{
    "Version": "2012-10-17",
    "Statement": [
        {
            "Effect": "Allow",
            "Principal": {
                "Service": "ec2.amazonaws.com"
            },
            "Action": "sts:AssumeRole"
        }
    ]
}
```

2. Deploy your Java application to AWS using WAS CLI: Elastic Beanstalk (PaaS)

    I.   Create the IAM role:

```
aws iam create-role --role-name aws-elasticbeanstalk-ec2-role
--assume-role-policy-document file://trust-policy.json
```

    II.   Attach the required policies to the role:

```
aws iam attach-role-policy --role-name aws-elasticbeanstalk-
ec2-role-java --policy-arn arn:aws:iam::aws:policy/
AWSElasticBeanstalkWebTier
aws iam attach-role-policy --role-name aws-elasticbeanstalk-
ec2-role-javaa --policy-arn arn:aws:iam::aws:policy/
AWSElasticBeanstalkMulticontainerDocker
aws iam attach-role-policy --role-name aws-elasticbeanstalk-
ec2-role-java --policy-arn arn:aws:iam::aws:policy/
AWSElasticBeanstalkWorkerTier
```

III.   Create the instance profile:

```
aws iam create-instance-profile --instance-profile-name
aws-elasticbeanstalk-ec2-role-java
```

IV.   Add the role to the instance profile:

```
aws iam add-role-to-instance-profile --instance-profile-
name aws-elasticbeanstalk-ec2-role-java --role-name
aws-elasticbeanstalk-ec2-role-java
```

3.   Deploying to Elastic Beanstalk

I.   Create an Elastic Beanstalk Application:

```
aws elasticbeanstalk create-application --application-name
MyJavaApp
```

II.   Create a new Elastic Beanstalk environment using the latest Corretto 21 version on Amazon Linux 2023.

```
aws elasticbeanstalk create-environment --application-name
MyJavaApp --environment-name my-env --solution-stack-name
"64bit Amazon Linux 2023 v4.2.6 running Corretto 21"
```

III.   Uploads my-application.jar to the deployments folder in the my-bucket S3 bucket. Adjust the parameters as needed for your specific use case.

```
aws s3 cp target/my-application.jar s3://my-bucket/
my-application.jar
```

IV.   Create a new application version in Elastic Beanstalk

```
aws elasticbeanstalk create-application-version --application-
name MyJavaApp --version-label my-app-v1 --description
"First version" --source-bundle S3Bucket= my-bucket,S3Key=
my-application.jar
```

V.   Update the Elastic Beanstalk environment to use the new application version

```
aws elasticbeanstalk update-environment --environment-name
my-env --version-label my-app-v1
```

VI.   Check the environment health

```
aws elasticbeanstalk describe-environment-health
--environment-name my-env --attribute-names All
```

4. Deploy your Java application: ECS (Containers)

    I. Push the Docker image to Amazon **Elastic Container Registry (ECR)**: First, create an
        ECR repository:

        ```
        aws ecr create-repository --repository-name my-application
        ```

    II. Authenticate Docker to your ECR:

        ```
        aws ecr get-login-password --region <your-region> | docker
        login --username AWS --password-stdin <account-id>.dkr.
        ecr.<region>.amazonaws.com
        ```

    III. Tag your Docker image:

        ```
        docker tag my-application:1.0
                                                        <account-id>.
        dkr.ecr.<region>.amazonaws.com/my-application:1.0
        ```

    IV. Push your Docker image to ECR:

        ```
        docker push <account-id>.dkr.ecr.<region>.amazonaws.com/
        my-application:1.0
        ```

5. Set up **Elastic Container Service** or **ECS** using AWS CLI

    I. **Create a task definition**: Create a JSON file named `task-definition.json` with
        the task definition configuration:

        ```
        {
            "family": "my-application-task",
            "networkMode": "awsvpc",
            "requiresCompatibilities": [
                "FARGATE"
            ],
            "cpu": "256",
            "memory": "512",
            "containerDefinitions": [
                {
                    "name": "my-application",
                    "image": "<account-id>.dkr.ecr.<
                    region>.amazonaws.com/my-application:1.0",
                    "portMappings": [
                        {
                            "containerPort": 8080,
                            "protocol": "tcp"
                        }
        ```

```
            ],
            "essential": true
        }
    ]
}
```

II.   Register the task definition:

```
aws ecs register-task-definition --cli-input-json file://task-
definition.json
```

III. Create a Cluster:

```
aws ecs create-cluster --cluster-name cloudapp-cluster
```

IV.   **Create a Service**: Create a service using the following command:

```
aws ecs create-service \
    --cluster cloudapp-cluster \
    --service-name cloudapp-service \
    --task-definition cloudapp-task \
    --desired-count 1 \
    --launch-type FARGATE \
    --network-configuration "awsvpcConfiguration={
        subnets=[subnet-XXXXXXXXXXXXXXXXX],securityGroups=[
            sg-XXXXXXXXXXXXXXXX],assignPublicIp=ENABLED}"
```

Important notes

Replace subnet-XXXXXXXXXXXXXXXXX with the actual ID of the subnet where you want to run your tasks.

Replace sg-XXXXXXXXXXXXXXXXX with the actual ID of the security group you want to associate with your tasks.

This command uses forward slashes (\) for line continuation, which is appropriate for Unix-like environments (Linux, macOS, Git Bash on Windows).

For Windows Command Prompt, replace the backslashes (\) with caret symbols (^) for line continuation.

For PowerShell, use backticks (`) at the end of each line instead of backslashes for line continuation.

The --desired-count 1 parameter specifies that you want one task running at all times.

The --launch-type FARGATE parameter specifies that this service will use AWS Fargate, which means you don't need to manage the underlying EC2 instances.

6. Deploy your Java serverless Lambda function

    I.   Create an AWS Lambda function role:

```
aws iam create-role --role-name lambda-role --assume-role-
policy-document file://trust-policy.json
```

    II.   Attach the `AWSLambdaBasicExecutionRole` policy to the role:

```
aws iam attach-role-policy --role-name lambda-role
--policy-arn arn:aws:iam::aws:policy/service-role/
AWSLambdaBasicExecutionRole
```

    III.   Create the Lambda function:

```
aws lambda create-function \
    --function-name java-lambda-example \
    --runtime java17 \
    --role arn:aws:iam::<account-id>:role/lambda-role \
    --handler com.example.LambdaHandler::handleRequest \
    --zip-file fileb://target/lambda-example-1.0-SNAPSHOT.jar
```

    IV.   **Invoke the Lambda function**: When using the aws lambda invoke command to test your AWS Lambda function, it's important to update the --payload parameter to match the expected input format of your specific Lambda function.

```
aws lambda invoke --function-name java-lambda-example
--payload '{"name": "World"}' response.json
```

    V.   Check the response:

```
cat response.json
```

Now that you have an understanding of how to set up a cloud-native Java environment and deploy your applications on various cloud platforms, you may want to dive deeper into specific cloud services. The following links provide additional resources and documentation to help you further your knowledge and skills in deploying and managing Java applications in the cloud.

Useful links for further information on AWS

- **Amazon EC2**: Getting Started with Amazon EC2 (https://docs.aws.amazon.com/AWSEC2/latest/UserGuide/EC2_GetStarted.html)

- **AWS Elastic Beanstalk**: Getting Started with AWS Elastic Beanstalk (https://docs.aws.amazon.com/elasticbeanstalk/latest/dg/GettingStarted.html)

- **Amazon ECS**: Getting Started with Amazon ECS (https://docs.aws.amazon.com/AmazonECS/latest/developerguide/ECS_GetStarted.html)

- **AWS Lambda**: Getting Started with AWS Lambda (`https://docs.aws.amazon.com/lambda/latest/dg/getting-started.html`)

- **Managing environment variables**: Best practices for managing environment variables in AWS Lambda (`https://docs.aws.amazon.com/lambda/latest/dg/configuration-envvars.html`)

Microsoft Azure

In this section, you will learn the steps required to deploy Java applications on Microsoft Azure. This includes setting up the Azure environment, deploying applications on virtual machines and containers, and utilizing **Azure Kubernetes Service** (**AKS**) for containerized applications. Additionally, you will explore how to deploy Java functions on Azure Functions, enabling you to leverage serverless computing for your Java applications.

1. Set up the Azure environment:

   I.   **Download and install Azure CLI**: Follow the official installation instructions for your operating system from the Azure CLI installation guide (`https://learn.microsoft.com/en-us/cli/azure/install-azure-cli`).

   II. **Configure Azure CLI**: Open your terminal or command prompt and run the following command to log in to your Azure account:

   ```
   az login
   ```

2. Follow the instructions to log in to your Azure account.

Next, you will learn how to deploy a regular Java application on Azure Virtual Machines.

Deploying a Regular Java Application on Azure Virtual Machines

1. Create a Resource Group:

   ```
   az group create --name myResourceGroup --location eastus
   ```

2. Create a Virtual Machine:

   ```
   az vm create --resource-group myResourceGroup --name myVM --image UbuntuLTS --admin-username azureuser --generate-ssh-keys
   ```

3. Open port 8080:

   ```
   az vm open-port --port 8080 --resource-group myResourceGroup --name myVM
   ```

4. SSH into the VM:

```
ssh azureuser@<vm-ip-address>
```

5. Install Java on the VM:

```
sudo apt update
sudo apt install openjdk-21-jre -y
```

6. Transfer and Run the JAR File:

```
scp target/myapp-1.0-SNAPSHOT.jar azureuser@<vm-ip-address>:/home/
azureuser
ssh azureuser@<vm-ip-address>
java -jar myapp-1.0-SNAPSHOT.jar
```

Once you have successfully deployed your Java application on an Azure Virtual Machine, you can manage and scale your application as needed using the Azure portal and CLI tools. This approach provides a solid foundation for running traditional Java applications in a cloud environment.

Next, you will learn how to deploy a Java application in containers using AKS, which offers a more flexible and scalable solution for containerized applications.

Deploying a Java Application in Containers on AKS

1. Create an **Azure Container Registry (ACR)**:

```
az acr create --resource-group myResourceGroup --name myACR --sku
Basic
```

2. Login to ACR:

```
az acr login --name myACR
```

3. Tag and push Docker image to ACR:

```
docker tag myapp:1.0 myacr.azurecr.io/myapp:1.0
docker push myacr.azurecr.io/myapp:1.0
```

4. Create AKS Cluster:

```
az aks create --resource-group myResourceGroup --name myAKSCluster
--node-count 1 --enable-addons monitoring --generate-ssh-keys
```

5. Get AKS Credentials:

```
az aks get-credentials --resource-group myResourceGroup --name
myAKSCluster
```

6. Deploy the application to AKS:

    I.   Create a Deployment YAML file (`deployment.yaml`):

```yaml
apiVersion: apps/v1
kind: Deployment
metadata:
    name: myapp-deployment
spec:
    replicas: 1
    selector:
        matchLabels:
            app: myapp
    template:
        metadata:
            labels:
                app: myapp
        spec:
            containers:
                - name: myapp
                image: myacr.azurecr.io/myapp:1.0
                ports:
                    - containerPort: 8080
```

    II. Apply the deployment:

```
kubectl apply -f deployment.yaml
```

    III. Expose the deployment:

```
kubectl expose deployment myapp-deployment--type=LoadBalancer
--port=8080
```

This completes the process of deploying a containerized Java application to AKS. However, for scenarios where you need more granular control over your application's execution or want to build serverless microservices, Azure Functions provides an excellent alternative. Next, you will learn how to deploy Java functions on Azure Functions, enabling you to take advantage of serverless computing for event-driven applications and microservices.

Deploying Java Functions on Azure Functions

1. Install Azure functions core tools:

   - **Windows**: Use MSI installer (`https://learn.microsoft.com/en-us/azure/azure-functions/functions-run-local?tabs=windows%2Cisolated-process%2Cnode-v4%2Cpython-v2%2Chttp-trigger%2Ccontainer-apps&pivots=programming-language-csharp#v2`)

   - **macOS**:

   ```
   brew tap azure/functions
   brew install azure-functions-core-tools@4
   ```

2. Create a new function app:

   ```
   func init MyFunctionApp --java
   cd MyFunctionApp
   func new
   ```

3. Build the function:

   ```
   mvn clean package
   ```

4. Deploy to Azure:

   ```
   func azure functionapp publish <FunctionAppName>
   ```

> **Notes**
> Replace placeholders like <vm-ip-address>, <your-region>, <FunctionAppName>, etc., with your actual values.

For detailed information on configuring environment variables and managing configurations specifically for Azure environments, you can refer to the official Azure documentation:

- **Azure App Service configuration**: Configure apps in Azure App Service (`https://learn.microsoft.com/en-us/azure/app-service/configure-common?tabs=portal`)

- **AKS configuration**: Best practices for cluster and node pool configuration in AKS (`https://learn.microsoft.com/en-us/azure/aks/operator-best-practices-scheduler`)

- **Azure functions configuration**: Configure function app settings in Azure functions (`https://learn.microsoft.com/en-us/azure/azure-functions/functions-how-to-use-azure-function-app-settings?tabs=azure-portal%2Cto-premium`)

Now that you've learned how to deploy Java applications on various Azure services, including virtual machines, AKS, and Azure Functions, let's explore another major cloud provider. The following section will guide you through similar deployment processes on GCP, allowing you to broaden your cloud deployment skills across different environments.

Google Cloud Platform

In this section, you'll learn how to deploy Java applications on **Google Cloud Platform** or **GCP**, one of the leading cloud service providers. GCP offers a wide range of services that cater to various deployment needs, from virtual machines to containerized environments and serverless functions. We'll cover the setup process for GCP and guide you through deploying Java applications using different GCP services, including **Google Compute Engine (GCE)**, **Google Kubernetes Engine (GKE)**, and Google Cloud Functions. This knowledge will empower you to leverage GCP's robust infrastructure and services for your Java applications.

Setting up the Google Cloud Environment

1. Create a Google Cloud account if you don't have one.

2. Install the Google Cloud SDK. Follow the instructions for your operating system from the official Google Cloud SDK documentation (`https://cloud.google.com/sdk/docs/install`)

3. Initialize the Google Cloud SDK:

```
gcloud init
```

4. Follow the prompts to log in and select your project.

5. Set your project ID:

```
gcloud config set project YOUR_PROJECT_ID
```

With your Google Cloud environment successfully set up, you are now prepared to deploy and manage Java applications using GCP's robust infrastructure. In the next sections, you will explore specific methods for deploying Java applications on GCE, GKE, and Google Cloud Functions.

Deploy your Java application to Google Cloud

GCE for regular Java Applications:

1. Create a VM instance:

```
gcloud compute instances create my-java-vm --zone=us-central1-a
--machine-type=e2-medium --image-family=ubuntu-2004-lts --image-
project=ubuntu-os-cloud
```

2. SSH into the VM:

```
gcloud compute ssh my-java-vm --zone=us-central1-a
```

3. Install Java on the VM:

```
sudo apt update
sudo apt install openjdk-17-jdk -y
```

4. Transfer your JAR file to the VM:

```
gcloud compute scp your-app.jar my-java-vm:~ --zone=us-central1-a
```

5. Run your Java application:

```
java -jar your-app.jar
```

By following these steps, you can efficiently deploy and manage your Java applications on Google Cloud, leveraging the various services and tools provided by GCP. With your Java application successfully deployed to Google Cloud, you are now ready to explore containerized deployments using GKE, which offers powerful orchestration capabilities for managing containers at scale.

GKE for containerized applications

In this section, you will learn how to deploy Java applications in containers using GKE. GKE provides a managed environment for deploying, managing, and scaling containerized applications using Kubernetes. You will be guided through setting up a GKE cluster, deploying your Docker images, and managing your containerized applications efficiently.

1. Create a GKE cluster:

```
gcloud container clusters create my-cluster --num-nodes=3
--zone=us-central1-a
```

2. Get credentials for the cluster:

```
gcloud container clusters get-credentials my-cluster --zone=us-
central1-a
```

3. Push your Docker image to **Google Container Registry (GCR)**:

```
docker tag your-app:latest gcr.io/YOUR_PROJECT_ID/your-app:latest
docker push gcr.io/YOUR_PROJECT_ID/your-app:latest
```

4. Create a Kubernetes deployment: Create a file named deployment.yaml:

```
apiVersion: apps/v1
kind: Deployment
metadata:
    name: your-app
spec:
    replicas: 3
    selector:
        matchLabels:
            app: your-app
    template:
        metadata:
            labels:
                app: your-app
    spec:
        containers:
            - name: your-app
            image: gcr.io/YOUR_PROJECT_ID/your-app:latest
            ports:
            - containerPort: 8080
```

5. Apply the deployment:

```
kubectl apply -f deployment.yaml
```

6. Expose the deployment:

```
kubectl expose deployment your-app --type=LoadBalancer --port 80
--target-port 8080
```

By leveraging GKE, you can take full advantage of Kubernetes' robust features to ensure your containerized Java applications are highly available, scalable, and easy to maintain.

Google Cloud Functions for serverless Java functions

In this section, you will learn how to deploy Java functions using Google Cloud Functions, enabling you to run event-driven code in a fully managed serverless environment. You will be guided through setting up your development environment, creating and deploying your Java functions, and managing them effectively using Google Cloud's powerful serverless tools.

1. Create a new directory for your function:

```
mkdir my-java-function
cd my-java-function
```

2. Initialize a new Maven project:

```
mvn archetype:generate -DgroupId=com.example -DartifactId=my-
java-function -DarchetypeArtifactId=maven-archetype-quickstart
-DinteractiveMode=false
```

3. Add the necessary dependencies to your pom.xml:

```
<dependency>
  <groupId>com.google.cloud.functions</groupId>
  <artifactId>functions-framework-api</artifactId>
  <version>1.0.4</version>
  <scope>provided</scope>
</dependency>
```

4. Create your function class:

```
package com.example;

import com.google.cloud.functions.HttpFunction;
import com.google.cloud.functions.HttpRequest;
import com.google.cloud.functions.HttpResponse;
import java.io.BufferedWriter;

public class Function implements HttpFunction {
    @Override
    public void service(HttpRequest request,
        HttpResponse response) throws Exception {
        BufferedWriter writer = response.getWriter();
        writer.write("Hello, World!");
        }
    }
```

5. Deploy the function:

```
gcloud functions deploy my-java-function --entry-point com.
example.Function --runtime java17 --trigger-http --allow-
unauthenticated
```

These instructions provide a basic setup for deploying Java applications to Google Cloud using different services. Remember to adjust the commands and configurations based on your specific application requirements and Google Cloud project settings.

Useful links for further information

- **Google Compute Engine**: Get started creating and managing virtual machines in GCP with the Compute Engine quickstart guide: `https://cloud.google.com/compute/docs/quickstart`

- **Google Kubernetes Engine**: Dive into container orchestration with GKE and deploy your first Kubernetes cluster using the GKE quickstart: `https://cloud.google.com/kubernetes-engine/docs/quickstart`

- **Google Cloud Functions**: Develop and deploy serverless functions that respond to events with the Cloud Functions deployment guide: `https://cloud.google.com/functions/docs/deploying`

- **Managing environment variables**: `https://cloud.google.com/run/docs/configuring/services/environment-variables`

Appendix B
Resources and Further Reading

Recommended books, articles, and online courses

Chapters 1–3

Books

- *Cloud Native Java: Designing Resilient Systems with Spring Boot, Spring Cloud, and Cloud Foundry* by Josh Long and Kenny Bastani. This comprehensive guide offers practical insights into building scalable, resilient Java applications for cloud environments, covering Spring Boot, Spring Cloud, and Cloud Foundry technologies. Link: `https://www.amazon.com/Cloud-Native-Java-Designing-Resilient/dp/1449374646`

- *Java Concurrency in Practice* by Brian Goetz et al. A seminal work on Java concurrency, this book provides in-depth coverage of concurrent programming techniques, best practices, and pitfalls to avoid when developing multi-threaded applications.

- *Parallel and Concurrent Programming in Haskell* by Simon Marlow. While focused on Haskell, this book offers valuable insights into parallel programming concepts that can be applied to Java, providing a broader perspective on concurrent and parallel application design. Link: `https://www.oreilly.com/library/view/parallel-and-concurrent/9781449335939/`

- *Designing Distributed Systems: Patterns and Paradigms for Scalable, Reliable Services* by Brendan Burns (Microsoft Azure). This explores essential patterns for building scalable and reliable distributed systems, offering insights from Microsoft Azure's experience in cloud computing. Link: `https://www.amazon.com/Designing-Distributed-Systems-Patterns-Paradigms/dp/1491983647`

Articles

- *Microservices Patterns* by Chris Richardson (microservices.io). A comprehensive guide to microservices architecture patterns, this article helps developers understand and implement effective microservices-based systems. Link: `https://microservices.io/patterns/index.html`

- *A Java Fork/Join Framework* by Doug Lea. An in-depth look at the Fork/Join framework by its creator, providing valuable insights into its design and implementation for parallel processing in Java. Link: `http://gee.cs.oswego.edu/dl/papers/fj.pdf`

- *Amdahl's Law in the Multicore Era* by Mark D. Hill and Michael R. Marty. This article offers a modern perspective on Amdahl's Law and its implications for parallel computing, helping developers understand the limits and potential of parallel processing in contemporary systems. Link: `https://research.cs.wisc.edu/multifacet/papers/ieeecomputer08_amdahl_multicore.pdf`

Online courses

- *Java Multithreading, Concurrency & Performance Optimization* by Udemy. This comprehensive course covers Java multithreading, concurrency, and performance optimization techniques, providing practical examples and hands-on exercises to master advanced Java programming concepts. Link:`https://www.udemy.com/course/java-multithreading-concurrency-performance-optimization/`

- *Concurrency in Java* by Coursera (offered by Rice University). Focusing on the foundational principles of concurrency in Java, this course offers practical exercises to solidify understanding of concurrent programming concepts and techniques. Link: `https://www.coursera.org/learn/concurrent-programming-in-java`

- *Reactive Programming in Modern Java using Project Reactor* by Udemy. A comprehensive course on reactive programming in Java using Project Reactor, teaching developers how to build reactive applications for better scalability and resilience in modern software architectures. Link: `https://www.udemy.com/course/reactive-programming-in-modern-java-using-project-reactor/`

- *Parallel, Concurrent, and Distributed Programming in Java Specialization* on Coursera by Rice University. This specialization offers a comprehensive coverage of advanced concurrency topics in Java, including parallel, concurrent, and distributed programming techniques for developing high-performance applications. Link: `https://www.coursera.org/specializations/pcdp`

Key blogs and websites

- *Baeldung* offers comprehensive tutorials and articles on Java, Spring, and related technologies, including in-depth content on concurrency and parallelism. Their concurrency section is particularly valuable for learning advanced Java threading concepts. Link: `https://www.baeldung.com/java-concurrency`

- *DZone Java Zone* is a community-driven platform, offering a wealth of articles, tutorials, and guides on Java and cloud-native development. The Java Zone is an excellent resource for staying up-to-date with the latest trends and best practices in Java development. Link: `https://dzone.com/java-jdk-development-tutorials-tools-news`

- *InfoQ Java* provides news, articles, and interviews on software development, with a strong focus on Java, concurrency, and cloud-native technologies. InfoQ is particularly useful for gaining insights into industry trends and emerging technologies in the Java ecosystem. Link: `https://www.infoq.com/java/`

Chapters 4–6

Books

- *Patterns for Distributed Systems* by Unmesh Joshi (InfoQ)

- This book provides an overview of common patterns used in distributed systems, offering practical advice for designing robust and scalable architectures. Link: `https://www.amazon.com/Patterns-Distributed-Systems-Addison-Wesley-Signature/dp/0138221987`

Articles and blogs

- Martin Fowler's blog on microservices and distributed systems. This blog is a treasure trove of information on microservices and distributed systems, offering in-depth articles and thought leadership on modern software architecture. Link: `https://martinfowler.com/articles/microservices.html`

- *LMAX Disruptor documentation and performance guide* is a high-performance inter-thread messaging library for Java. This resource provides documentation and performance guides for implementing low-latency, high-throughput systems. Link: `https://lmax-exchange.github.io/disruptor/`

Online courses

- *Microservices Architecture* by the University of Alberta. This course provides a comprehensive introduction to microservices architecture, covering design principles, implementation strategies, and best practices for building scalable and maintainable systems. Link: `https://www.coursera.org/specializations/software-design-architecture`

- *Building Scalable Java Microservices with Spring Boot and Spring Cloud* on Coursera by Google Cloud Offered by Google Cloud, this course teaches how to build scalable Java microservices using Spring Boot and Spring Cloud, with a focus on cloud-native development practices. Link: `https://www.coursera.org/learn/google-cloud-java-spring`

Chapters 7–9

Books

- *Serverless Architectures on AWS* by Peter Sbarski This book provides comprehensive coverage of serverless concepts and practical implementations on AWS, offering valuable insights for developers looking to build scalable and cost-effective applications. Link: `https://www.amazon.com/Serverless-Architectures-AWS-Peter-Sbarski/dp/1617295426`

Articles

- *Serverless Computing: One Step Forward, Two Steps Back* by Joseph M. Hellerstein et al. This article provides a critical analysis of serverless computing, discussing its advantages and limitations, and offering a balanced perspective on its place in modern architecture. Link: `https://arxiv.org/abs/1812.03651`

- *Serverless Architecture Patterns and Best Practices* by freeCodeCamp

- This article provides an overview of key serverless patterns like messaging, function focus, and event-driven architecture, emphasizing the benefits of decoupling and scalability. Link: `https://www.freecodecamp.org/news/serverless-architecture-patterns-and-best-practices/`

Online courses

- *Developing Serverless Solutions on AWS* by the AWS training team. This includes comprehensive coverage of AWS Lambda, best practices, frameworks, and hands-on labs. Link: `https://aws.amazon.com/training/classroom/developing-serverless-solutions-on-aws/`

Technical papers

- *Serverless Computing: Current Trends and Open Problems* by Ioana Baldini et al. This academic paper provides a thorough examination of serverless computing, discussing current trends, challenges, and future directions in this rapidly evolving field. Link: `https://arxiv.org/abs/1706.03178`

Online resources

- AWS Lambda Developer Guide: `https://docs.aws.amazon.com/lambda/latest/dg/welcome.html`

- Azure Functions Java developer guide: `https://docs.microsoft.com/en-us/azure/azure-functions/functions-reference-java`

- Google Cloud Functions Java Tutorials: `https://codelabs.developers.google.com/codelabs/cloud-starting-cloudfunctions#`

Chapters 10–12

Books

- *Quantum Computing for Developers* by Johan Vos. This groundbreaking book offers a developer-friendly introduction to quantum computing, bridging the gap between theoretical concepts and practical implementation. It provides clear explanations of quantum principles and includes hands-on examples using Java-based frameworks, preparing software developers for the emerging quantum computing landscape. The author, Johan Vos, expertly guides readers through quantum algorithms, quantum gates, and quantum circuits, demonstrating how to leverage existing programming skills in this cutting-edge field. Link: `https://www.manning.com/books/quantum-computing-for-developers`

Articles

- *Transitioning your service or application* by Amazon Web Services

- This article explores optimizing Java applications to provide more details regarding the individual steps involved in transitioning an application to Graviton2. Link: `https://docs.aws.amazon.com/whitepapers/latest/aws-graviton2-for-isv/transitioning-your-service-or-application.html`

- *Java in the Era of Cloud Computing* by Cogent University. This article focuses on the cloud-native advancements in Java, with frameworks like Spring Boot and Quarkus facilitating cloud-based development. It also mentions tools like Maven, Gradle, and JUnit for enhancing productivity and ensuring code quality. Link: `https://www.cogentuniversity.com/post/java-in-the-era-of-cloud-computing`

Online courses

- *Parallel, Concurrent, and Distributed Programming in Java Specialization* by Rice University on Coursera. This specialization covers advanced concurrency topics in Java, which apply to cloud computing environments and auto-scaling scenarios. Link: `https://www.coursera.org/specializations/pcdp`

- *Serverless Machine Learning with Tensorflow on Google Cloud Platform* by Google Cloud on Coursera. This course explores the intersection of serverless computing and machine learning, aligning with discussions on AI/ML integration in cloud environments and future trends in cloud computing. Link: `https://www.coursera.org/learn/serverless-machine-learning-gcp-br`

This appendix provides a curated selection of resources, including books, articles, and online courses, to deepen your understanding of concurrency, parallelism, and cloud-native development in Java. Leveraging these materials will enhance your knowledge and skills, enabling you to build robust, scalable, and efficient cloud-native Java applications.

Answers to the end-of-chapter multiple-choice questions

Chapter 1: Concurrency, Parallelism, and the Cloud: Navigating the Cloud-Native Landscape

1. B) Easier to scale and maintain individual services
2. B) Synchronization
3. D) Stream API
4. C) Automatic scaling and management of resources
5. B) Data consistency and synchronization

Chapter 2: Introduction to Java's Concurrency Foundations: Threads, Processes, and Beyond

1. C) Threads share a memory space, while processes are independent and have their own memory.
2. B) It offers a set of classes and interfaces for managing threads and processes efficiently.
3. B) Allowing multiple threads to read a resource concurrently but requiring exclusive access for writing.
4. B) It allows a set of threads to wait for a series of events to occur.
5. B) It allows for lock-free thread-safe operations on a single integer value.

Chapter 3: Mastering Parallelism in Java

1. B) To enhance parallel processing by recursively splitting and executing tasks
2. A) `RecursiveTask` returns a value, while `RecursiveAction` does not
3. B) It allows idle threads to take over tasks from busy threads
4. B) Balancing task granularity and parallelism level
5. B) The task's nature, resource availability, and team expertise

Chapter 4: Java Concurrency Utilities and Testing in the Cloud Era

1. C) To efficiently manage thread execution and resource allocation
2. B) `CopyOnWriteArrayList`
3. B) Enables asynchronous programming and non-blocking operations
4. B) They enable efficient data handling and reduce locking overhead in concurrent access scenarios
5. C) By offering more control over lock management and reducing lock contention

Chapter 5: Mastering Concurrency Patterns in Cloud Computing

1. C) To prevent failures in one service from affecting other services
2. B) Employing a lock-free ring buffer to minimize contention
3. C) It isolates services to prevent failures in one from cascading to others.
4. B) Scatter-Gather pattern
5. D) Resilience and data flow management

Chapter 6: Java and Big Data – a Collaborative Odyssey

1. B) Volume, velocity, and variety
2. A) **Hadoop Distributed File System (HDFS)**
3. C) Spark offers faster in-memory data processing capabilities.
4. A) Spark can only process structured data.
5. C) It helps to break down large datasets into smaller, manageable chunks for processing.

Chapter 7: Concurrency in Java for Machine Learning

1. C) To optimize computational efficiency
2. C) Parallel Streams

3. C) They improve scalability and manage large-scale computations.

4. B) To perform data preprocessing and model training more efficiently

5. B) Combining Java concurrency with generative AI

Chapter 8: Microservices in the Cloud and Java's Concurrency

1. C) Independent deployment and scalability

2. C) CompletableFuture

3. B) Distributing incoming network traffic across multiple instances

4. C) Circuit breaker pattern.

5. C) Assigning a separate managed database instance for each microservice

Chapter 9: Serverless Computing and Java's Concurrent Capabilities

1. C) Automatic scaling and reduced operational overhead.

2. B) CompletableFuture.

3. C) Managing recursive tasks by dividing them into smaller subtasks.

4. B) Optimize function size and use provisioned concurrency.

5. B) Improved performance through concurrent data processing.

Chapter 10: Synchronizing Java's Concurrency with Cloud Auto-Scaling Dynamics

1. C) Dynamic resource allocation based on demand

2. B) `CompletableFuture`

3. A) Managing a fixed number of threads

4. C) Implementing stateless services

5. C) Improving performance through concurrent data processing

Chapter 11: Advanced Java Concurrency Practices in Cloud Computing

1. D) User interface design

2. C) Improved performance for parallel tasks

3. B) VisualVM

4. B) Minimize data loss and improve availability

5. C) Difficulty in obtaining a cohesive view of distributed operations

Chapter 12: The Horizon Ahead

1. A) Improved startup time and reduced memory usage

2. D) All of the above

3. B) Google Cloud AutoML

4. C) Quantum superposition

5. A) Reduced operational overhead for managing infrastructure

Index

A

J

‹packt›

www.packtpub.com

Subscribe to our online digital library for full access to over 7,000 books and videos, as well as industry leading tools to help you plan your personal development and advance your career. For more information, please visit our website.

Why subscribe?

- Spend less time learning and more time coding with practical eBooks and Videos from over 4,000 industry professionals

- Improve your learning with Skill Plans built especially for you

- Get a free eBook or video every month

- Fully searchable for easy access to vital information

- Copy and paste, print, and bookmark content

Did you know that Packt offers eBook versions of every book published, with PDF and ePub files available? You can upgrade to the eBook version at packtpub.com and as a print book customer, you are entitled to a discount on the eBook copy. Get in touch with us at customercare@packtpub.com for more details.

At www.packtpub.com, you can also read a collection of free technical articles, sign up for a range of free newsletters, and receive exclusive discounts and offers on Packt books and eBooks.

Other Books You May Enjoy

If you enjoyed this book, you may be interested in these other books by Packt:

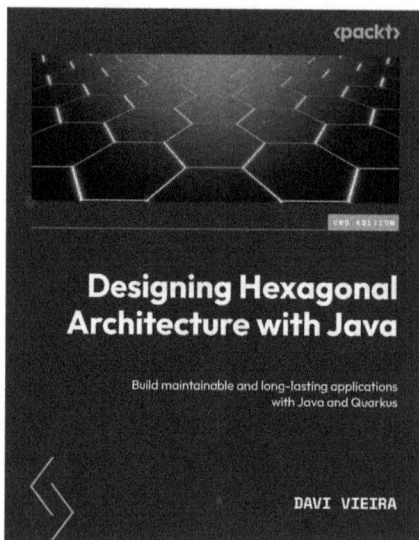

Designing Hexagonal Architecture with Java

Davi Vieira

ISBN: 978-1-83763-511-5

- Apply SOLID principles to the hexagonal architecture
- Assemble business rules algorithms using the specified design pattern
- Combine domain-driven design techniques with hexagonal principles to create powerful domain models
- Employ adapters to enable system compatibility with various protocols such as REST, gRPC, and WebSocket
- Create a module and package structure based on hexagonal principles
- Use Java modules to enforce dependency inversion and ensure software component isolation
- Implement Quarkus DI to manage the life cycle of input and output ports

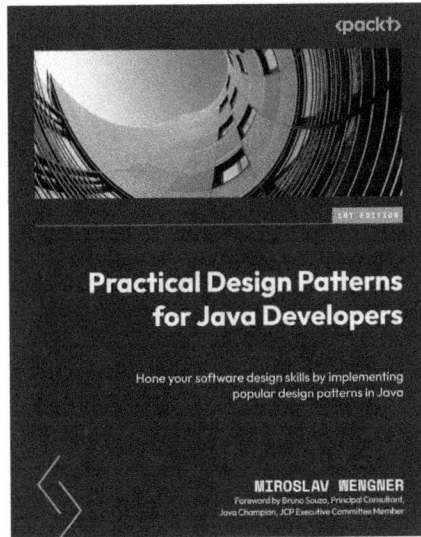

Practical Design Patterns for Java Developers

Miroslav Wengner

ISBN: 978-1-80461-467-9

- Understand the most common problems that can be solved using Java design patterns
- Uncover Java building elements, their usages, and concurrency possibilities
- Optimize a vehicle memory footprint with the Flyweight Pattern
- Explore one-to-many relations between instances with the observer pattern
- Discover how to route vehicle messages by using the visitor pattern
- Utilize and control vehicle resources with the thread-pool pattern
- Understand the penalties caused by anti-patterns in software design

Packt is searching for authors like you

If you're interested in becoming an author for Packt, please visit authors.packtpub.com and apply today. We have worked with thousands of developers and tech professionals, just like you, to help them share their insight with the global tech community. You can make a general application, apply for a specific hot topic that we are recruiting an author for, or submit your own idea.

Share Your Thoughts

Now you've finished *Java Concurrency and Parallelism*, we'd love to hear your thoughts! Scan the QR code below to go straight to the Amazon review page for this book and share your feedback or leave a review on the site that you purchased it from.

https://packt.link/r/1805129260

Your review is important to us and the tech community and will help us make sure we're delivering excellent quality content.

Download a free PDF copy of this book

Thanks for purchasing this book!

Do you like to read on the go but are unable to carry your print books everywhere?

Is your eBook purchase not compatible with the device of your choice?

Don't worry, now with every Packt book you get a DRM-free PDF version of that book at no cost.

Read anywhere, any place, on any device. Search, copy, and paste code from your favorite technical books directly into your application.

The perks don't stop there, you can get exclusive access to discounts, newsletters, and great free content in your inbox daily

Follow these simple steps to get the benefits:

1. Scan the QR code or visit the link below

https://packt.link/free-ebook/9781805129264

2. Submit your proof of purchase
3. That's it! We'll send your free PDF and other benefits to your email directly

www.ingramcontent.com/pod-product-compliance
Lightning Source LLC
Chambersburg PA
CBHW061739210326
41599CB00034B/6728